AE-11

Spring Design Manual

All SAE standards are abstracted and indexed in the SAE STANDARDS SEARCH Database.

Prepared under the auspices
of the SAE Spring Committee

Published by:
Society of Automotive Engineers, Inc.
400 Commonwealth Drive
Warrendale, PA 15096-0001

Library of Congress Cataloging-in-Publication Data

Spring design manual / prepared under the auspices of the SAE
Spring Committee.
      p. cm. -- (AE ; 11)
    Includes bibliographical references.
    ISBN 0-89883-777-4 : $80.00
    1. Springs (Mechanism)--Design and construction--
Handbooks, manuals, etc. 2. Motor vehicles--Springs and
suspension--Design and construction--Handbooks, manuals,
etc. I. Society of Automotive Engineers. Spring Committee.
II. Series: AE (Series) ; 11.
TJ210.S67 1989
629.24'3--dc20   89-26130
                                             CIP

Preface

It was in 1678 that an English scientist, Robert Hooke, stated that, within certain limits, deflection is proportional to load. His work on flexible members is still the basis for spring design today. The wide use of springs to store and release energy can easily be observed by looking at the tools, appliances, and vehicles that we use in our daily lives. While there are several basic types of spring forms, the variations within each category are endless. The processes which are used to manufacture these items may also vary, further compounding and enlarging the subject.

In 1943, at the behest of the U.S. Ordinance Department, the first spring manual was published under the title "Manual on Design and Application of Helical Springs for Ordinance." The manual found wide distribution, and revised editions were issued in 1945, 1958, 1962, 1973, 1980, and 1989. In 1944 the manual "Design and Application of Leaf Springs" was published with revised editions being issued in 1962, 1970, and 1982. The manual "Design and Manufacture of Torsion Bar Springs" was first published in 1947 and revisions were made in 1966 and 1981. The manual "Design and Manufacture of Coned Disk Springs and Spring Washers" was first published in 1950 with revisions in 1955 and 1989. The newest manual "Incorporating Pneumatic Springs in Vehicle Suspension Designs" was published in 1989. The publication of this manual on pneumatic springs, along with all the recent revisions to the other manuals, reflect the addition of new technologies and industry practices for automotive applications.

The incorporation of these five manuals into one volume represents the most comprehensive reference work available today. The Spring Committee wishes to note that these manuals should not be regarded as a compilation of design or manufacturing specifications; instead, they should be considered as reference works which contain essential information which may be helpful to the engineer and designer on a broad range of topics — material selection, tolerances, end configurations, fatigue life, load and stress calculations, and processing information. All of the manuals employ SI Units in accordance with SAE 916, "Rules For Use of SI (Metric) Units."

The Spring Committee recognizes the generous measure of time, effort, and dedication which the respective Subcommittees put forth in the preparation of these manuals for publication.

E. H. Judd

Table of Contents

Part 1

Design and Application of Leaf Springs

SAE HS 788

SPRING COMMITTEE

H. M. Reigner (Sponsor), Eaton Corp., Engineering & Research Center

J. F. Kelly (Chairman), Detroit Steel Products, Div. of Marmon Group

K. Campbell (Vice-Chairman), Rockwell International Corp., Suspension Components Div.

J. A. Alfes, Pontiac Motor Div., General Motors Corp.

T. A. Bank, Firestone Industrial Products Co.

J. J. Bozyk, Chrysler Corp.—Product Planning & Development

G. W. Folland, Rockwell International Corp., Suspension Components Div.

L. A. Habrle, Engineering Consultant

R. E. Hanslip, Toledo Spring Co.

D. J. Hayes, United States Steel Corp.

E. H. Judd, Associated Spring—Barnes Group, Inc.

W. Mayers, Peterson American Corp.

M. W. Mericle, Caterpillar Tractor Co., Materials Div.

G. W. Myrick, XM1 Tank System

E. C. Oldfield, Burton Auto Spring Corp.

W. Platko, Chevrolet Motor Div., General Motors Corp.

G. L. Radamaker, Eaton Corp., Suspension Div.

F. T. Rowland, Registered Professional Engineer

H. L. Schmedt, Caterpillar Tractor Co.

G. Schremmer, Schnorr-Neise Disc Spring Corp.

K. E. Siler, Ford Motor Co., Chassis Engineering

J. E. Silvis, Winamac Steel Products Div.—Norris Industries

B. Sterne, Bernhard Sterne Associates

W. M. Wood, Associated Spring—Barnes Group, Inc.

LEAF SPRING SUBCOMMITTEE

F. T. Rowland (Chairman), Registered Professional Engineer

G. W. Folland (Vice-Chairman), Rockwell International Corp., Suspension Components Div.

K. Campbell, Rockwell International Corp., Suspension Components Div.

R. E. Hanslip, Toledo Spring Co.

J. F. Kelly, Detroit Steel Products, Div. of Marmon Group, Inc.

E. C. Oldfield, Burton Auto Spring Corp.

G. L. Radamaker, Eaton Corp., Suspension Div.

K. E. Siler, Ford Motor Co., Chassis Engineering

B. Sterne, Bernhard Sterne Associates

E. J. Streichert, General Motors Corp.

Past members of the Spring Committee who have contributed materially to the original (1944) and several of the revised issues of the Manual:

H. H. Clark	E. H. Lindeman	Robert Schilling*
Tore Franzen*	Maurice Olley*	C. A. Tea
H. O. Fuchs	J. W. Rosenkrands	F. P. Zimmerli*
N. E. Hendrickson*	Max Ruegg*	M. C. Turkish

*Deceased

TABLE OF CONTENTS

Chapter 1
General Data

1. Introduction

This Manual is written as a guide for the designer of leaf spring installations. It contains information which will make it possible to calculate the space required for a leaf spring, to provide suitable attachments, and to determine the elastic and geometric properties of the assembly.

The detail design of the spring itself also is described, but it was not the intention of the Committee to lay down fixed rules for this. The choice of leaf lengths, leaf thicknesses, and leaf curvatures depends upon the type of installation and upon the kind of service. Only an experienced spring engineer can make the best choice of these factors. It is therefore recommended that the designer of a leaf spring installation consult a spring maker before the design is finalized.

For standards and practices not covered in this Manual, see the current SAE Handbook.

No attempt has been made to investigate or consider patents which may apply to subject matter presented in this Manual. Those who intend to use any of the constructions described herein should make their own investigations and arrangements in order to avoid liability for infringements.

The term multi-leaf has generally been applied to springs of constant width and with stepped leaves, each of constant thickness, except where leaf ends may be tapered in thickness. More recently, the term has been extended to include an assembly of stacked "single" leaves, each of which is characterized by tapering either in width or in thickness or by a combination of both.

Chapter 10 includes design data for single leaf springs which may be of variable width and constant thickness, constant width and variable thickness, or a combination of variable width and variable thickness.

2. General Characteristics of Leaf Springs

The leaf spring, like all other springs, serves to absorb and store energy and then to release it. During this cycle the stress in the spring must not exceed a certain maximum in order to avoid settling or premature failure. This consideration limits the amount of energy which can be stored in any spring.

For leaf springs based on a maximum stress of 1100 MPa, the energy listed in Table 1.1 may be stored in the active part of the spring. If consideration of the inactive part of the spring required for axle anchorage, spring eyes,

TABLE 1.1—ENERGY STORAGE OF STEEL SPRINGS AT 1100 MPa

Type*	Spring Design		Energy J/kg
F-1	Single leaf or all leaves full length		43
F-2	Properly stepped multi-leaf with B_e = 0.20		94
F-4	Single leaf	with H = 0.20	122
P-2	Single leaf	with H = 0.16	
		therefore J_c = 0.40	121
T-1	Single leaf	with J_e = 0.36	108
T-2	Single leaf	with J_e = 0.40	
		H = 0.16	
		therefore J_c = 0.496	105

*For description of Type see Chapter 10.

etc., is included, the energy per kg of the total spring mass will be less than shown.

For comparison, the stored energy in the active material of a helical spring of round bar section is 510 J/kg at 1100 MPa, and for a torsion bar of round section is 390 J/kg at 965 MPa.

This comparison shows that a leaf spring is heavier in mass than other types of springs.

Balancing this disadvantage of mass, the leaf spring possesses the advantage that it can also be used as an attaching linkage or structural member. In order to be economically competitive, the leaf spring must therefore be so designed that this advantage is fully utilized.

Also, a leaf spring made entirely of full length leaves of constant thickness (see type F-1) is very much heavier and less efficient than a leaf spring made of properly stepped leaves (see type F-2) or single leaf springs (see types F-4, P-2, T-1, and T-2).

The maximum permissible leaf thickness for a given deflection is proportional to the square of the spring length. By choosing too short a length, the designer often makes it impractical for the spring maker to build a satisfactory spring, although the requirements for normal load, deflection, and stress can be fulfilled.

For example: A cube of steel, weighing 44 kg and measuring about 178 mm on each side, can be made into a spring carrying a load of 16 000 N at 125 mm deflection with a stress of 480 MPa.

If 1500 mm is allowed for the length, the spring will look like Fig. 1.1. It will consist of 10 leaves, each 75.0 mm wide and 10.00 mm thick.

If only 750 mm is allowed for the length, the spring will look like Fig. 1.2. It will consist of 80 leaves, each 75.0 mm wide and only 2.50 mm thick.

When springs are made with stepped leaf lengths of type F-2, it is desirable to choose a length so that the spring will have no less than three leaves. Springs with many leaves

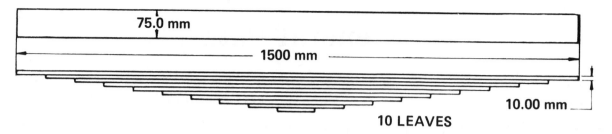

Fig. 1.1—Leaf spring of type F-2: Practical design with adequate length

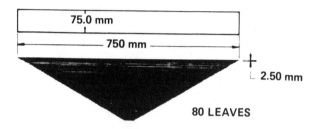

Fig. 1.2—Leaf spring of type F-2: Impractical design with inadequate length

are sometimes used for heavy loads, but they are economical only where the shortening of the spring leads to definite savings in the supporting structure. In addition, allowance will have to be made for increased spring rate and greater eye stress, assuming the same load and width are used.

In most installations the spring is also subject to windup loads. A typical example is that of the suspension spring (in a vehicle with Hotchkiss drive) which must withstand both driving and braking torque. The stresses under such loads are inversely proportional to the spring length; and the windup stiffness is proportional to the square of the length for the spring of given load rate (see Chapter 6). This is another reason why it is important to make the spring long enough and to check the resulting stresses and deflections.

When a leaf spring is used as an attaching linkage, it will tend to guide the supported members in a certain geometrical path (see Chapter 4). If no other guiding members are used, the desired geometry must be obtained by properly placing the supporting parts on the structure which carries the spring. If other guiding members are used, their geometry must fit that of the spring, or forces may be set up that will cause failure.

3. Leaf Springs for Vehicle Suspension

Leaf springs are most frequently used in suspensions. This Manual, therefore, contains information which is most useful in the design of suspension springs, but it is also applicable to leaf springs for other installations.

The characteristics of a spring suspension are affected chiefly by the spring rate and the static deflection of the spring.

The *rate* of a spring is the change of load per unit of deflection (N/mm). This is not the same amount at all positions of the spring, and is different for the spring as installed. *Static deflection* of a spring equals the static load divided by the rate at static load; it determines the stiffness of the suspension and the ride frequency of the vehicle. In most cases the static deflection differs from the actual deflection of the spring between zero load and static load, due to influences of spring camber and shackle effect.

A soft ride generally requires a large static deflection of the suspension. There are, however, other considerations and limits, among them the following:

1. A more flexible spring will have a larger total deflection and will be heavier.

2. In most applications a more flexible spring will cause more severe striking through or will require a larger "ride clearance" (the spring travel on the vehicle from the design load position to the metal-to-metal contact position), disregarding rubber bumpers.

3. The change of standing height of the vehicle due to a variation of load is larger with a more flexible spring.

The static deflection to be used also depends upon the available ride clearance. Further, the permissible static deflection depends upon the size of the vehicle because of considerations of stability in braking, accelerating, cornering, etc.

Table 1.2 shows typical static deflections and ride clearances for various types of vehicles. These values are approximate and are meant to be used only as a general indication of current practice in suspension system design.

The mass of a spring subject to a given maximum stress is determined by the energy which is to be stored (see Table 1.1). This energy is represented by the area under the load-deflection diagram, which therefore is also a measure of the required spring mass. The following consideration will indicate what effect some changes in either rate or clearance will have on the required spring mass and therefore on the load-deflection diagram.

	Static Deflection, mm	Ride Clearance, mm
Passenger automobiles (at design load)	100–300	75–125
Motor coaches (at maximum load)	100–200	50–125
Trucks (at rated load)		
For highway operation	75–200	75–125
For "off the road" operation	25–175	50–125

Concerning changes in rate, Fig. 1.3 shows a theoretical load-deflection diagram of a stiff (high rate) spring, and Fig. 1.4 that of a very flexible (low rate) spring, both for the same design load and clearance. The energy stored in each, when fully deflected, is the same (1125 J), and the two springs will have almost the same mass if made of the same kind of material.

In the case of the stiff spring, energy and mass will be decreased by making the spring more flexible. In the case of the very flexible spring, energy and mass will be decreased by making the spring stiffer. The dividing point between these two cases is defined by "static deflection = clearance". The load-deflection diagram of this "minimum energy" or "minimum mass" spring is shown by a dashed line in both figures; it indicates a stored energy of 1000 J.

Concerning changes in clearance, Figs. 1.3 and 1.4 bring out the fact that a change in clearance by a given distance will affect the stored energy and therefore the required mass of the stiff spring, much more than that of the very flexible spring.

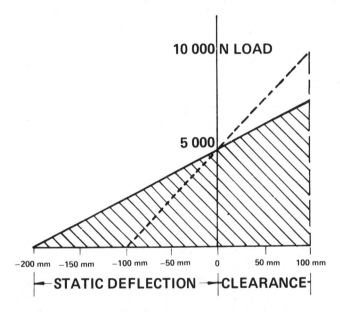

Fig. 1.4—Theoretical load-deflection diagram of a low rate spring

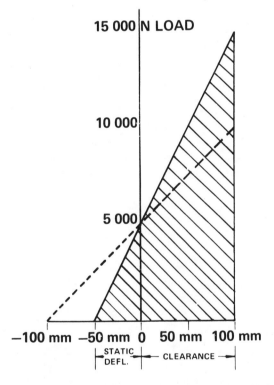

Fig. 1.3—Theoretical load-deflection diagram of a high rate spring

1.3

Chapter 2
Nomenclature And Specifications

1. Nomenclature

The following terms are recommended for use on drawings and in specifications to avoid misunderstandings. The terms apply mainly to semi-elliptic springs.

Datum Line—Most of the dimensions defined here refer to a datum line. In Figs. 2.1 and 2.2 (where the springs are shown inverted as in a machine for load and rate checking) it is shown as the line X-X. On springs with eyes, the datum line passes through the centers of the eyes. On other springs it passes through the points where the load is applied near the ends of the spring. These points must be indicated on the drawing.

Seat Angle Base Line—(see Figs. 2.1 and 2.2.) Reference line drawn through the terminal points of the active spring length at each eye, taken along the tension surface of the main leaf. For a Berlin type eye (see Fig. 3.3 E, F, G in Chapter 3), the terminal point is the intersection of an extension to the contour of the tension surface with a perpendicular line through the center of the eye. On springs without eyes, the seat angle base line is coincident with the datum line.

Loaded Length—(see Figs. 2.4–2.8.) Distance between the spring eye centers when the spring is deflected to the specified load position. On springs without eyes, it is the distance between the lines where load is applied under the specified conditions. Tolerance, $\pm$ 3.0 mm.

Loaded Fixed End Length—(see Figs. 2.4–2.8.) Distance from the center of the fixed end eye to the projection on the datum line of the point where the centerline of the center bolt intersects the spring surface in contact with the spring seat. Tolerance, $\pm$ 1.5 mm.

Straight Length—Distance between spring eye centers when the tension surface of the main leaf at the center bolt centerline is in the plane of the seat angle base line. The distance is measured parallel to the seat angle base line. Tolerance, $\pm$ 3.0 mm.

Seat Length—Length of spring that is in actual engagement with the spring seat when installed on a vehicle at design height. It is always greater than the inactive length.

Inactive Length—Length of the spring rendered inactive by the action of the U-bolts or clamping bolts. For metal-to-metal type spring seats, this length is usually assumed to be equal to the distance between the insides of the U-bolts, except for some curved seats where it is apt to be slightly shorter. For soft seats (using rubber type isolation, as in many passenger car installations) the inactive length may approach zero.

Seat Angle—(see Figs. 2.1 and 2.2.) Angle between the tangent to the center of the spring seat and the seat angle base line. When the spring is viewed with the fixed end of the spring to the left as shown, and the load is applied to the shortest leaf from above, the seat angle may be specified as either positive (counterclockwise) or negative (clockwise), depending upon the angular direction in which the tangent to the center of the spring seat is disposed from the seat angle base line.

Consequently, with the spring in normal vehicle position so that the load is applied to the shortest leaf from below as shown in Figs. 2.4, 2.6, 2.7, 2.8, and 2.9, and again with the fixed end of the spring to the left of the drawing, the seat angle is defined as positive when that tangent is disposed clockwise, and as negative when the tangent is disposed counterclockwise.

For suspension layout purposes, the seat angle is usually established with the main leaf *straight* (see *Straight Length*). In this position (in which the center of the spring seat lies on the seat angle base line), the suspension layout specifies the contour of the main leaf, namely either:

- flat, so that the seat angle is zero; or else
- with front and rear segments being approximately circular arcs tangent to each other at the spring seat ("S-shaped" main leaf). This is then the tangent of a seat angle defined as other-than-zero.

Tolerance is usually held within $\pm$ 0.5 deg, or as required for a particular application.

For production checking purposes it is sometimes convenient to deflect the spring to the position specified for load checking, and there measure the angle between the tangent to the spring seat and the datum line (instead of the seat angle base line). For correct angle evaluation, the following relations between this "checked angle" and the seat angle in the straight-main-leaf (or any other) position must be considered:

1. the two angles differ according to the distance by which the spring is deflected between any two positions, and according to the spring control (Φ, see Chapter 4).

2. The angles also differ according to the distance at each end between the seat angle base line and the datum line. With the spring inverted as shown in Fig. 2.1, the terminal point of the seat angle base line is:

    a) Higher by ID/2 than the datum line of the upturned eye.

    b) Lower by t/2 than the datum line of the Berlin eye.

    c) Lower by (t + ID/2) than the datum line of the downturned eye.

    d) Identical with the datum line when there are no eyes.

Finished Width—Width to which the spring leaves are ground or milled to give the edges a flat bearing surface.

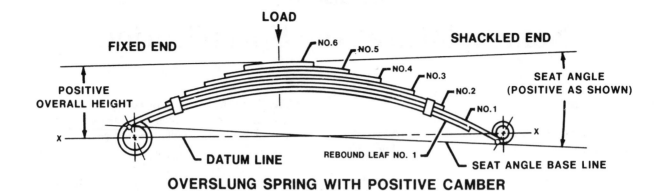

OVERSLUNG SPRING WITH POSITIVE CAMBER

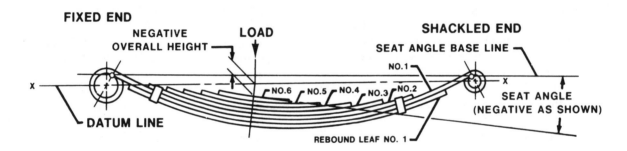

UNDERSLUNG SPRING WITH POSITIVE CAMBER

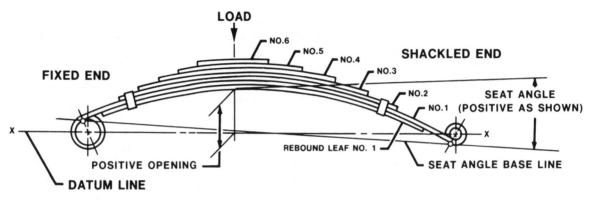

OVERSLUNG SPRING WITH NEGATIVE CAMBER

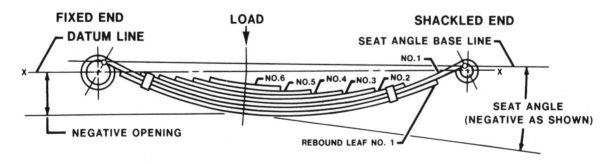

UNDERSLUNG SPRING WITH NEGATIVE CAMBER

Fig. 2.1—Measurement of opening, overall height, and seat angle

1.6

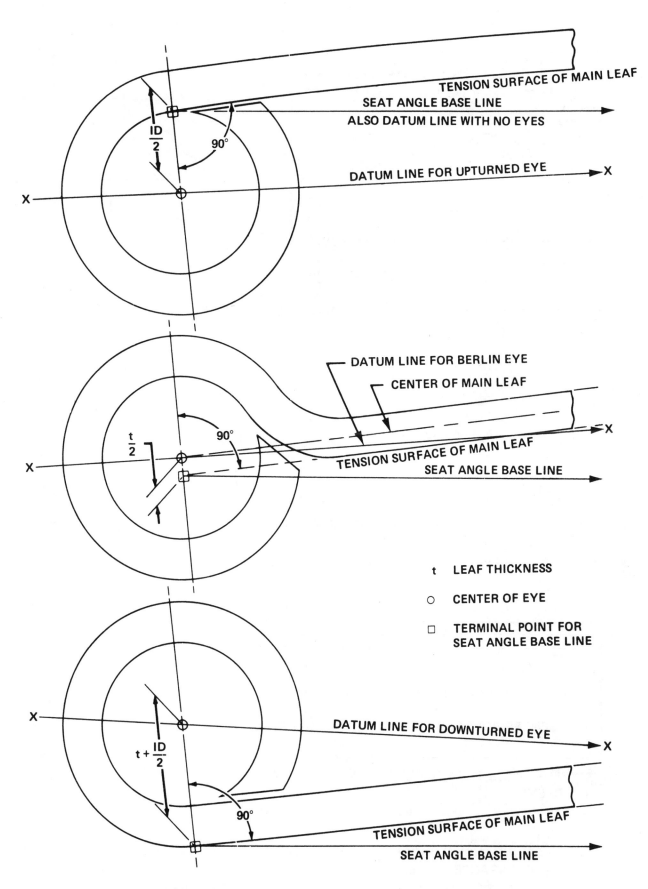

Fig. 2.2—Datum line and seat angle base line for upturned, Berlin, downturned, and no eyes

1.7

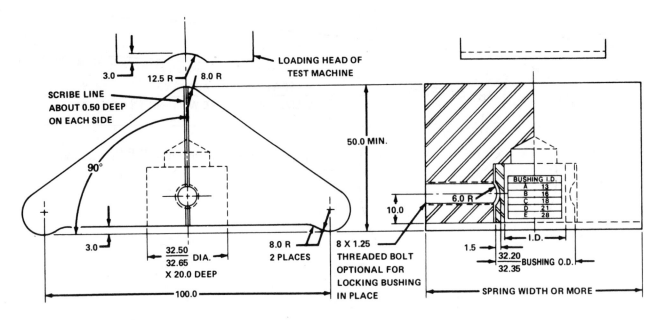

Fig. 2.3—Spring loading block

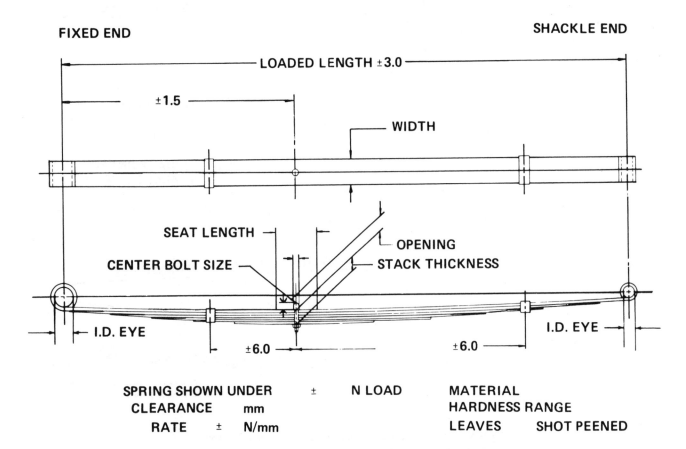

FIXED END

SHACKLE END

Fig. 2.4—Minimum specification requirements for underslung
springs with positive opening

1.8

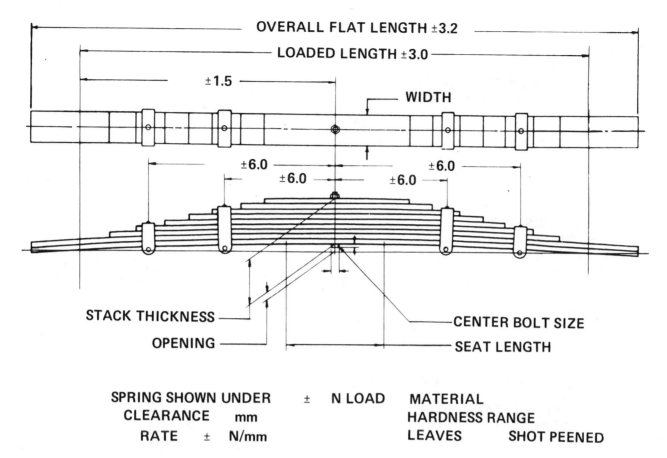

SPRING SHOWN UNDER ± **N LOAD** **MATERIAL**
CLEARANCE mm   **HARDNESS RANGE**
RATE ± **N/mm**   **LEAVES** **SHOT PEENED**

Fig. 2.5—Minimum specification requirements for springs with plain ends

If the spring ends have a finished width, the required length of the finished edge must also be indicated (see distance A, Fig. 2.6.) The usual tolerances for finished width are:

Leaf Width		Tolerance from
Over	To and Including	Nominal Width +0.00
0	50	−0.25
50	63	−0.35
63	150	−0.50

Assembled Spring Width—Where more than one leaf constitutes a spring assembly, the overall width tolerance of the assembly within the spring seat length shall be as follows:

Leaf Width		Tolerance from
Over	To and Including	Nominal Width −0.0
0	63	+2.5
63	100	+3.0
100	125	+3.5
125	150	+4.5

Stack Thickness—Aggregate of the nominal thicknesses of all leaves of the spring including any liners and spacer plates which are part of the spring at the seat.

Leaf Numbers—(see Fig. 2.1.) Leaves are designated by numbers, starting with the main leaf which is No. 1. The adjoining leaf is No. 2, and so on. If rebound leaves are used, the rebound leaf adjoining the main leaf is rebound leaf No. 1, the next one is rebound leaf No. 2, and so on. (Rebound leaves are assembled adjacent to the main leaf on the side opposite the load bearing leaves.) Helper springs are considered as separate units.

Opening And Overall Height—(see Fig. 2.1.) Distance from the datum line to the point where the center bolt centerline intersects the surface of the spring that is in contact with the spring seat.

If the surface in contact with the seat is on the main leaf or a rebound leaf (as on underslung springs), this distance is called "opening."

If the surface in contact with the seat is on the shortest leaf (as on overslung springs), this distance is called "overall height."

"Opening" and "overall height" may be positive or negative (see Fig. 2.1.) They are specified dimensions not subject to a tolerance (see paragraph on Load in this chapter.)

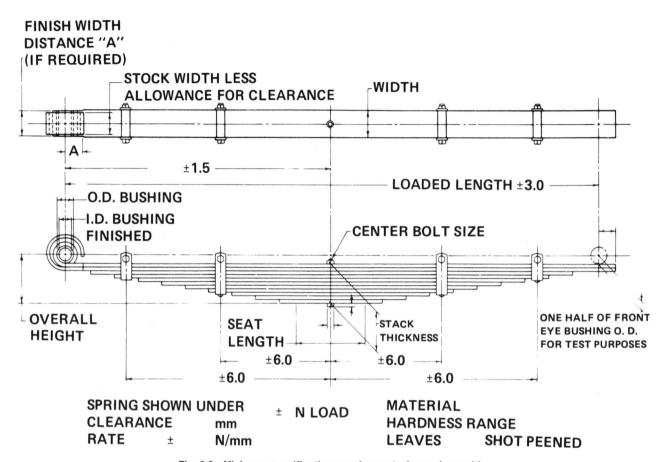

FINISH WIDTH
DISTANCE "A"
(IF REQUIRED)

STOCK WIDTH LESS
ALLOWANCE FOR CLEARANCE

WIDTH

A

±1.5

LOADED LENGTH ±3.0

O.D. BUSHING

I.D. BUSHING
FINISHED

CENTER BOLT SIZE

OVERALL
HEIGHT

SEAT
LENGTH

STACK
THICKNESS

±6.0

±6.0

ONE HALF OF FRONT
EYE BUSHING O. D.
FOR TEST PURPOSES

±6.0

±6.0

SPRING SHOWN UNDER   ±   N LOAD   MATERIAL
CLEARANCE   mm   HARDNESS RANGE
RATE   ±   N/mm   LEAVES   SHOT PEENED

Fig. 2.6—Minimum specification requirements for springs with
one eye and one plain end

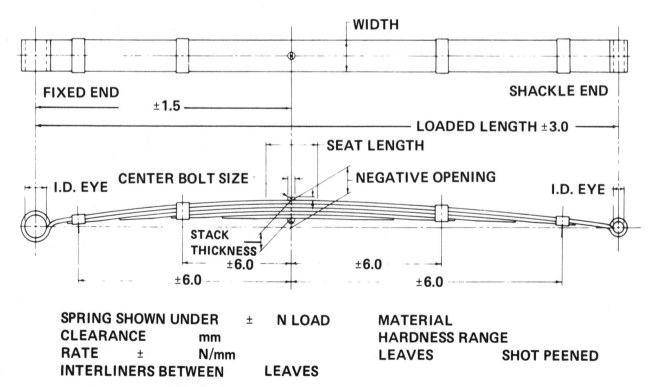

WIDTH

FIXED END

±1.5

SHACKLE END

LOADED LENGTH ±3.0

SEAT LENGTH

I.D. EYE

CENTER BOLT SIZE

NEGATIVE OPENING

I.D. EYE

STACK
THICKNESS

±6.0

±6.0

±6.0

±6.0

SPRING SHOWN UNDER   ±   N LOAD   MATERIAL
CLEARANCE   mm   HARDNESS RANGE
RATE   ±   N/mm   LEAVES   SHOT PEENED
INTERLINERS BETWEEN   LEAVES

Fig. 2.7—Minimum specification requirements for underslung
springs with negative opening

1.10

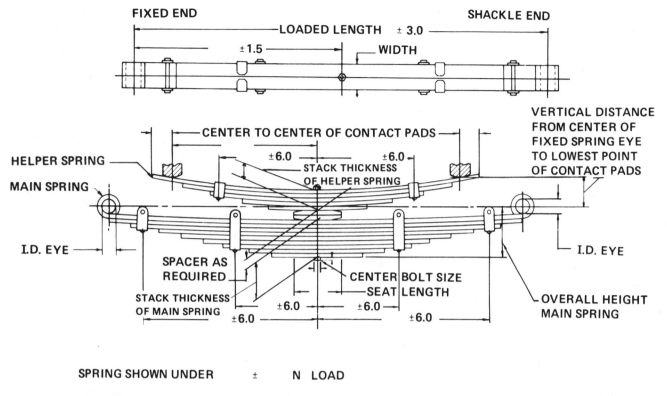

SPRING SHOWN UNDER   ±   N   LOAD

HELPER TO CONTACT AT   N   LOAD ON MAIN SPRING

CLEARANCE   mm   MATERIAL

RATE OF MAIN SPRING   ±   N/mm   HARDNESS RANGE

RATE OF HELPER SPRING   ±   N/mm   LEAVES   SHOT PEENED

Fig. 2.8—Minimum specification requirements for overslung commercial vehicle springs

Clearance—Difference in opening, or overall height, between the design load position and the extreme position (of maximum stress) to which the spring can be deflected on the vehicle.

Camber—Camber is not strictly defined and should therefore not be used in specifications, but it is sometimes convenient as a descriptive term. It is usually defined as the arc height of the main leaf. Camber is positive or negative analogous to opening, but this should not be confused with the fact that zero camber may be equivalent to either positive or negative opening, depending upon whether the spring has upturned or downturned eyes.

Curvature—Curvature (1/R) is the reciprocal of radius (R). The curvature of a flat leaf is zero. Curvature is considered positive in the direction in which it increases with added load. Positive curvature corresponds with negative camber.

Load and Rate—Terms which are usually employed to describe the basic characteristics of a leaf spring. As specified on the spring drawing, they refer to quantities measured on the spring without center clamp and without

shackles. They are not the same as those of the installed spring. If it is necessary to specify load and rate as clamped, this should be clearly shown on the drawing with full particulars of the clamp.

Load is the force in newtons (N) exerted by the spring at the specified opening or overall height. This force is greater during loading of the spring ("compression load") than during unloading of the spring ("release load"). The specified load shall be the average of the compression load and the release load. For practical reasons, load as well as rate shall be measured in terms of compression loads only, but the compression load in any position shall be read only after the spring has been thoroughly rapped in that position with a plastic or soft metal hammer.

The tolerance on load at the specified overall height or opening is usually expressed as a load range (N) which is equivalent to a deflection (mm) at the nominal rate (N/mm). This deflection may be as small as 6.0 mm for a passenger car spring and as large as 13.0 mm for a heavy truck spring.

Rate is the change of load per unit of spring deflection

(N/mm). For leaf springs it is determined as one fiftieth (2%) of the difference between the loads measured 25 mm above and 25 mm below the specified position, unless otherwise specified (see Fig. 2.9.) The tolerance is usually held within $\pm$ 5% on low rate springs and within $\pm$ 7% on high rate springs.

Measuring Methods—Instead of measuring loads at the specified position and 25 mm above and below, some users measure loads at more than three positions during compression and release and plot a complete load-deflection diagram with a friction loop, similar to Fig. 7.1. This method requires more time but provides additional information. Such a diagram is preferably obtained by loading and unloading continuously and recording the data with an X-Y plotter. Load and rate are then obtained from the diagram.

When the load is measured, the spring ends shall be free to move in the direction of the datum line. The ends are usually mounted on carriages with rollers. The spring shall be supported on its ends, and the load shall be applied from above to the shortest leaf.

It shall be transmitted from the testing machine head through a standard SAE loading block shown in Fig. 2.3. The loading block shall be centered over the center bolt with the legs of the V resting on the spring. It is understood that the load specified on the spring drawing does not include the force of gravity (usually called "weight" and equalling mass times acceleration of gravity) on either the spring or the loading block.

Just before the spring is checked for load or rate, it shall undergo a preloading operation. During the initial preloading by the spring maker, the spring shall be deflected at least to the position defined under the paragraph on *Clearance*. During any subsequent preloading, the spring shall be deflected only to and not beyond this "clearance position" in order to remove any temporary recovery from the set incurred during the initial preloading. After the spring has been preloaded, it shall be released to the free position before the load is applied for load and rate checking.

2. Specification Requirements

Minimum specification requirements are given in Figs. 2.4-2.9. They illustrate what information should be given to the spring maker for working out the detail design of the spring.

The spring shown in Fig. 2.4 is designed for underslung mounting. Therefore, the center bolt head is on the main leaf side and the height is dimensioned by specifying the opening. On an overslung spring, the center bolt head would be on the opposite side and the overall height would be specified.

The type of spring shown in Fig. 2.5 is often used on truck rear suspensions mounted in the position shown and with the center bolt head located as shown. In this case the opening should be specified.

Fig. 2.6 shows a spring which has a main leaf constructed with an eye at one end and the other end plain; a construction frequently used for truck suspensions.

Fig. 2.7 shows the details of an underslung rear spring designed with considerable negative opening.

The combination of main spring and helper spring shown in Fig. 2.8 is frequently used for truck rear suspensions and is mounted as an overslung spring.

Fig. 2.9 shows an overslung variable rate spring of the multistage type. The graph indicates the method of measuring rates of such springs, where rate (1) is usually measured at "curb load" (that is, at the load on the spring which is due to the mass of the vehicle without any payload), while rate (2) is measured at "design load" (that is, at the load on the spring which is due to the mass of the vehicle plus the payload).

3. Spring Eye Tolerances

Spring Eyes and Bushings

For round eyes with specified inside diameter, the size and roundness of the eye should be checked by means of a round plug gage from which two opposite segments of 60 deg have been removed. The gage shall be tapered by 0.05 mm in diameter per 25.0 mm of length (see Fig. 2.10.)

The gage shall be inserted into the eye three times from each side at angular positions differing by about 60 deg. The eye is acceptable only if the gage reading on the side of the eye from which the gage is inserted is within the specified diametral limits at each of the six checks.

Also, the round eye should be checked with a round plug, GO/NO-GO gage, to determine if the eye is cone shaped or tapered. The GO diameter must pass completely through the eye, and the NO-GO diameter must not enter the eye from either side.

The total tolerance shall be 1% of the nominal diameter of the eye, except for large diameter eyes (40 mm or more), where bushing retention may require a smaller tolerance of 0.75% of the nominal eye diameter. For eye diameters of less than 25 mm, the minimum tolerance is 0.25 mm.

Where the ID of a bushing may have been affected by pressing into the spring eye, it should be checked with a round plug gage. Total tolerance is to be 0.13 mm unless otherwise specified.

Oval eyes (see Fig. 3.3H) consist of two half circle ends joined by flat sections. One method of checking their sizes is by using a GO/NO-GO plug gage system. This consists of:

1. An oblong GO gage to the minimum inside dimensions.

2. An oblong NO-GO gage to the maximum inside dimensions for the half circle ends only, with the flat sides of the gage undercut.

3. A rectangular NO-GO gage for the inside dimensions between the flat sides only.

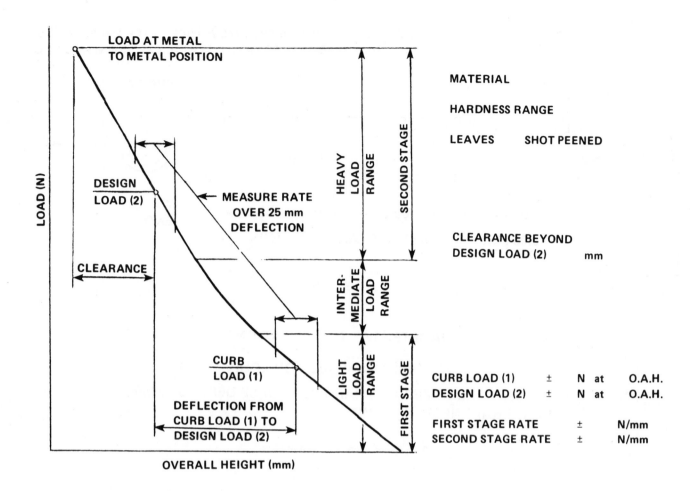

LOAD AT METAL
TO METAL POSITION

LOAD (N)

DESIGN
LOAD (2)

MEASURE RATE
OVER 25 mm
DEFLECTION

CLEARANCE

CURB
LOAD (1)

DEFLECTION FROM
CURB LOAD (1) TO
DESIGN LOAD (2)

OVERALL HEIGHT (mm)

HEAVY LOAD RANGE

SECOND STAGE

INTER-MEDIATE LOAD RANGE

LIGHT LOAD RANGE

FIRST STAGE

MATERIAL

HARDNESS RANGE

LEAVES   SHOT PEENED

CLEARANCE BEYOND
DESIGN LOAD (2)   mm

CURB LOAD (1)	±	N at	O.A.H.
DESIGN LOAD (2)	±	N at	O.A.H.
FIRST STAGE RATE	±	N/mm	
SECOND STAGE RATE	±	N/mm	

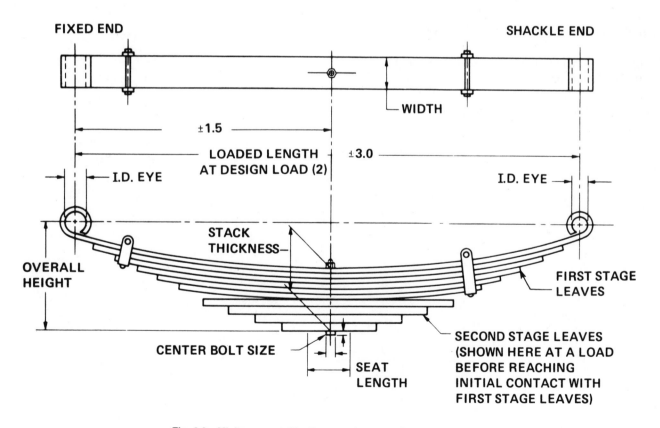

FIXED END

SHACKLE END

WIDTH

±1.5

±3.0

LOADED LENGTH
AT DESIGN LOAD (2)

I.D. EYE

I.D. EYE

OVERALL
HEIGHT

STACK
THICKNESS

FIRST STAGE
LEAVES

CENTER BOLT SIZE

SEAT
LENGTH

SECOND STAGE LEAVES
(SHOWN HERE AT A LOAD
BEFORE REACHING
INITIAL CONTACT WITH
FIRST STAGE LEAVES)

**Fig. 2.9—Minimum specification requirements for variable rate
or progressive rate springs (overslung type shown)**

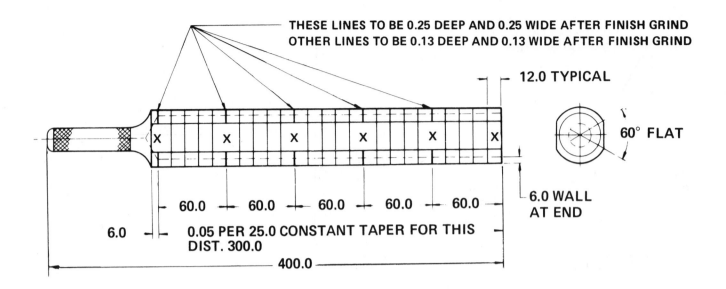

THESE LINES TO BE 0.25 DEEP AND 0.25 WIDE AFTER FINISH GRIND
OTHER LINES TO BE 0.13 DEEP AND 0.13 WIDE AFTER FINISH GRIND

12.0 TYPICAL

60° FLAT

6.0 WALL AT END

X X X X X X

60.0 — 60.0 — 60.0 — 60.0 — 60.0

6.0 — 0.05 PER 25.0 CONSTANT TAPER FOR THIS DIST. 300.0

400.0

X-STAMP GAGE DIAMETERS AT THESE STATIONS

MATERIAL: STEEL - G40270 (SAE 4027) OR EQUIVALENT

PROCESS: CARBURIZED AND HARDENED; CASE DEPTH 0.50 MIN.
SURFACE HARDNESS: R_c 58 MIN.

Fig. 2.10—Gage—leaf spring eye plug

Parallelism and Squareness of Spring Eyes

Eyes of the main leaf in the assembled spring, measured in the unloaded condition, shall be parallel to the surface at the spring seat, and square with a tangent to either edge of the main leaf at the spring seat, within $\pm$ 1 deg.

Chapter 3
Design Elements

1. Leaf Sections[1]

For automotive springs, round edge flat steel was adopted as the SAE standard in 1938. The bars shall be of flat rolled steel having two flat surfaces and two rounded (convex) edges. The cross section tolerances permit the two flat surfaces to be slightly concave. When that occurs, the radii of the arcs of the two concave surfaces shall be of approximately equal length.

The rounding of the convex edges shall be an arc with a radius of curvature that may vary from 65–85% of the thickness of the bar.

Bars shall be substantially straight and free from physical characteristics known as "kinks" or "twists" which render them unsatisfactory for spring manufacturing purposes.

Distortions due to a bar being bent about either major axis of section shall be measured with the bar against a flat checking surface so as to make contact with this surface near both bar ends. Gaps between the bar and the checking surface shall not exceed 4.0 mm/1 m of bar length out of contact with the checking surface when this bar length is greater than 1 m. Also, a gap between the bar and a straight edge 1 m long applied along any portion of the surface or edge of the bar shall not exceed 4.0 mm.

It is recommended that all leaf spring bars which have been cold straightened be identified by the steel mill so that the spring manufacturer can use them selectively.

The bar sections, which are generally provided in alloy steel, shall be specified and rolled in the widths and thicknesses shown in Table 3.1. These sections are subject to the tolerances given in Table 3.2.

TABLE 3.1 (mm)

Widths		Thicknesses					
40.0	75.0	5.00	7.10	10.00	14.00	20.00	28.00
45.0	90.0	5.30	7.50	10.60	15.00	21.20	30.00
50.0	100.0	5.60	8.00	11.20	16.00	22.40	31.50
56.0	125.0	6.00	8.50	11.80	17.00	23.60	33.50
63.0	150.0	6.30	9.00	12.50	18.00	25.00	35.50
		6.70	9.50	13.20	19.00	26.50	37.50

It should be noted that all the widths and thicknesses are "Preferred Numbers" in accordance with American National Standard ANSI Z17.1.

Tables showing the mass per meter length and the actual moment of inertia for each size of these bars are provided in Chapter 5.

TABLE 3.2—CROSS SECTION TOLERANCES (mm)

Width	Width Tolerance	Tolerance In Thickness ($+$)[a] And In Flatness ($-$)[b]			Maximum Difference In Thickness[c]		
		for Thickness			for Thickness		
	Minus 0.00	5.00 to 9.50	10.00 to 21.20	22.40 to 37.50	5.00 to 9.50	10.00 to 21.20	22.40 to 37.50
40.0	+0.75	0.13	0.15	—	0.05	0.05	—
45.0	+0.75	0.13	0.15	—	0.05	0.05	—
50.0	+0.75	0.13	0.15	—	0.05	0.05	—
56.0	+0.75	0.13	0.15	—	0.05	0.05	—
63.0	+0.75	0.13	0.15	—	0.05	0.05	—
75.0	+1.15	0.15	0.20	0.30	0.08	0.10	0.15
90.0	+1.15	0.15	0.20	0.30	0.08	0.10	0.15
100.0	+1.15	0.15	0.20	0.30	0.08	0.10	0.15
125.0	+1.65	0.18	0.25	0.40	0.10	0.13	0.20
150.0	+2.30	—	0.30	0.50	—	0.15	0.25

[a]Thickness measurements shall be taken at the edge of the bar where the flat surfaces intersect the rounded edge.
[b]This tolerance represents the maximum amount by which the thickness at the center of the bar may be less than the thickness at the edges. Thickness at the center may never exceed the thickness at the edges.
[c]Maximum difference in thickness between the two edges of each bar.

It is well known that fatigue failures in spring leaves usually start on the tension side of the leaf. Taking advantage of this fact, special sections shown in Fig 3.1. have been developed which place the neutral axis nearer to the

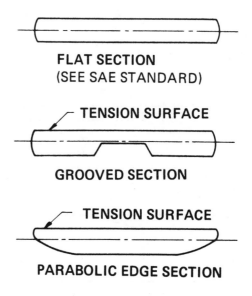

FLAT SECTION
(SEE SAE STANDARD)

TENSION SURFACE

GROOVED SECTION

TENSION SURFACE

PARABOLIC EDGE SECTION

Fig. 3.1—Sections of spring steel

[1]Ref. SAE J1123.

tension surfaces. Springs made of such sections are 5–10% lower in mass than those made of conventional section.

When grooved sections are used, special precautions should be taken to prevent corrosion caused by the moisture which tends to become trapped within the grooves. It is suggested that spring users interested in these special sections contact the manufacturers who produce such springs.

2. Leaf Ends

Square End (Blunt End) (Fig. 3.2A)

This is the cheapest end to produce but is often unsatisfactory. It causes concentration of interleaf pressure, resulting in more friction and galling than tapered ends. It is a very poor approximation of the theoretical triangular leaf uniform stress spring, and is therefore heavier than necessary.

Diamond Point (Spear End) (Fig. 3.2B)

This end makes a better approximation of the uniform stress spring by omitting excess material. The pressure distribution between leaves is slightly improved.

Tapered End (Fig. 3.2C)

This end can be formed to approximate very closely the ideal uniform stress shape. The plan view contour is con-

trolled by trimming or edge squeezing as part of the tapering operation. Due to the flexibility of the leaf end, the pressure distribution in the bearing area is improved and interleaf friction is generally reduced.

Tapered And Trimmed End (Fig. 3.2D)

This end is similar to Fig 3.2C, except that the plan view contour is controlled by trimming after the tapering operation, and thus has the added advantage of the maximum obtainable area of contact.

3. Spring Eyes and Spring Ends

Upturned Eye (Fig. 3.3A)

This construction is most commonly used. If required, the second leaf can be extended to give support to the eye.

Military Wrapper (Fig. 3.3B)

In this design no attempt is made to use the second leaf wrapper as an eye under design loads; but it may come into action on rebound and thus assist the main leaf. It also provides an emergency support if the main leaf breaks. The design has been widely used on military vehicles and trucks where the service is severe.

A. END SQUARE AS SHEARED

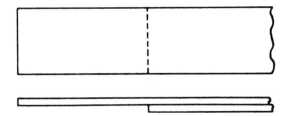

C. END TAPERED

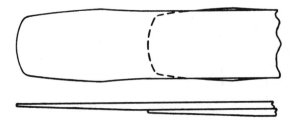

B. END TRIMMED WITH DIAMOND POINT

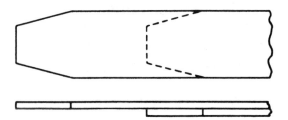

D. END TAPERED, THEN TRIMMED
BURRS AWAY FROM BEARING SURFACE

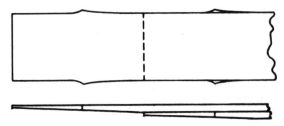

Fig. 3.2—Leaf ends

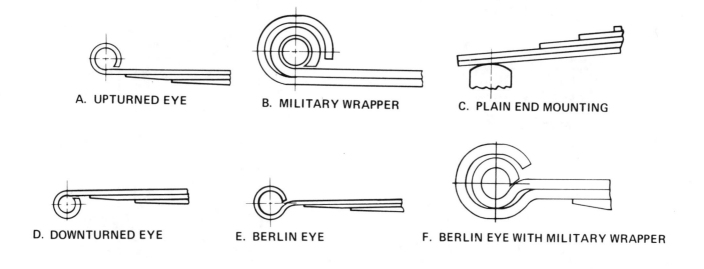

Fig. 3.3—Spring eyes and spring ends

Plain End Mounting (Fig. 3.3C)

This construction can be built as a flat leaf on a curved spring pad, or as a curved leaf on either a flat or a curved spring pad. The leaf ends used with the rubber insulators shown in Fig. 3.11 are similar, but are sometimes provided with a T end or a circular hole for the transmission of lengthwise forces.

Downturned Eye (Fig. 3.3D)

This is sometimes used because it produces a desired spring geometry (or suspension motion) which may improve steering or axle control. If support to the eye by the second leaf is required, this construction is not recommended.

Berlin Eye (Fig. 3.3E)

Longitudinal loads are applied centrally to the main leaf, thereby reducing the tendency of the eye to unwrap.

Berlin Eye With Military Wrapper (Fig. 3.3F)

This construction is a variation of Figure 3.3B.

Welded Eye (Fig. 3.3G)

This construction is used predominantly in applications such as torque rods where the horizontal force is high. The welding must be performed before heat treatment, using appropriate technique.

Oval Eye (Fig. 3.3H)

This eye construction permits the use of rubber bushings which have different rates in the vertical and horizontal directions. This eye was developed specifically to reduce the magnitude of the horizontal force inputs in suspension applications.

4. Spring Eye Bearings

Threaded Bushings (Figs. 3.4, 3.5, and 3.6)

This type of construction has the following advantages: It takes side thrust as well as vertical load, retains lubricant, and excludes dirt better than a plain bushing, thus requiring less frequent lubrication. Spring eyes need not be finished in width.

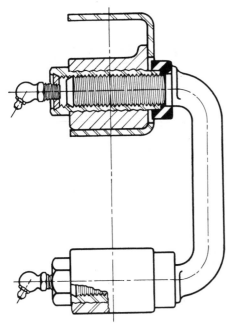

Fig. 3.4—Spring eye bearing: Threaded bushings in one piece C shackle

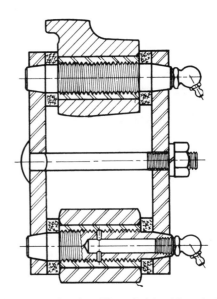

Fig. 3.6—Spring eye bearing: Threaded bushings in taper pin shackle

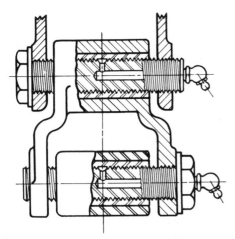

Fig. 3.5—Spring eye bearing: Threaded bushings and pins in one piece Y shackle

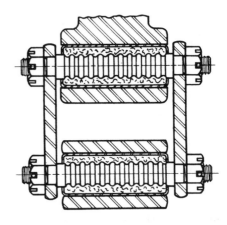

Fig. 3.7—Spring eye bearing: Self-lubricated bushings in double bolted shackle

The bushing has a 60 deg V thread on the inside which fits the pin loosely. The outside is either plain or provided with a very flat angle thread. It is forced into the spring eye or bracket.

The bushings and pins are made of carbon or alloy steel, carburized and hardened.

Thread sizes generally used are M14×2 to M36×4. Load pressures up to 7.00 MPa on the projected area at normal load are used. Figs. 3.4 and 3.6 show seals.

Self-Lubricated Bushings (Fig. 3.7)

Various designs and materials have been introduced on passenger cars and light trucks. They do not require lu-

brication and are noiseless. The design shown in Fig. 3.7 takes side thrust on rounded circular grooves and ridges. Spring eyes are not finished in width. Some types will stand pressures up to 8.40 MPa on the projected area at normal load.

Plain Bushings (Fig. 3.8)

This type bearing, usually bronze, is used on heavy trucks. It is simple to manufacture and service, and will give satisfactory life if it is regularly lubricated. Side thrust is taken on the finished faces of the spring eyes. The wall thickness is usually 3.0 mm. Load pressures used are between 3.50–7.00 MPa on the projected area at normal load.

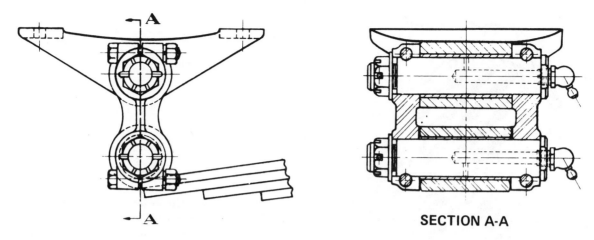

SECTION A-A

Fig. 3.8—Spring eye bearing: Plain bushings, periodically lubricated, in double bolted shackle

Rubber Bushings (Figs. 3.9 and 3.10)

Various types are used successfully. Their flexibility is an added insulation against noise, but the effect of the flexibility on road holding ability, steering control, and increase in spring rate must be considered.

Rubber Cushion ("Shock Insulator") (Fig. 3.11)

This bearing is used on heavy vehicles. The design permits a limited amount of longitudinal motion of the spring ends. It is, therefore, successful only with fairly long springs which are approximately flat at design load.

One Piece Y-Shackle (Fig. 3.5)

Has more load capacity than one piece C-shackle.

Taper Pin Shackle (Fig. 3.6)

Double Bolted Shackle (Figs. 3.7 and 3.8)

These constructions have been used where they must resist forces transverse to the spring (that is, in the direction of the spring eye axis). Careful design with close fits must be used to avoid loosening in service.

5. Shackles

One Piece C-Shackle (Fig. 3.4)

Used on passenger cars and light trucks.

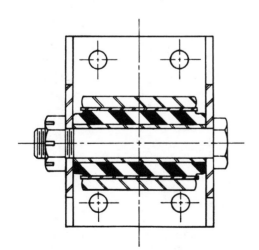

Fig. 3.9—Spring eye bearing: Rubber bushing in fixed eye pivot

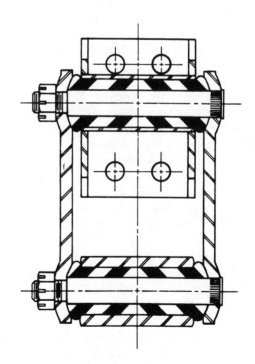

Fig. 3.10—Spring eye bearing: Rubber bushings in shackle

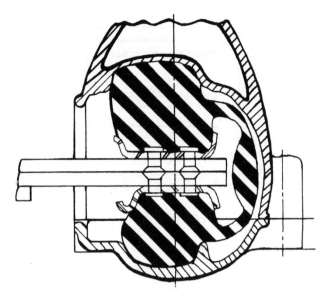

Fig. 3.11—Spring end bearing: Rubber cushion ("Shock insulator")

Riveted Bolted Shackle (Fig. 3.10)

Well suited to rubber bushings, and therefore, in general use on passenger cars and light trucks.

6. Center Bolt And Cup Center

The center bolt is required to hold the spring leaves together, and the center bolt head is used as a locating dowel during installation to the vehicle. For underslung springs, the head should be adjacent to the main leaf; and for overslung springs, the head should be adjacent to the short leaf.

In most cases, center bolts are highly stressed in the handling of the spring and in service. Therefore, it is necessary to use bolts and nuts of high mechanical properties.

The diameter of the center bolt hole in the spring leaves should be at least equal to the thickness of the heaviest leaf in order to permit cold punching. If the diameter of the center bolt hole should be less than the thickness of the leaf, it may require heating the leaf in the area to be punched. However, it is not recommended to cold punch leaves which are thicker than 14 mm. (See Table 3.3 for sizes.)

Generally, the spring leaf material at the center bolt area is inoperative when assembled to the vehicle. However, the diameter of the center bolt hole should not be too large in relation to the width of the leaf.

When it is not desirable to use a center bolt and hole in the spring leaves, a nib or cup, for nesting the adjacent leaves, is forged from the leaf material itself by forming a depression on one side and a corresponding projection on

TABLE 3.3—RECOMMENDED CENTER BOLT AND NUT DIMENSIONS (mm)

Nominal Bolt Diameter	Threads		Bolt Head Size		Nut Size Style 1		
	Pitch	Minimum Length	Diameter	Height	Width Across Flats (Max.)	Width Across Corners (Max.)	Thickness (Max.)
8	1.25	25	12.00	6.00	13.00	15.01	6.6
10	1.5	25	15.00	7.00	15.00	17.32	9.0
12	1.75	30	17.00	8.00	18.00	20.78	10.7
16	2	35	20.00	10.00	24.00	27.71	14.5

the other side. The leaves are then held together with clamps.

Cup centers are often used in heavy duty springs which may not safely depend on clamps and center bolts to prevent shifting of the spring on the axle seat due to driving and braking forces (See Fig. 3.14.)

When the main leaf is assembled adjacent to the axle seat as in underslung springs, the cup is hot forged in the main leaf only (away from the No. 2 leaf). When the shortest leaf is mounted above the axle seat as in overslung springs, all the leaves must be cupped toward the shortest leaf.

This method of cupping locks the main leaf to the axle seat. The horizontal forces which are applied to the main leaf will be resisted by the cup rather than the clamp and the center bolt.

There are many types of cup centers in general use, one of which is shown in Fig. 3.12. The cup dimensions are listed according to center bolt diameter; however, the cup diameter should not exceed one-half the leaf width and the cup depth should not exceed one-half the leaf thickness.

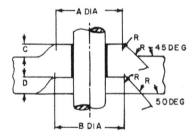

Fig. 3.12—Dimensioning of typical cup center

Dimension (mm)	Tolerance + 0.0	For Use With Centerbolt Diameters	
		10, 12	16
Diameter A	− 0.5	21.3	31.5
Diameter B	− 0.5	22.4	33.0
Height C	− 0.5	3.6	5.1
Depth D	− 0.5	4.6	6.1
Radius R	− 0.3	2.5	3.0

7. Center Clamp

The center clamp provides the permanent tie between the leaves, and between the spring and the spring seat. Figs. 3.13, 3.14, and 3.15 show some typical designs.

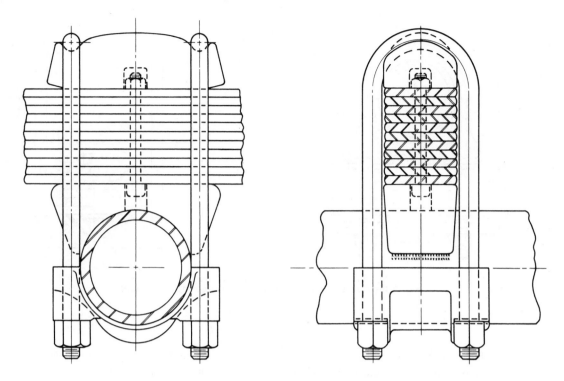

Fig. 3.13—Typical center clamping of overslung spring

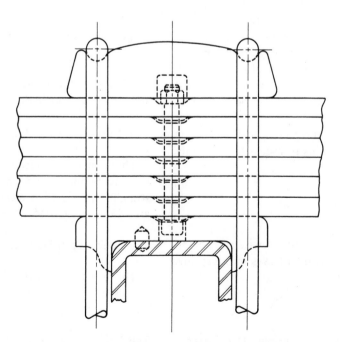

Fig. 3.14—Center bolt assisted by cup centers

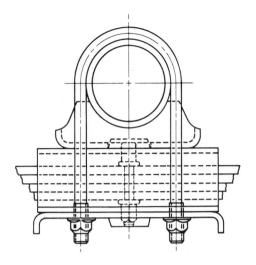 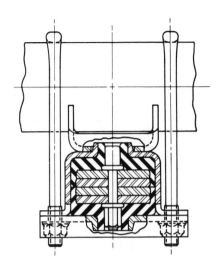

Fig. 3.15—Center clamping of underslung spring with rubber pads

The functions of the clamp are to attach the spring firmly to its seat to prevent leaf breakage through the center bolt section, and to prevent center bolt breakage due to horizontal forces. It therefore must remain tight in service.

Excessive clamp length reduces the active length of the spring and wastes metal. The clamp length is usually between 8–15% of the spring length. The ends of steel clamps should be well rounded to avoid sharp edges in contact with the spring leaves.

Clamps with rubber pads (Fig. 3.15) are frequently used on passenger car rear axles because of their important influence on reduction of noise transmission.

Similar to the rubber shackle bushings mentioned in Section 4 of this chapter, considerations of road holding, axle control, and steering control set a limit to the amount of softness which it is permissible to use at the center clamp.

The center clamp has an effect on load rate and on loaded height of the spring. This is discussed in Chapter 5, Section 2.

8. Alignment Clips

Alignment clips are used to limit sidewise spread and vertical separation of the individual leaves in the spring.

Bolt Clip (Fig. 3.16A)

This clip is used for most heavy springs. The clearance between the bolt and the main leaf must be sufficient to permit the main leaf to twist longitudinally so that this twist will not be concentrated in the free ends near the eyes. A spacer tube is recommended to prevent the sides

of the clip from binding the main leaf. For heavy duty applications, a double rivet construction may be used in springs 100 mm wide and over. Material is hot rolled steel strip of the following sizes: 4.5×20, 6.0×25, 6.0×30, 8.0×35, 10×40.

Clinch Clip (Fig. 3.16B)

This is used on springs where the clearances are limited. The material is usually hot rolled steel strip 4.5×20 and 6.0×25 size.

Single Piece Box Clip (Fig. 3.16C)

This clip is manufactured from hot rolled steel strip 2.5×25. It is used with and without a rubber liner. Bolt sizes are $M8 \times 1.25$ and $M10 \times 1.5$.

Two Piece Box Clip (Fig. 3.16D)

This clip is also made from 2.5×25 hot rolled steel strip. Note that the sides of this clip, as in Fig 3.16C, are straight and provide a clearance for all leaves. Bolt sizes are $M8 \times 1.25$ and $M10 \times 1.5$.

Tab Lock Clip (Fig. 3.16E)

This clip is also made of 2.5×25 hot rolled steel strip. It is used with and without a rubber liner on the main leaf only.

Tab Lock Clip With Locating Tang (Fig. 3.16F)

This clip is similar to Fig. 3.16E except that it has a tang for retaining it on the spring leaf.

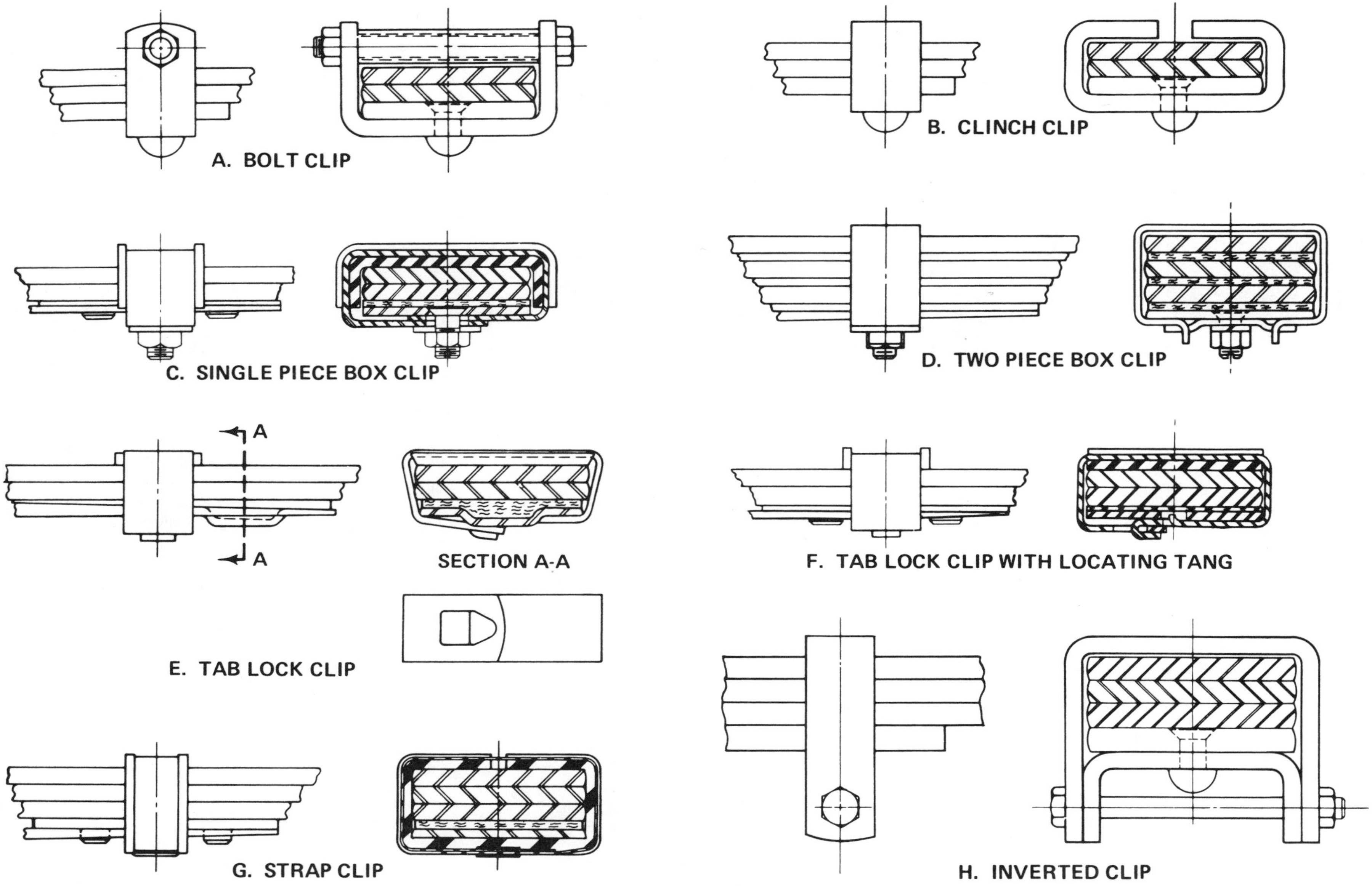

B. CLINCH CLIP

D. TWO PIECE BOX CLIP

F. TAB LOCK CLIP WITH LOCATING TANG

H. INVERTED CLIP

A. BOLT CLIP

C. SINGLE PIECE BOX CLIP

SECTION A-A

E. TAB LOCK CLIP

G. STRAP CLIP

Fig. 3.16—Alignment clips

1.23

Strap Clip (Fig. 3.16G)

This clip has a rubber liner on all four sides of the spring. The strap is 0.5 × 16 stainless steel. The ends are overlapped and secured by means of a fastener which is crimped at assembly.

Inverted Clip (Fig. 3.16H)

This clip is used on heavy springs where there is a clearance problem. It is usually made from 30 or 35 × 8.0 stock.

9. Rebound Leaves

Figure 2.1 shows an example of a spring with rebound leaf. There are two principal objectives for the application of one or several rebound leaves.

One of these concerns the spring subject to very high and frequent windup loads which tend to distort the main leaf by separating it from the shorter leaves. The rebound leaf or leaves serving to prevent these excessive windup stresses in the main leaf act substantially like the other leaves and should be treated in the same manner as to length, free curvature, etc. They are loaded through the alignment clips, which must be properly placed and designed so as to maintain tip contact between rebound leaves and main leaf.

The other case is that of the truck spring which in extreme rebound will have to support the weight of the axle and of other unsprung components, thereby becoming subject to detrimental reverse bending stresses, partic-

ularly in the main leaf. In this case the rebound leaves are usually formed with less free curvature than the other leaves. When the leaves are bolted together in the spring assembly operation, assembly stresses are set up in the various leaves including rebound leaves, as described in Chapter 5, Section 4. The rebound leaf will protect the main leaf in the region between the eye and the nearest alignment clip from distortion in extreme rebound.

As long as rebound leaves are under load, they contribute to the load rate in the same manner as other leaves, regardless of the free camber in the rebound leaves.

10. Variable Rate Springs

Variable rate springs are used primarily on vehicles which operate with large variations in load, such as trucks and buses. Variable spring rates are generally required to provide desirable ride and handling characteristics under these conditions. There are several ways to obtain variable rates, some of which may be combined with others.

The helper spring is one method of obtaining increased rate with deflection. As shown in Fig. 2.8, the helper is mounted above the main spring and has its own bearing pads. The helper spring does not support any load until contact is made with the bearing pads. The change in rate at contact is necessarily abrupt.

Shackles may be used to obtain some variation in rate as described in Chapter 6, Section 1.

Curved bearing pads or cams which shorten the effective length of the spring as it is deflected will provide a variable rate. Such a configuration is shown in Fig. 3.17A.

Another method to obtain variable rate is by means of

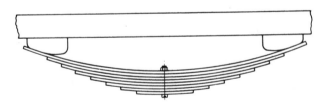

A. VARIABLE EFFECTIVE LENGTH SPRING

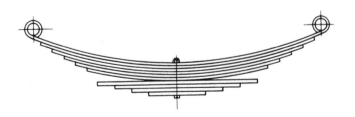

B. MULTI-STAGE SPRING

Fig. 3.17—Variable rate springs

the multi-stage spring, shown in Fig. 2.9, and Fig. 3.17B. This spring has one or more leaves called "second stage" leaves, mounted adjacent to the shortest leaf of the main or "first stage" portion of the spring. This spring gradually increases in rate with deflection as the contact between the stages increases. Load and rate for each stage are usually specified as shown in Fig. 2.9. They are generally checked in the same manner as single stage multi-leaf springs.

Combinations, such as the use of curved bearing pads in conjunction with a multi-stage spring, are sometimes used to provide a greater change of spring rate.

Chapter 4
Geometry

1. Deflection Theory

As a spring with leaves of constant cross section properly stepped to approach the condition of uniform strength is deflected, it will assume the shape of a circular arc at all loads between zero and maximum, provided it has a circular arc shape or is flat at no load or at any given load.

Most springs approximate these conditions closely enough so that the circular arc shape can be used to calculate their geometric properties. The following relations have been derived analytically and found to agree closely with a number of actual springs checked. (However, see Chapter 10, Section 6 concerning the contour of single leaf springs in the free camber.)

2. Cantilever Spring

For a spring of this type the center of the eye of the Berlin type moves in a path with a radius of 0.75ℓ central to the main leaf, as shown in Fig. 4.1A. If the eye center is offset the distance "e" from the center of the main leaf, the center of arc will be offset by $0.5e$ in the opposite direction, as shown in Figs. 4.1B and 4.1C. This construc-

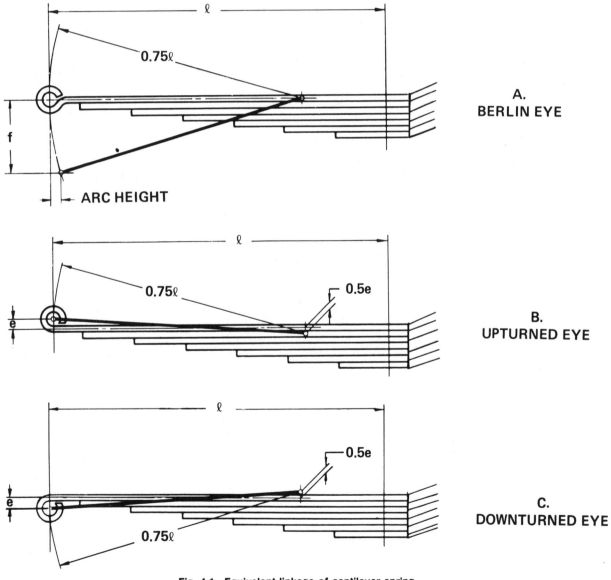

Fig. 4.1—Equivalent linkage of cantilever spring

tion reproduces the change of arc height with an accuracy of 1% up to deflections $f = 0.6\ell$.

3. Semi-Elliptic Spring

This type of spring can be considered as two cantilever springs, and the resulting spring action can be determined by considering the spring as a three-link mechanism, as shown in Figs. 4.2 and 4.3. These layouts can be drawn even if the spring is unsymmetrical and if the eye offsets are opposed. The three-link equivalent layouts are useful in determining the geometry of spring action, including the path of the axle attached to the spring seat, and the axle control which is defined as the seat angle change in degree per millimeter of deflection. They also permit establishing the axle path and control corrected for the shackle effects, as explained later.

Two different methods for determining the spring geometry are described, and their constructions are shown in Figs. 4.2 and 4.3. These methods may be used whether the spring is a conventional or unconventional design. A conventional design has the same number of leaves and leaf spacing of uniform strength in both cantilevers, and their rates are inversely proportional to their lengths cubed. If the leaf spacing of either cantilever is not of uniform strength or if the number of leaves of the two cantilevers differ, or both, the design is called unconventional, and the cantilever deflections no longer bear a simple relationship to the cantilever lengths (see formula B in Table 4.1). The unconventional design has certain advantages in geometry, particularly when specific axle control requirements are to be met with the predetermined front and rear cantilever lengths. However, the unconventional design may engender a loss in efficiency.

4. Center Link Extension Method (Fig. 4.2)

The basic principle used in this construction is that every extension of the center link for any position of linkage will intersect at a common point. The center link tilts in such a manner that its extension always passes through the common point O, although the center link does not rotate about this point. In further explanation, it might be said that the foregoing extension to the center link would reciprocate back and forth through an imaginary slot located at point O as the center link travels from its positions of compression to rebound.

In making a layout by the Center Link Extension Method, the position of the center link DE for different values of spring deflection establishes the axle control. A succession of triangles DEH establishes the axle path. The three-link equivalent layout is most conveniently made by starting from the position of the spring where the main leaf is flat. In this position the three links do not lie in a straight line except when both eyes are of the Berlin type.

The linkage is dependent upon the eye offset and the spring unsymmetry. The motion of the center link for different deflections of point M depends upon:

1. Path of point D at radius R_a.
2. Path of point M at radius R_M.
3. Position of center link DE so that its extension intersects point O.
4. Correction for shackle effects, if required.

Procedure

1. Start layout with main leaf in flat position with lengths a, b, and L measured along the main leaf and axle center H at distance h from center of main leaf. Axle center is above main leaf in an underslung spring, below main leaf in an overslung spring.
2. Establish lengths m and n, which represent the inactive material. These are considered equal for most springs. They can be neglected in relatively long flexible springs without serious error.
3. Draw arc R_a and at the intersection with $0.5e_a$ locate point D.
4. Draw arc R_b and at the intersection with $0.5e_b$ locate point E.
5. Construct the three links AD, DE, and EB.
6. Locate point M at the intersection of centerline of center bolt and link DE.
7. Locate point O on extension of center link DE at computed distance Q from point M (see formulae in Table 4.1).
8. Draw arc R_M where $R_M = \lambda \cdot L$. Its center is located on extension of line OA.
9. For a given deflection into rebound or compression, new position of center link DE is established by locating point M_r or M_c and then drawing a line through point M_r or M_c and point O.
Note: An alternate, and frequently a more convenient method is to work with an overlay drawing to determine the new position of center link DE by locating points D_r/D_c or M_r/M_c for given deflections in rebound and in compression so that extension of center link DE intersects at point O.
10. For each position of the center link DE, the axle position can be located by constructing the triangle DEH. When three or more such positions have been located, the approximate radius R_H of the axle can be established by geometric construction.
11. The "control" or "tilt" of the center link, and thus of the spring seat, is the change which the angle θ undergoes during a displacement x. It is known as both "spring control" and "axle control" and is labeled $\Phi = \theta/x$ (deg/mm).
12. In the symmetrical spring the control is zero, with the center link moving parallel to itself throughout the compression and rebound range. Actually, however, the center link undergoes a small angular change due to the vertical displacement of the shackled spring eye.

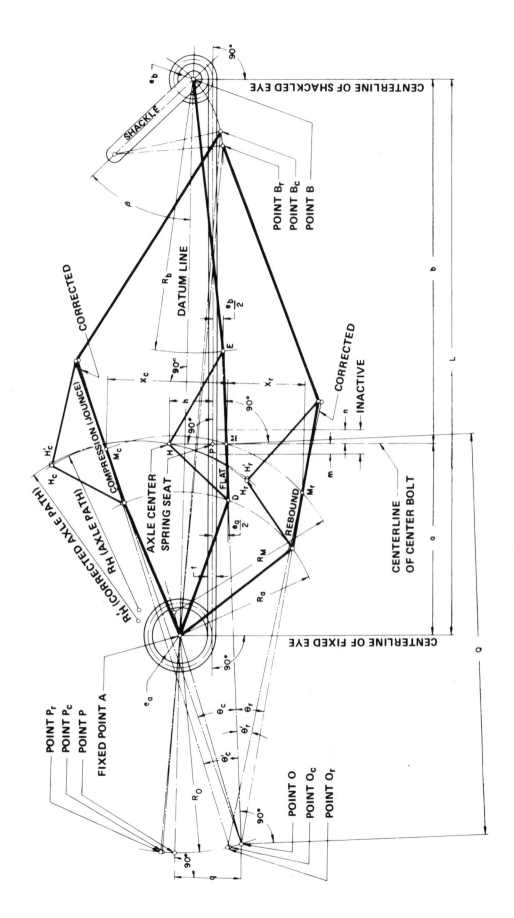

Fig. 4.2—Layout by center link extension method using three link mechanism

1.29

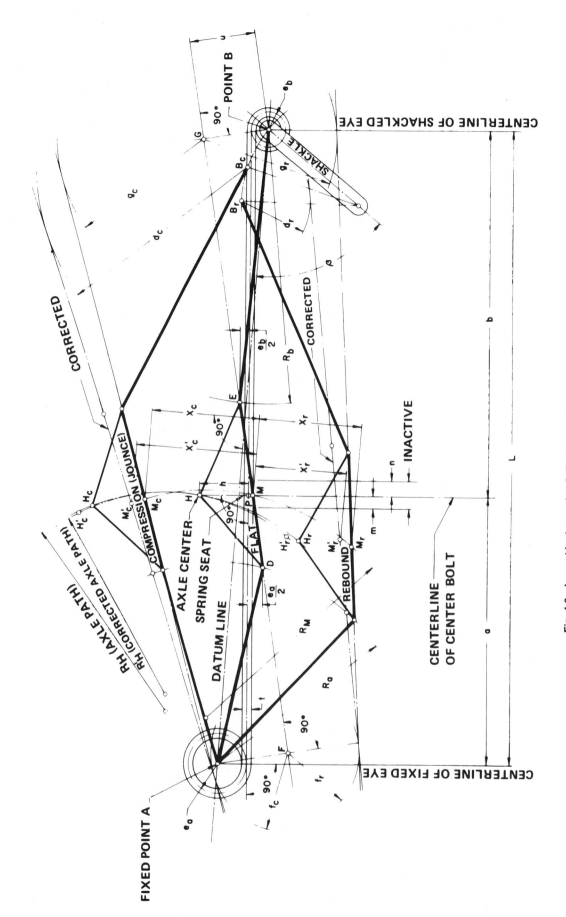

Fig. 4.3—Layout by two point deflection method using three link mechanism

1.30

TABLE 4.1—GEOMETRY FORMULAE FOR SEMI-ELLIPTIC SPRINGS

A. For conventional spring where $W = Z/Y^3$ equals one

$$f = x/Y$$

$$g = x \cdot Y = f \cdot Y^2$$

$$q = \frac{3[(b \cdot e_a + a \cdot e_b) + 2Q(e_a - e_b)]}{2L}$$

$$Q = \frac{a \cdot b}{b - a} = \frac{L \cdot Y}{(Y^2 - 1)}$$

$$\lambda = \frac{3Y^2}{(3Y^2 + 1)(Y + 1)}$$

$$V = \frac{Y^2}{Y + 1}$$

$$Z = \frac{b(57.3 + \Phi \cdot b)}{a(57.3 - \Phi \cdot a)} = Y \cdot \frac{57.3(Y + 1) + \Phi \cdot L \cdot Y}{57.3(Y + 1) - \Phi \cdot L}$$

$$\Phi = \frac{57.3}{Q} = \frac{57.3(b - a)}{a \cdot b} = \frac{57.3(Y^2 - 1)}{L \cdot Y}$$

B. For unconventional spring where $W = Z/Y^3$ does not equal one

$$f = x \cdot \frac{Y(Y + 1)}{Z + Y^2}$$

$$g = x \cdot \frac{Z(Y + 1)}{Z + Y^2} = f \cdot \frac{Z}{Y}$$

$$q = \text{Same as in A.}$$

$$Q = \frac{Z \cdot a^2 + b^2}{Z \cdot a - b} = \frac{L(Z + Y^2)}{(Z - Y)(Y + 1)}$$

(See Fig. 4.5)

$$\lambda = \frac{3(Z + Y^2)^2}{(Y + 1)[3(Z + Y^2)^2 + Y^2(Y + 1)^2]}$$

(See Fig. 4.4)

$$V = \frac{Z + Y^2}{(Y + 1)^2}$$

$$Z = \text{Same as in A.}$$

$$\Phi = \frac{57.3}{Q} = \frac{57.3(Z \cdot a - b)}{Z \cdot a^2 + b^2} = \frac{57.3(Z - Y)(Y + 1)}{L(Z + Y^2)}$$

13. Depending upon the accuracy demanded of the layout, a correction for the effect of the shackle may be necessary, particularly when the shackle angle is exceptionally small (β less than 60 deg in the flat main leaf position) and the shackle is exceptionally long. The correction may be made in the following manner:

Locate point P at intersection of datum line and R_o, where R_o is equal to distance OA.

After determining linkage layout for a given deflection, such as for rebound, point B_r is located.

Locate point P_r on extension of line $B_r A$.

Locate point O_r where chordal distance PO equals $P_r O_r$.

Draw line $O_r M_r$ to give corrected tilt to center link DE in rebound.

In like manner, establish line $O_c M_c$ to give corrected tilt to center link DE in compression.

These corrected positions of center link DE determine the corrected control in degrees per millimeter (equal to θ'/x) and can be used to establish the approximate radius R'_H for the corrected axle path.

5. Two-Point Deflection Method (Fig. 4.3)

This method has the advantage that all of the layout work can be done within the overall length of the spring. In cases where the unsymmetry factor is small and the O point is far from the axle center, it is the only known procedure which permits construction within the confines of the standard layout board and straight edge.

The principle of this method is based upon the use of the two cantilever deflections corresponding to a given deflection at the center of the spring seat. These deflections may be computed for two vertical positions of the spring seat, for example maximum compression (metal-to-metal) and maximum rebound. When they are applied to the three-link equivalent of the spring with the main leaf in the flat position, the path of the axle and the angles of the spring seat can be determined entirely by construction.

Procedure

1. Start layout with main leaf in flat position with

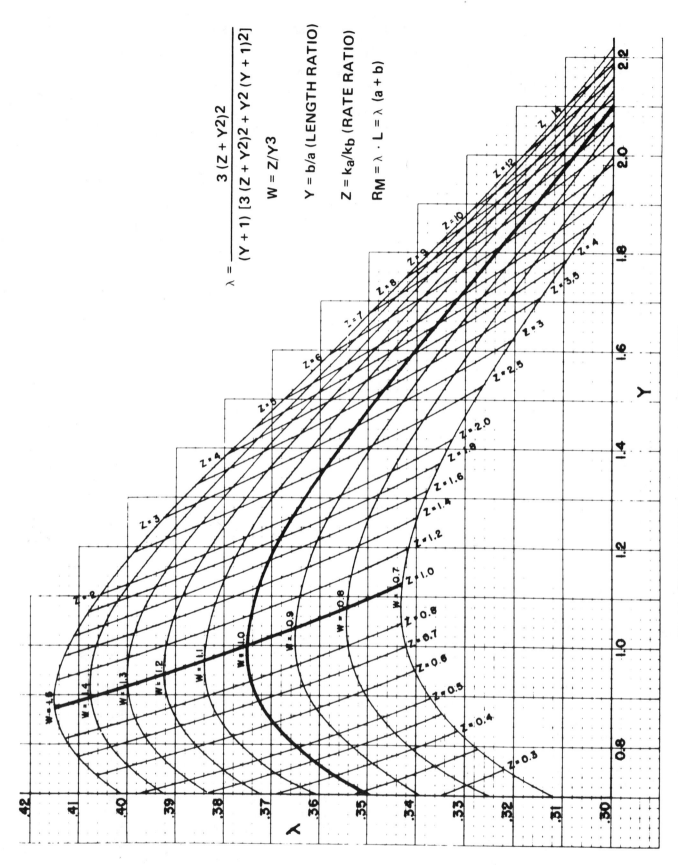

$$\lambda = \frac{3(Z + Y2)2}{(Y + 1)[3(Z + Y2)2 + Y2(Y + 1)2]}$$

$W = Z/Y3$

$Y = b/a$ (LENGTH RATIO)

$Z = k_a/k_b$ (RATE RATIO)

$R_M = \lambda \cdot L = \lambda (a + b)$

Fig. 4.4—Chart for parameter λ to determine path of point M in three link mechanism

1.32

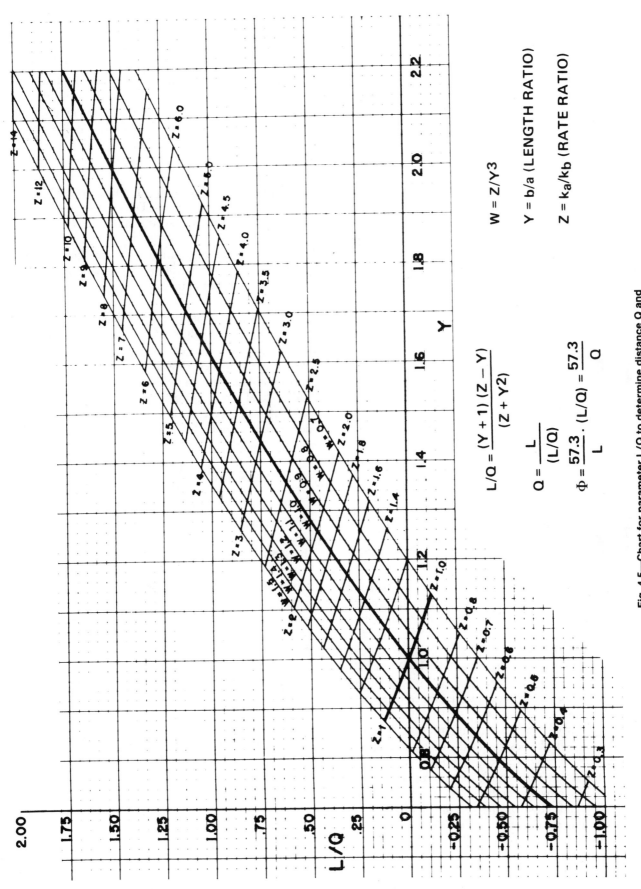

$$\frac{L}{Q} = \frac{(Y+1)(Z-Y)}{(Z+Y2)}$$

$$Q = \frac{L}{(L/Q)}$$

$$\Phi = \frac{57.3}{L} \cdot (L/Q) = \frac{57.3}{Q}$$

W = Z/Y3

Y = b/a (LENGTH RATIO)

Z = ka/kb (RATE RATIO)

Fig. 4.5—Chart for parameter L/Q to determine distance Q and control Φ in three link mechanism

lengths a, b, and L measured along the main leaf and axle center H at distance h from center of main leaf.

2. Establish lengths m and n, which represent the inactive material.

3. Draw arc R_a and at the intersection with $0.5e_a$ locate point D.

4. Draw arc R_b and at the intersection with $0.5e_b$ locate point E.

5. Construct the three links AD, DE, and EB.

6. Locate point M at intersection of centerline of center bolt and link DE.

7. Draw reference lines AF and BG through the eye centers and perpendicular to the extension of the center link DE.

8. For any given deflections such as x_r and x_c, compute f_r and f_c from the following formulae (see also Table 4.1) and draw arcs about point F.

For a conventional spring:

$$f = \frac{x}{Y} = x\left(\frac{a}{b}\right)$$

For an unconventional spring:

$$f = x \cdot \frac{Y(Y + 1)}{Z + Y^2}$$

9. Similarly, for given deflections x_r and x_c, compute g_r and g_c from the following formulae (see also Table 4.1) and draw arcs about point G.

For a conventional spring:

$$g = x\, Y = x\left(\frac{b}{a}\right)$$

For an unconventional spring:

$$g = f\left(\frac{Z}{Y}\right)$$

10. Tangent line to arcs f_r and g_r establishes the position of center link DE in rebound, and tangent line to arcs f_c and g_c establishes the position of center link DE in compression.

11. For each position of the center link DE the axle position can be located by the triangle DEH. When three or more such positions have been located, the approximate radius R_H for the axle path can be established by geometric construction.

12. The control in degrees per millimeter is equal to the angular change in the position of the center link divided by the deflection x: $\Phi = \theta/x$ (deg/mm).

13. In the symmetrical spring the control is zero, with the center link moving parallel to itself throughout the compression and rebound range. Actually, however, the center link undergoes a small angular change due to the vertical displacement of the shackled spring eye.

14. Depending upon the accuracy demanded of the layout, a correction for the effect of the shackle may be necessary, particularly when the shackle angle is exceptionally small (β less than 60 deg in the flat main leaf

position) and the shackle is exceptionally long. The correction may be made in the following manner:

After determining linkage layout for a given deflection, such as for rebound, point B_r is located.

Draw arc d_r where $d_r = (g_r - u)$.

Draw tangent line to arcs f_r and d_r to establish corrected position of center link DE in rebound. (Note that the layout deflection x_r is changed to x'_r in applying this correction.)

Similarly, after locating point B_c, draw arc d_c where $d_c = (g_c + u)$.

Draw tangent line to arcs f_c and d_c to establish corrected position of center link DE in compression. (Note that the layout deflection x_c is changed to x'_c in applying this correction.)

These corrected positions of center link DE determine the corrected control in degrees per millimeter and can be used to establish the approximate radius R'_H for the correct axle path.

The difference between deflections x_c, x_r and x'_c, x'_r respectively is so small in a full size spring layout that it usually can be neglected. However, when required, an overlay of triangle DEH positioned with point M lying on horizontal line through deflection points M_r and M_c and with points D and H lying on arcs R_a and R'_H respectively, will determine exactly the angle of seat at the original deflections, x_r and x_c.

6. Layouts and Nomenclature

Figures 4.2 and 4.3 were constructed with exaggerated eye diameters in order to illustrate the mechanics of construction. Also, the relatively small shackle angle and long shackle used in these layouts results in a large vertical displacement to the shackled spring eye. This was done to better illustrate the effect of the shackle correction. Under these conditions the compression shackle (Fig. 4.2.) results in a control increase during rebound and a control decrease during compression, while the reverse is true for the tension shackle (Fig. 4.3.)

Nomenclature for Figs. 4.2 and 4.3 and for related formulae in Table 4.1.

a	=	Fixed cantilever length (including inactive length m) called "front" length
b	=	Shackled cantilever length (including inactive length n) called "rear" length
c	=	Subscript for compression (jounce) position of linkage
d	=	Tangent arc radius at B where d = (g $\pm$ u)
e	=	Eccentricity = 0.5 (eye I.D. + t) (for Berlin eye, e equals zero)
f	=	Deflection of point F
g	=	Deflection of point G
h	=	Axle distance from center of main leaf
k	=	Spring rate, N/mm (clamped). See Chapter 5
k_a	=	Front cantilever rate, N/mm (clamped)
k_b	=	Rear cantilever rate, N/mm (clamped)
L	=	Total spring length—measured along flat main leaf

m   = Front inactive length

n   = Rear inactive length

q   = Distance of point O below datum line

Q   = Distance between points O and M (see Figs. 4.2 and 4.5), the "equivalent torque arm" for vertical load—see Chapter 6.

r   = Subscript for rebound position of linkage

R_a = 0.75 (a-m)

R_b = 0.75 (b-n)

R_H = Radius of axle path = path of point H (approx.)

R_M = Radius of path of point M (approx. $R_M = \lambda \cdot L$)

R_O = Radius of points P and O

t   = Thickness of main leaf

u   = Distance between points B and G

V   = Front cantilever rate proportion (k_a/k)

W   = Ratio Z/Y^3 (see Fig. 4.4 for λ and Fig. 4.5 for L/Q)

x   = Deflection of point M

Y   = Length ratio (b/a)

Z   = Cantilever rate ratio (k_a/k_b)

β = Shackle angle with main leaf flat, deg

θ = Angular displacement of center link, deg

λ = Parameter to determine R_M (see Fig. 4.4)

Φ = Control of center link (θ/x, deg/mm), known as both "spring control" and "axle control"

(')   = Primes denote shackle effect corrections

Chapter 5
Design Calculations

1. Rate, Load and Stress

A leaf spring may be considered as a beam of uniform strength composed of leaves of equal thickness where the fiber stress is the same throughout the length of the beam.

This approximation is justified for most springs within the accuracy necessary for layout work and—with certain correction factors—for estimates of the necessary length, thickness, width, and number of leaves. It also serves as a base for more detailed calculations.

Figure 5.1 shows a cantilever of six leaves and the same cantilever rearranged with the leaves split and laid side by side for comparison with the triangular beam of uniform strength which is shown in broken lines. It must be appreciated that the straight line contour shown in Fig. 5.1 applies only when all the leaves are of the same thickness. See the remarks under the "Stepping" paragraphs in Section 4 of this chapter.

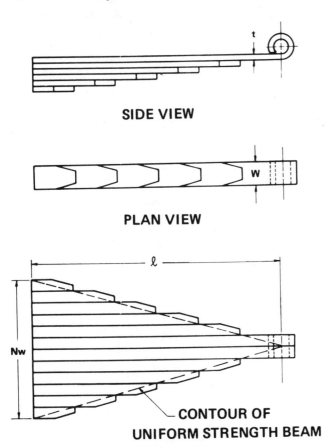

SIDE VIEW

PLAN VIEW

CONTOUR OF UNIFORM STRENGTH BEAM

LEAVES SPLIT AND LAID SIDE BY SIDE

Fig. 5.1—Description of multi-leaf spring

The formulae in Fig. 5.2 are given for leaf springs which approximate beams of uniform strength (except for the last column) and are derived from the following fundamental facts:

1. Stress is proportional to leaf thickness multiplied by change of curvature.

2. Change of curvature is proportional to change of bending moment divided by moment of inertia.

3. Stress is proportional to leaf thickness multiplied by bending moment divided by moment of inertia.

For comparison, the formulae for the cantilever spring of one leaf or of several full length leaves (so-called "uniform section beam") are also given in Fig. 5.2. In this spring the stress is highest at the clamping area, and the stress formulae refer to this highest value. It will be noted that with the same load, length, thickness, and stress, the uniform section spring produces only two-thirds of the deflection and weighs twice as much as the uniform strength spring, or in other words, the uniform strength spring is three times as efficient as the uniform section spring.

As shown by the different formulae in Fig. 5.2, the stress can be calculated from strain, deflection, or load, depending on what information is known.

It is evident from the "stress from strain" formula that for the same change in curvature $(1/R - 1/R_o)$, the stress will vary directly with the leaf thickness.

Again, it will be seen from the "stress from deflection" formula that the stress will vary directly with the leaf thickness and inversely with the square of the effective spring length.

The "stress from load" formula is the standard beam formula for stress where, for a given load, the stress will vary directly with effective length and inversely with the square of the leaf thickness. This may appear paradoxical when comparing the "stress from load" formula with the "stress from deflection" formula. However, in the "stress from load" formula the deflection is not considered. If the expression for load (P) is replaced by the product of rate (k) and deflection (f), the "stress from load" formula will reduce to the "stress from deflection" formula.

Although the foregoing substitution marks the two formulae as merely two different ways of expressing stress, the "stress from deflection" formula is particularly significant for this reason. It shows that for a given stress and deflection the leaf thickness varies as the square of the effective spring length, and since thin leaves will not supply sufficient strength for the spring eyes, the formula emphasizes the desirability of long springs. Another important consideration for the use of long springs is the fact

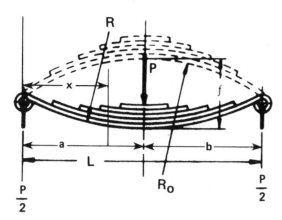

Symmetrical (a = b) Semi-elliptic

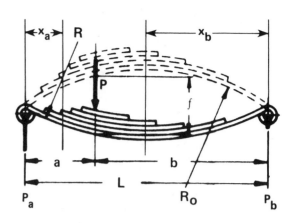

Unsymmetrical Semi-elliptic

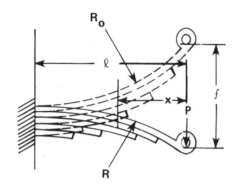

Cantilever

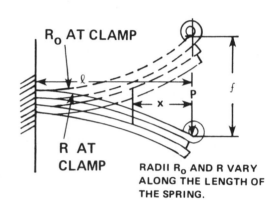

RADII R_o AND R VARY ALONG THE LENGTH OF THE SPRING.

Uniform Section Cantilever

Type	Symmetrical (a = b) Semi-elliptic	Unsymmetrical Semi-elliptic		Cantilever	Uniform Section Cantilever
Length Ratio	$Y = b/a = 1$	$Y = b/a$		——	——
Deflection From Geometry	$f = \dfrac{L^2}{8}\left(\dfrac{1}{R} - \dfrac{1}{R_o}\right)$	$f = \dfrac{ab}{2}\left(\dfrac{1}{R} - \dfrac{1}{R_o}\right) = \dfrac{YL^2}{2(Y+1)^2}\left(\dfrac{1}{R} - \dfrac{1}{R_o}\right)$		$f = \dfrac{\ell^2}{2}\left(\dfrac{1}{R} - \dfrac{1}{R_o}\right)$	$f = \dfrac{\ell^2}{3}\left(\dfrac{1}{R} - \dfrac{1}{R_o}\right)$
Stress From Strain	$S = \dfrac{E}{2}t\left(\dfrac{1}{R} - \dfrac{1}{R_o}\right)$	$S = \dfrac{E}{2}\cdot t\left(\dfrac{1}{R} - \dfrac{1}{R_o}\right)$		$S = \dfrac{E}{2}t\left(\dfrac{1}{R} - \dfrac{1}{R_o}\right)$	$S = \dfrac{E}{2}t\left(\dfrac{1}{R} - \dfrac{1}{R_o}\right)$
Stress From Deflection	$S = \dfrac{4Et}{L^2}\cdot f \cdot SF$	$S = \dfrac{Et}{ab}\cdot f \cdot SF = \dfrac{Et}{L^2}\cdot\dfrac{(Y+1)^2}{Y}\cdot f \cdot SF$		$S = \dfrac{Et}{\ell^2}\cdot f \cdot SF$	$S = \dfrac{Et}{\ell^2}\cdot f \cdot 1.5$
Stress From Load At Any Section	$S_x = \dfrac{xt}{4\Sigma I_x}\cdot P$	$S_{xa} = \dfrac{x_a t}{2\Sigma I_{xa}}\cdot P_a$	$S_{xb} = \dfrac{x_b t}{2\Sigma I_{xb}}\cdot P_b$	$S_x = \dfrac{xt}{2\Sigma I_x}\cdot P$	$S_x = \dfrac{xt}{2\Sigma I}\cdot P$
Stress From Load At Spring Seat	$S = \dfrac{at}{4\Sigma I}\cdot P = \dfrac{Lt}{8\Sigma I}\cdot P$	$S = \dfrac{abt}{2\Sigma IL}\cdot P = \dfrac{Lt}{2\Sigma I}\cdot\dfrac{Y}{(Y+1)^2}\cdot P$		$S = \dfrac{\ell t}{2\Sigma I}\cdot P$	$S = \dfrac{\ell t}{2\Sigma I}\cdot P$
Load Rate	$k = \dfrac{P}{f} = \dfrac{4E\Sigma I}{a^3}\cdot SF = \dfrac{32E\Sigma I}{L^3}\cdot SF$	$k = \dfrac{P}{f} = \dfrac{2E\Sigma IL}{a^2 b^2}\cdot SF = \dfrac{2E\Sigma I}{L^3}\cdot\dfrac{(Y+1)^4}{Y^2}\cdot SF$		$k = \dfrac{P}{f} = \dfrac{2E\Sigma I}{\ell^3}\cdot SF$	$k = \dfrac{P}{f} = \dfrac{2E\Sigma I}{\ell^3}\cdot 1.5$
Volume: Efficiency mJ/mm³	$\dfrac{S^2 peak}{6E}$	$\dfrac{S^2 peak}{6E}$		$\dfrac{S^2 peak}{6E}$	$\dfrac{S^2 peak}{18E}$

Fig. 5.2—Design formulae for leaf springs: Unloaded springs have a radius of curvature R_o which is considered negative when camber is positive

that windup stiffness also varies as the square of the length. (See Chapter 6, Section 2.)

In considering the load (P) and the rate (k), it must be kept in mind that the test load and rate are not the same as the desired installed load and rate. The effect of the installation (shackles, etc.) is discussed in Chapter 6.

2. Stiffening Factor

Actual leaf springs are not truly beams of uniform strength. How closely they approach such a beam depends chiefly on the following factors:

1. Length of Leaves—Two or more full length leaves are sometimes used. The shorter leaves may be longer than they would be for uniform strength. This is to reduce the main leaf stresses in the area of the eyes. The spring will therefore be stiffer, and it will be intermediate between a uniform strength spring and a uniform section spring.

2. Leaf Ends—Fig. 5.1 shows how the ends of the leaves exceed the outline of the triangular beam. This makes the spring stiffer. The various leaf ends and their effects are discussed in Chapter 3, Section 2.

3. Center Clamp—Standard procedure demands that semi-elliptic springs be tested without center clamp, and the formulae are given for this condition. Since the springs are used with the clamp, the leaf lengths are designed for the clamped springs. Theoretically, this requires subtracting the length of the clamped part from the total length, designing two cantilevers for this reduced length, and adding a uniform section in the middle between the two cantilevers. Actually, the effect of the clamp can be allowed for by using an "active length" in the formulae instead of the full length.

The amount of active length within the seat area depends on the design of the clamping parts and also on any liner or insulator material which may be used between the clamping plates and the spring itself. In semi-elliptic and cantilever springs, the active length extends into the seat area and is therefore longer than the distance from the edge of the clamp to the point of applied load. For springs without any liner material, the active length generally extends to the inside edge of the clamp bolt. Spring seat liners will increase the active length farther into the clamped area, the amount depending on the compressibility of the liner, when the clamping bolts have been tightened.

At one particular deflection the addition of the clamp does not change the load carried at that height. This is the position at which the curvature of the spring without clamp fits the spring seat.

The influences of the lengths of leaves and the types of leaf ends are taken into consideration by a "stiffening factor" which is designated as "SF" in Fig. 5.2 and Table 5.1. Note that the stress from deflection and the load rate formulae are multiplied by SF, except for the uniform section cantilever for which SF always equals 1.5.

The value of SF is exactly 1.00 when the leaf lengths and leaf thicknesses are selected to produce a uniform strength beam (spring), and the resultant curvature of the spring in bending is a circular arc with radius R.

The farther the design deviates from uniform strength, the farther will the elastic curve deviate from a circular arc. "R" will be only an approximate parameter for the elastic curve, and SF will have to be a higher value.

SF will have its maximum value of 1.50 when all the leaves of a multi-leaf spring are full length, or when there is a single leaf of constant thickness and width. A spring of this type is known as a uniform section spring (column 4, Fig. 5.2), where "R" is the smallest radius for the elastic curve, namely at the line of clamp or encasement, and "S" is the highest stress (at that line), as explained in Section 1 of this chapter.

Selection of the correct SF value in the final spring design is predicated on factors gained from experience, since the value may vary from less than 1.10 to 1.50, depending on the design specifications.

For the preliminary design calculation, the following SF values may be applied in formula I of Table 5.1 to provide a moment of inertia (ΣI) for a selection of the approximate number of leaves and gages.

For passenger car and light truck springs with tapered leaf ends and more or less "uniform" stress design:
$$SF = 1.10$$
For passenger car and light truck springs with tapered leaf ends and extended leaf lengths:
$$SF = 1.15$$
For truck springs with untapered leaf ends and more or less "uniform" stress design:
$$SF = 1.15$$
For truck springs with untapered leaf ends and two full length leaves:
$$SF = 1.20$$
For truck springs with untapered leaf ends and three full length leaves:
$$SF = 1.25$$
For the first stage of variable or progressive rate springs, before contact with the second stage:
$$SF = 1.40$$
For the second stage of variable or progressive rate springs, when all leaves are operable:
$$SF = 1.10 \text{ with tapered leaf ends,}$$
$$SF = 1.15 \text{ with untapered leaf ends}$$
For springs with all leaves full length:
$$SF = 1.50$$

Table 5.1—FORMULAE FOR CALCULATIONS ON LEAF SPRINGS

Formula		Symmetrical Semi-Elliptic	Unsymmetrical Semi-Elliptic		Multi-Leaf Cantilever
I	Total Moment of Inertia — mm^4	$\Sigma I = \dfrac{k \cdot L^3}{32 \cdot E \cdot SF}$	$\Sigma I = \dfrac{k \cdot a^2 \cdot b^2}{2 \cdot E \cdot SF \cdot L}$	$= \dfrac{k \cdot L^3}{2 \cdot E \cdot SF} \cdot \dfrac{Y^2}{(Y+1)^4}$	$\Sigma I = \dfrac{k \cdot \ell^3}{2 \cdot E \cdot SF}$
II	Maximum Leaf Thickness t_{max} — mm	$t_{max} = \dfrac{8 \cdot \Sigma I}{L} \cdot \dfrac{S}{P}$	$t_{max} = \dfrac{2 \cdot \Sigma I \cdot L}{a \cdot b} \cdot \dfrac{S}{P}$	$= \dfrac{2 \cdot \Sigma I}{L} \cdot \dfrac{S}{P} \cdot \dfrac{(Y+1)^2}{Y}$	$t_{max} = \dfrac{2 \cdot \Sigma I}{\ell} \cdot \dfrac{S}{P}$
III	Stress with standard gage (t) Leaf — MPa	$S = \dfrac{L \cdot t}{8 \cdot \Sigma I} \cdot P$	$S = \dfrac{a \cdot b \cdot t}{2 \cdot \Sigma I \cdot L} \cdot P$	$= \dfrac{L \cdot t}{2 \cdot \Sigma I} \cdot P \cdot \dfrac{Y}{(Y+1)^2}$	$S = \dfrac{\ell \cdot t}{2 \cdot \Sigma I} \cdot P$
IV	Approx. Mass of Spring Steel — kg	(The number of leaves of the same gage*) $\cdot \dfrac{\text{Unit mass from Table 5.3}}{1000} \cdot$ one half of spring length $\cdot$ SF			

*Where the grading of the spring has several gages, the several products are additive before multiplying by (one half of spring length) and (SF). The denominator of 1000 reduces the unit mass values to kg per mm.

NOMENCLATURE FOR FIG. 5.2 AND TABLE 5.1

L	=	length of semi-elliptic spring	— mm
ℓ	=	length of cantilever spring	— mm
a	=	front length of semi-elliptic spring	— mm
b	=	rear length of semi-elliptic spring	— mm
Y	=	cantilever length ratio in semi-elliptic spring (=b/a)	
P	=	load on spring	— N
f	=	spring deflection	— mm
k	=	load rate (change in load / change in deflection)	— N/mm
ΣI	=	total moment of inertia (summation for all the leaves in the spring)	— mm^4
ΣI_x	=	summation of the moments of inertia of those leaves comprising the section for which the stress is to be calculated	— mm^4
t_{max}	=	maximum leaf thickness for maximum stress	— mm
t	=	thickness of the leaf selected from standard gages (t < t_{max}) for which the stress is to be calculated; usually this is the main leaf and/or the leaf of the greatest thickness in the spring	— mm
S_{max}	=	maximum specified stress (N/mm^2 = MN/m^2 =)	— MPa
S	=	stress with selected standard gage	— MPa
E	=	modulus of elasticity (for steel: 200 $\cdot$ 10^3)	— MPa
R_o	=	radius of curvature in the unloaded spring (shown in dash lines), considered negative	— mm
SF	=	Stiffening Factor	

3. Preliminary Calculations

Table 5.1 lists the formulae (I, II, III) for the three essential steps in calculating the spring design.

Formula I is used to establish the total moment of inertia for the specified rate and length.

Formula II is used to establish the maximum permissible leaf thickness within the specified maximum stress limit at a corresponding load. The grading of the leaves in the spring is established by selecting the number of leaves, the leaf width, and the combination of leaf gages to provide the calculated total moment of inertia obtained with formula I. The moments of inertia for the individual leaf sections are shown in Table 5.2.

Formula III is used to calculate the spring stress.

For preliminary calculations of unsymmetrical springs in which the length ratio Y does not exceed 1.30, the symmetrical formulae in Table 5.1 may be used with assurance that the results will be within 3% of those obtainable by use of the more complicated unsymmetrical formulae.

Formula IV is used to estimate an approximate mass of steel for comparing optional designs from preliminary calculations.

The exact mass of spring steel is calculated after the eye diameters and the leaf end constructions have been established. Generally, the exact mass is heavier than the approximate mass from formula IV.

TABLE 5.2—MOMENTS OF INERTIA (mm⁴) FOR LEAF SPRING BAR CROSS SECTIONS OF MEAN DIMENSIONS PER TABLE 3.1 IN CHAPTER 3 (TABLE 1 IN SAE STANDARD J1123)

Thick-ness	Moments Of Inertia For Width Of									
	40.0	45.0	50.0	56.0	63.0	75.0	90.0	100.0	125.0	150.0
5.00	398.8	449.6	500.3	561.2	632.3					
5.30	474.9	535.4	596.0	668.6	753.3					
5.60	560.0	631.5	703.0	788.8	888.9					
6.00	688.3	776.4	864.5	970.1	1093.4	1304	1567			
6.30	796.3	898.4	1000	1123	1266	1510	1815			
6.70	956.8	1079.7	1203	1350	1522	1817	2184			
7.10	1137.3	1283.7	1430	1606	1811	2162	2600	2892		
7.50	1338.9	1511.6	1684	1892	2134	2548	3065	3410		
8.00	1622.0	1831.9	2042	2294	2588	3092	3720	4139		
8.50	1942	2194	2446	2748	3101	3707	4462	4964	6212	
9.00	2301	2600	2899	3259	3678	4399	5295	5892	7376	
9.50	2700	3052	3405	3828	4321	5170	6225	6928	8677	
10.00	3136	3547	3957	4450	5024	5996	7222	8038	10050	12050
10.60	3726	4215	4704	5291	5976	7137	8598	9572	11980	14360
11.20	4383	4961	5538	6231	7040	8412	10137	11287	14130	16950
11.80	5111	5787	6463	7274	8221	9829	11848	13195	16520	19840
12.50	6056	6860	7664	8629	9755	11670	14073	15675	19640	23590
13.20			9002	10140	11460	13730	16560	18450	23130	27780
14.00			10710	12060	13650	16350	19730	21990	27580	33150
15.00			13120	14790	16740	20070	24230	27010	33900	40770
16.00			15850	17880	20250	24300	29360	32730	41120	49470
17.00				21370	24210	29080	35150	39200	49270	59320
18.00				25280	28650	34440	41650	46460	58430	70380
19.00				29620	33590	40410	48890	54540	68650	82720
20.00				34410	39050	47010	56910	63510	80000	96410
21.20						55810	67610	75470	95110	114700
22.40						65340	79210	88450	111300	134000
23.60						76190	92410	103220	130000	156600
25.00						90230	109530	122400	154300	186000
26.50								145400	183400	221300
28.00								171000	215900	260700
30.00								209500	264900	320100
31.50									306000	370000
33.50									367000	444100
35.50										527300
37.50										620100

For derivation of formulae for these tabulated values, see Appendix B

First Example

Design a symmetrical semi-elliptic passenger car spring with tapered leaf ends to meet the following specifications:

Rate, as tested without center clamp	k	= 17.5 N/mm
Design load	P	= 3500 N
Metal-to-metal clearance	x_c	= 110 mm
Length	L	= 1320 mm
Maximum stress	S_{max}	= 1000 MPa

Step One—Find the required total moment of inertia from formula I in Table 5.1:

$$\Sigma I = \frac{17.5 \cdot 1320^3}{32 \cdot 200 \cdot 10^3 \cdot 1.10} = 5717 \text{ mm}^4$$

Step Two—Find the maximum permissible leaf thickness from formula II in Table 5.1:

with $P_{max} = 3500 + 17.5 \cdot 110 = 5425$ N

$$t_{max} = \frac{8 \cdot 5717 \cdot 1000}{1320 \cdot 5425} = 6.39 \text{ mm}$$

Thinner leaves would give a lower stress, thicker leaves a higher stress. The nearest standard gage size is 6.30 mm, producing a 1½% lower stress.

Step Three—Find the Number of leaves and the leaf width (w) which will produce the required rate; using the moment of inertia values in Table 5.2:

First option: assume $w = 50$ mm

Number of leaves = 6

4 leaves gage 6.30	I =	4000
2 leaves gage 6.00	I =	1729
	ΣI =	5729 mm⁴

$$k = 17.5 \cdot \frac{5729}{5717} = 17.54 \text{ N/mm}$$

With this selection of leaf gages, the rate k is within less than 1% of the specified rate.

The stress at P_{max} (5429N) from formula III

$$S_{max} = \frac{1320 \cdot 6.30}{8 \cdot 5729} \cdot 5429 = 985 \text{ MPa}$$

$$\text{Approx. Mass} = \frac{4 \cdot 2.438 + 2 \cdot 2.322}{1000}$$
$$\cdot \frac{1320}{2} \cdot 1.10 = 10.45 \text{ kg}$$

Second option: assume w = 63 mm
Number of leaves = 5

2 leaves gage 6.30   I = 2532
3 leaves gage 6.00   I = 3280
                   ΣI = 5812 mm^4

$$k = 17.5 \cdot \frac{5812}{5717} = 17.79 \text{ N/mm}$$

With this selection of gages in the 63 mm width the rate k is within 2% of the specified rate.

The stress at P_{max} (5457 N)

$$S_{max} = \frac{1320 \cdot 6.30}{8 \cdot 5812} \cdot 5457 = 976 \text{ MPa}$$

$$\text{Approx. Mass} = \frac{2 \cdot 3.076 + 3 \cdot 2.930}{1000}$$
$$\cdot \frac{1320}{2} \cdot 1.10 = 10.85 \text{ kg}$$

Conclusions—Both options appear acceptable. The choice may depend on some of the following considerations:

1. Limited available space for the spring width will favor the first option (50 versus 63 mm width);

2. Restrictions on mass will favor the first option (10.45 versus 10.85 kg);

3. Mandate on low fabrication cost (6 versus 5 leaves) will favor the second option;

4. Demand for the greater lateral stiffness will favor the second option (63^3 versus 50^3 in the lateral rate formula—see Table 10.2).

Unconventional Springs (See Table 4.1)

Some semi-elliptic springs may require that one cantilever end is made of more leaves than the other, or that its leaf lengths are extended substantially beyond the uniform strength beam requirements. In such cases it is more convenient to consider the complete semi-elliptic spring as made of two cantilevers. The rates or deflections of both cantilevers are calculated separately and combined by use of the following formulae:

$$k = \frac{PL}{bf_a + af_b} \text{ or}$$

$$k = \frac{k_a k_b L^2}{k_a a^2 + k_b b^2} = \frac{(Y+1)^2}{Z+Y^2} \cdot k_a = \frac{L^2}{\frac{a^2}{k_b} + \frac{b^2}{k_a}}$$

where:

P = Load on spring
L = Total spring length
a = Fixed ("front") cantilever length (including clamp length)
b = Shackled ("rear") cantilever length (including clamp length)
f_a = Deflection of front cantilever
f_b = Deflection of rear cantilever
k_a = Front cantilever rate
k_b = Rear cantilever rate
Y = b/a
Z = k_a/k_b

Second Example

Redesign the second option (w=63mm) of the first example with the additional provision to provide $\Phi = 0.030$ deg/mm control. This will require an unsymmetrical spring design where $W = Z/Y^3$ does not equal one. (See Table 4.1B)

Try front end length a = 580mm
Rear end length b = 740 mm

Length ratio $Y = \dfrac{b}{a} = \dfrac{740}{580} = 1.276$

Cantilever rate ratio =

$$Z = \frac{k_a}{k_b} = Y \cdot \frac{57.3 \cdot (Y+1) + \Phi L \, Y}{57.3 \cdot (Y+1) - \Phi L}$$

$$Z = 1.276 \cdot \frac{57.3(1.276+1) + 0.030 \cdot 1320 \cdot 1.276}{57.3(1.276+1) - 0.030 \cdot 1320}$$

$$Z = 2.542$$

$$Q = \frac{L(Z+Y^2)}{(Z-Y) \cdot (Y+1)} = \frac{1320(2.542+1.276^2)}{(2.542-1.276) \cdot (1.276+1)}$$

$$Q = 1910$$

Control $\Phi = \dfrac{57.3}{Q} = \dfrac{57.3}{1910} = 0.030$ deg/mm as required.

$$P_a \text{ max} = \frac{5425 \cdot 740}{1320} = 3041 \text{ N}$$

$$P_b \max = \frac{5425 \cdot 580}{1320} = 2384 \text{ N}$$

$$k_a = \frac{k(Z + Y^2)}{(Y+1)^2} = \frac{17.5(2.542 + 1.276^2)}{(1.276 + 1)^2} =$$

$$k_a = 14.09 \text{N/mm}$$

$$k_b = \frac{k_a}{Z} = \frac{14.09}{2.542} = 5.54 \text{N/mm}$$

Step One—From formula I for cantilever

$$\Sigma I_a = \frac{k_a \cdot a^3}{2E \cdot SF}$$

$$\Sigma I_a = \frac{14.09 \cdot 580^3}{2 \cdot 200 \cdot 10^3} \cdot \frac{1}{1.10} = 6248 \text{mm}^4$$

$$\Sigma I_b = \frac{5.54 \cdot 740^3}{2 \cdot 200 \cdot 10^3} \cdot \frac{1}{1.10} = 5102 \text{ mm}^4$$

Step Two—Under a given load the rear cantilever will have a higher stress than the front cantilever, because of its smaller ΣI. Therefore, $t_b$max will decide the selection of the standard gage to be used.

From formula II for rear cantilever

$$t_b \max = \frac{2 \cdot 5102}{740} \cdot \frac{1000}{2384} = 5.78 \text{ mm}$$

Step Three—Due to eye strength requirements it is not advisable to use a gage thinner than 6.00 mm in the main leaf. The maximum stress will then slightly exceed 1000 MPa in the rear (shackled) cantilever. This should be acceptable as long as the stress in the more critical front (fixed) cantilever will remain below 1000 MPa.

Establish the number of leaves and their gages in the specified width of 63 mm

Leaf	Gage	Front Cantilever I	Rear Cantilever I
1	6.00	1093.4	1093.4
2	6.00	1093.4	1093.4
3	6.00	1093.4	1093.4
4	6.00	1093.4	1093.4
5	5.60	888.9	888.9
6	5.60	888.9	none
		$\Sigma I = 6151.4$mm^4	5262.5mm^4

If Leaf #5 were of gage 6.00 the front cantilever$\Sigma I = 6355.9$ and the rear cantilever$\Sigma I = 5467.0$ and since the requiredΣI for both front cantilever$\Sigma I_a = 6248$ and rear cantilever$\Sigma I_b = 5102$ is between these two selections of leaf gages, actual tests for spring rates will establish the gages to be used.

For this analysis, it will be assumed that the checked rates with the 5.60 gage in No. 5 leaf will be satisfactorily close to the specified rates.

$$S_a \max = \frac{580 \cdot 6.00}{2 \cdot 6151.4} \cdot 3041 = 860 \text{ MPa}$$

$$S_b \max = \frac{740 \cdot 6.00}{2 \cdot 5262.5} \cdot 2384 = 1006 \text{ MPa}$$

Estimate of approximate mass

$$\text{Front Cantilever} = \frac{4 \cdot 2.930 + 2 \cdot 2.736}{1000} \cdot$$

$$\frac{580}{2} \cdot 1.10 = 5.484 \text{ kg}$$

$$\text{Rear Cantilever} = \frac{4 \cdot 2.930 + 1 \cdot 2.736}{1000} \cdot$$

$$\frac{740}{2} \cdot 1.10 = 5.884 \text{ kg}$$

$$\text{Total} = 11.37 \text{ kg}$$

A spring in which the specified control Φ requires an unconventional design will have more mass than a spring in which Z/Y^3 equals or nearly equals 1. This is verified by comparing the approximate mass of the Second Example (11.37 kg) with the mass of the First Example second option (10.85 kg).

It is always advisable to make sample springs in order to verify by test results the grading (gages) selected before issuing the spring design.

Third Example

Design a symmetrical semi-elliptic truck spring with 3 full length leaves and no tapered leaf ends.

Rate as tested with an
SAE loading block $k = 110$ N/mm
Design load $P = 14700$ N
Stress at P (not to exceed) $S_p = 540$ MPa
Length $L = 1460$ mm
Width $w = 75$ mm

Step One—From formula I calculate the required moment of inertia with the heavy truck SF value = 1.25

$$\Sigma I = \frac{kL^3}{32E} \cdot \frac{1}{SF}$$

$$\Sigma I = \frac{110 \cdot 1460^3}{32 \cdot 200 \cdot 10^3 \cdot 1.25} = 42792 \text{ mm}^4$$

Step Two—Find the maximum permissible gage thickness from formula II

$$t_{max} = \frac{8\Sigma I}{L} \cdot \frac{S}{P} = \frac{8 \cdot 42792}{1460} \cdot \frac{540}{14700} = 8.61 \text{ mm}$$

Step Three—The nearest thinner standard gage is 8.50. Since 12 leaves gage 8.50 in the 75 mm width has aΣI = 44484 which is too large, a combination of gages 8.50 and 8.00 will be selected.

9 leaves gage 8.50,	I =	33363	
3 leaves gage 8.00,	I =	9276	
	ΣI =	42639 mm^4	

With this selection of gages the rate k should be very close to the specified rate.

Stress at design load 14700N

$$S = \frac{1460 \cdot 8.50}{8 \cdot 42639} \cdot 14700 = 535 \text{ MPa}$$

Approx. Mass

$$= \frac{9 \cdot 4.946 + 3 \cdot 4.657}{1000} \cdot \frac{1460}{2} \cdot 1.25 = 53.4 \text{ kg}$$

Fourth Example

It is interesting to note the difference in calculated data which would result in the Third Example, if the specifications had not required the three full length leaves.

The stiffening factor would be SF = 1.15

Step One—

$$\Sigma I = \frac{110 \cdot 1460^3}{32 \cdot 200 \cdot 10^3} \cdot \frac{1}{1.15} = 46513 \text{ mm}^4$$

Step Two—

$$t_{max} = \frac{8 \cdot 46513}{1460} \cdot \frac{540}{14700} = 9.36 \text{ mm}$$

Step Three—The nearest thinner gage is 9.00 mm

8 leaves gage 9.00	I =	35192	
3 leaves gage 8.50	I =	11121	
	ΣI =	46313 mm^4	

With this selection of gages, the rate k should be very close to the specified rate.

Stress at design load 14700 N

$$S = \frac{1460 \cdot 9.00}{8 \cdot 46313} \cdot 14700 = 521 \text{ MPa}$$

Approx. Mass

$$= \frac{8 \cdot 5.235 + 3 \cdot 4.946}{1000} \cdot \frac{1460}{2} \cdot 1.15 =$$

47.6 kg

In comparing the Third and Fourth Examples, it is evident that due to the more uniform stress distribution throughout the length of the spring in the Fourth Example, the mass of the Fourth Example is less (47.6 kg versus 53.4 kg). However, where safety requires lower stresses at the bearing ends (eyes) with more than one full length leaf, the greater mass and increased cost of manufacture (12 versus 11 leaves) must be accepted.

Summary

The four examples show that the formulae in Table 5.1 are sufficient for preliminary calculation. The total moment of inertia required for a specified rate (the most important characteristic of any suspension spring) at a specified spring length is determined in formula I. It may be satisfied by a variety of combinations of width, thickness, and number of leaves, provided the leaf thickness does not exceed the maximum calculated in formula II for a specified maximum stress. This is checked in formula III in which the total moment of inertia for the chosen standard leaf size is incorporated.

4. Stress Distribution

When the types of leaf sections, ends, clamp, etc., (Chapter 3) have been chosen and the approximate spring dimensions have been determined as shown in preceding sections of this chapter, further specifications are generally developed in cooperation with the spring manufacturer. These specifications include thickness, length, free radius and peening of each leaf, and the amount of cold set to be given to the assembled spring.

Spring manufacturers use various methods to arrive at these data, and their results may be different because they involve judgment in the compromise of contradictory requirements.

In view of the compromises and assumptions involved in the detail design, service experience must remain the final test. Where sufficient background is available, service experience can be represented by the results of suitably arranged life tests (see Chapter 8), and in turn, life test results may lead to definite design rules for a particular type of service.

The basic considerations which are presented in this section will enable the user to follow the spring manufacturer's analysis. They will also help the interpretation of service troubles in terms of spring design.

Leaf Thickness

In automotive practice, springs are usually "graded," that is, composed of leaves of two or three different gage thicknesses. The main leaf, often together with adjacent leaves, is made one gage thicker, and several short leaves are made one gage thinner than the intermediate leaves.

This is done for a number of reasons: to give the main leaf more strength to resist eye forces; to allow more tolerance of quench radius on the short leaves; to compensate for the difference in free leaf radii, and because desired rates can be approached more closely by combinations of standard gages than by using the same gage for all the leaves.

Leaf Radii

The curvature is not the same for all the unassembled leaves of a spring. It becomes more and more negative or less and less positive from main leaf towards the shorter leaves. When the spring is assembled, the leaves are pulled up against each other and a common curvature is established (which may, of course, vary along the spring). The leaves of an assembled spring in the free position are therefore under some stress. In the main leaf, this assembly stress is subtractive from load stress; in the short leaves, it is additive to load stress. This is done to reduce the main leaf stress and to insure that the leaf ends have bearing on adjacent leaves.

The quench radii or curvatures of the quenching forms are obtained from the desired curvatures of the individual leaves by allowing for springback, for the effect of shot peening, and for cold setting of the spring (See paragraphs on "Free Radii" in Section 5 of this Chapter).

Stepping

The lengths of the leaves of a spring, together with the thicknesses and the individual leaf radii, determine the distribution of stresses along each leaf. They also control the shape of the spring under load and its rate.

The leaves of a spring bear on each other mainly on a relatively small area near the leaf tip. The center of pressure of this area (the "bedding centerline") is some distance behind the actual tip of the leaf. This distance ranges from about 10mm for blunt (full, thick) ends to about 50mm for tapered leaf ends. The distance from one center of pressure to that of the next shorter leaf is the step or overhang. To determine the overhang of the shortest leaf from the edge of the clamp, it is necessary to consider the design of the clamping parts as explained under "Center Clamp" in Section 2 of this chapter. The sum of all the steps equals the "active" length of the spring.

To discuss the effect of stepping, a hypothetical impractical spring in which all leaves are made with the same free radius will be considered first. If all leaves of this spring are of equal thickness, equal steps will give the closest approach to the beam of uniform strength. That is the condition shown in Fig. 5.1.

If the spring has a mixed grading, that is, composed of more than one size of gage, the steps should be made proportional to t^3. The stresses will be approximately uniform along each leaf, but the various leaves will be stressed in proportion to their thickness, which may not be the most desirable for fatigue life endurance.

If several leaves are made full length, the spring will be stiffened (deflection reduced under a given load). For a given load, this will reduce the stress adjacent to the spring ends, while the stress near the seat area will remain the same. However, if the applied load is increased to produce a given deflection, the stress near the seat area will be increased.

If the balance of the leaves are not extended in length with the full length leaves, the disturbance of efficiency will be confined to the full length leaves. However, if the balance of the leaves are also extended, the stresses will be less toward the leaf ends. The spring will be stiffened still more, and it will deviate from the uniform strength beam in the direction of the uniform section spring.

If the individual leaves have different curvatures, then assembly stresses are set up. These are desirable for various reasons. With assembly stresses it is impossible to have uniform stress along a leaf at all loads. By a suitable combination of assembly stresses and stepping, it is possible to distribute the stress in a desirable manner among the leaves and to make it uniform along each leaf at one particular load.

Though the stress distribution can be improved by a proper combination of leaf thicknesses, free radii, and steps, the rate is the same as for a spring made up of the same leaves fitted "dead," that is, without assembly stresses.

Fig. 5.3 shows the stress distribution along the main leaf on a spring with and without assembly stresses, with three different leaf steppings, and at various loads. It is intended to show only the principles discussed here. The spring itself is impractical. It is made up of two leaves of equal thickness, the main leaf rectangular in plan view, the short leaf triangular to simulate the effect of a series of stepped leaves.

Without assembly stresses, the 250-250 stepping (equal steps) is obviously most efficient. Also, it has the lowest rate and deflects with uniform change of curvature in the part made up of two leaves. The 400-100 stepping approaches a rate and shape of the uniform section spring.

With 300 MPa assembly stress at the clamp, the rate is not changed, but the stress distribution and shape are changed.

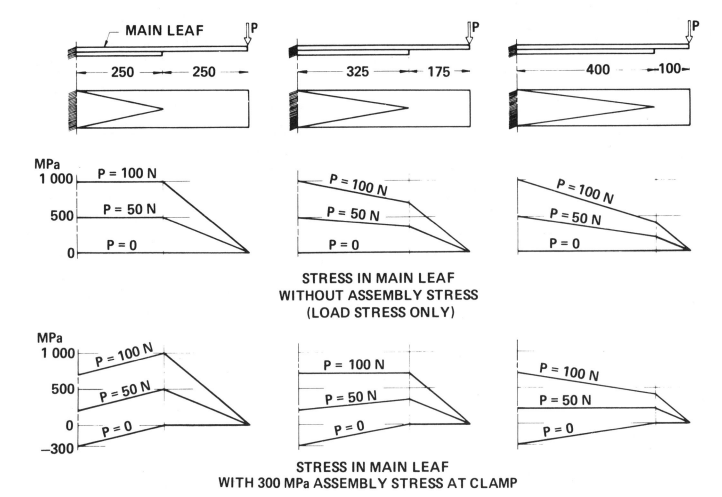

Fig. 5.3—Stress distribution along the main leaf with and without assembly stresses

The 250-250 stepping is now obviously inefficient, because the main leaf has a peak stress at the tip of the second leaf. If the unassembled leaves are circular, the shape under load will correspond to that shown in Fig. 5.4A. This condition corresponds to t^3 stepping.

With 325-175 stepping, the main leaf will have uniform stress within the contact area with the second leaf under 100 N load and at that load will have a circular shape. At lower loads the shape will be intermediate between Fig. 5.4B and 5.4C.

With 400-100 stepping, the main leaf will have uniform stress within the contact area with the second leaf under 50 N load and at that load will have a circular shape. At lower loads the shape will be more like Fig. 5.4B.

In Fig. 5.4C the spring is shown flat under load in order to bring out the deviations from the desired shape more clearly when the backing leaves are too short as in Fig. 5.4A and when the backing leaves are too long as in Fig. 5.4B.

In the following discussion the subscript "n" is used to denote stress (S), thickness (t), overhang (ℓ), curvature (q), etc., of "different" or "various" leaves.

To obtain uniform stress along the length of each leaf at a given load and to achieve circular or straight shape of a spring made up of leaves which are circular when free, it is necessary that each leaf be subjected only to pure couples applied to its ends. This condition can be fulfilled only if the loads between the various leaves are all equal to the load (P) on the end of the spring. In order to achieve this, the stresses (S_n) in the different leaves are adjusted by proper choice of free leaf radii. At the same time the overhangs (ℓ_n) of the various leaves are so chosen that at the particular load which is selected in advance as the load at which the spring shall have circular or straight shape,

$$S_n = 6\,P\,\ell_n/w\,t_n^2$$

Since P and w are constants, the overhangs ℓ_n are made proportional to $S_n t_n^2$.

The stresses S_n can be chosen, subject to the condition that the sum of the moments in the individual leaves equals the total bending moment of the load. It has been found that lower stresses in the thicker main leaf and higher stresses in the thinner leaves give good results.

1.46

A – BACKING LEAVES TOO SHORT

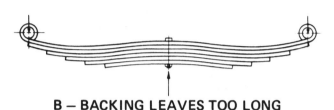

B – BACKING LEAVES TOO LONG

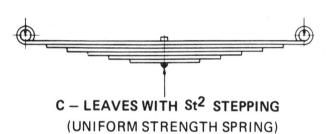

C – LEAVES WITH St^2 STEPPING
(UNIFORM STRENGTH SPRING)

Fig. 5.4—Effect of backing leaf lengths on spring shape

For "over-the-road" vehicles with a static deflection to design load of about half the total deflection, experience has shown that best service is obtained if the stresses are made uniform along the leaves under normal load. For "off-the-road" vehicles the stress at total deflection may be more important. Service experience and life tests must decide for which condition the spring should be designed.

When the length of a leaf has to be different from the value obtained from stress considerations (for example, to make room for a clip or to support the eye), the performance and efficiency of the spring are affected least if all the other leaves are made to the lengths originally obtained from stress considerations.

The change in spring rate due to changes in stepping can be estimated by comparing the overhangs as designed to overhangs proportional to t^3. The spring designed with overhangs proportional to t^3 would have a rate calculated from the formulae in Table 5.1. A spring with all leaves full length would have a 50% stiffer rate.

Tools Of Analysis

Two different assumptions about the action of leaf springs are reasonable. They shall be called "point pressure" and "common curvature".

"Point pressure" means that the leaves touch each other only at the bearing points and at the center clamp. It is based on the observation of worn-in areas near the leaf tips. With this assumption, it is possible to calculate the tip loads of successive leaves from the condition that two contacting leaves must have a common load and a common deflection. After the tip loads have been calculated, each leaf can be considered as a simple beam. This method is used by some spring designers, but unless the volume of work warrants the preparation of special curve sheets and leads to familiarity with the formulae, it is somewhat difficult. The assumption of "point pressure" is justified when the leaves are free to take the shape which corresponds to this load distribution. This is usually not possible for all leaves of a spring unless spacers are provided between them.

"Common curvature" means that all the leaves of a spring touch their neighbors all along their length. This assumption leads to simpler calculations. It is in general justified at all points where a leaf is surrounded by other leaves, that is, everywhere except for the bottom and top leaves and except for the overhanging leaf ends.

In some cases "point pressure" and "common curvature" lead to the same result. For a really thorough analysis, it would be necessary to use each assumption only where it is compatible with its results. "Common curvature" alone is usually justified in the major part of the spring. Calculations on this basis will be very nearly correct and are used here because of their relative simplicity. The following discussion, therefore, is based on "common curvature."

A bending moment M will produce at any cross section a change of curvature $q = M/E \Sigma I_n$, where the summation includes the leaves at that section. The relation between curvature free and loaded is given by

$$q_{free} + q = q_{loaded}$$

or

$$\frac{1}{R_{free}} + q = \frac{1}{R_{loaded}}$$

The correct signs for curvature must be introduced. Curvature is zero when the leaf is flat, and is positive under heavier loads (see the "Curvature" paragraph in Chapter 2, Section 1.) Therefore, on most springs the free curvature is negative. Numerically curvature is the inverse of radius, but the radius usually carries no plus or minus sign.

Deflections can be calculated from q, just as for beams of variable cross section, by any of the methods explained in textbooks. Where only the deflection at the point of load application is wanted, as in checking for rate, the strain energy method is the easiest of these. It is based on the equality between external work 0.5 P f and internal energy in bending $0.5 \int M q \, d\ell$ so that $f = (\int M q \, d\ell)/P$.

The load stress in any leaf at any cross section is $S_p = q E y$, where y is the distance from the neutral axis to the remotest fiber in stress. For SAE section leaves, y equals t/2 and the load stress then becomes $S_p = Mt/2\Sigma I_n$ where the summation includes the leaves at that section. The expression $(2\Sigma I/t)$ is the section modulus.

The load stress S_p would equal the total stress S (where $S = S_p + S_a$) only in springs which are fitted "dead," that is, where all leaves have the same unassembled curvature, because in that case S_a = zero. (S_a = assembly stress in leaf)

In almost all leaf springs the unassembled curvatures q_n are different in the individual leaves. In assembly a common (unloaded) curvature q_o is established which is variable along the spring even if the leaves are made to circular arcs. The common curvature q_o can be calculated from the condition that the internal bending moments of all the leaves must cancel when the spring is assembled but not loaded.

$$\Sigma S_{a_n} t_n^2 = 0$$
$$q_o = \Sigma q_n t_n^3 / \Sigma t_n^3 \text{ (weighted average)}$$

and the assembly stress S_a can be calculated for each leaf at any section in the usual manner.

$$S_a = E y (q_o - q_n)$$

The total stress S is the sum of the two stresses S_p and S_a. The sum of the internal moments produced by the stress S equals, of course, the external moment M at any section.

$$S = S_p + S_a$$
$$M = \Sigma S_n Z_n$$

where Z_n is the section modulus of a leaf. This last relation is implied in the previous ones, but may be used for re-checking.

The design and analysis of a leaf spring may therefore be considered in four stages:

1. Approximate estimate, as shown in Section 3 of this Chapter.

2. Determination of leaf lengths and individual leaf radii from a desired stress distribution at one particular load (see Section 5 in this chapter).

3. Analysis of stresses and stress ranges at various points in the spring by common curvature or point pressure.

4. Check of the soundness of the assumptions involved in number 3 and analysis by a combination of "common curvature" and "point pressure," or by strain gage measurements on the spring.

5. Sample Calculation

To design a spring for the following conditions:

Rate, as tested without center clamp	k =	17.5N/mm
Design load	P =	3500N
Opening at design load (negative)	=	−20mm
Eye diameter, (front) upturned	=	40mm
Eye diameter, (rear) upturned	=	40mm
Clearance (metal-to-metal)	x_c =	110mm
Length center to center of eyes	L =	1400mm
Length front cantilever	a =	570mm
Length rear cantilever	b =	830mm
Width	w =	63mm
Maximum stress	S_{max} =	1000 MPa
Length of seat clamp	=	100mm
Distance between inside edges of clamp bolts	=	75mm

Design the spring as a passenger car spring with tapered leaf ends except the main leaf, plus shot peening and cold setting.

Thickness of Leaves And ΣI

Since the cantilever length ratio $Y = 1.456$ exceeds 1.30, unsymmetrical formulae of Table 5.1 must be used.

$$P_{max} \text{ (at metal to metal position)} = 3500 + (17.5 \times 110)$$
$$= 5425N$$

Formula I

$$\Sigma I = \frac{kL^3}{2E} \cdot \frac{Y^2}{(y+1)^4} \cdot \frac{1}{SF}$$

$$\Sigma I = \frac{17.5 \cdot 1400^3}{2 \cdot 200 \cdot 10^3} \cdot \frac{1.456^2}{2.456^4} \cdot \frac{1}{1.10} = 6360mm^4$$

Formula II

$$t_{max} = \frac{2S\Sigma I}{P_{max}L} \cdot \frac{(Y+1)^2}{Y}$$

$$t_{max} = \frac{2 \cdot 1000 \cdot 6360}{5425 \cdot 1400} \cdot \frac{(1.456+1)^2}{1.456} = 6.94mm$$

Grading With Standard Gages

The nearest thinner standard gage is 6.70mm. Find required ΣI from Table 5.2 in gages of 6.70 or thinner in the specified 63.0 width.

$$\begin{array}{lll}
\text{1 leaf of 6.70} & \text{I} = 1522 & \text{thickness} = 6.70 \text{ mm} \\
\text{3 leaves of 6.30} & \text{I} = 3798 & \text{thickness} = 18.90mm \\
\underline{\text{1 leaf of 6.00}} & \underline{\text{I} = 1093} & \underline{\text{thickness} = 6.00 \text{ mm}} \\
& \Sigma I = 6413 & \text{thickness} = 31.60mm
\end{array}$$

1 leaf of 6.70 I = 1522 thickness = 6.70mm
2 leaves of 6.30 I = 2532 thickness = 12.60mm
<u>2 leaves of 6.00 I = 2187 thickness = 12.00mm</u>
ΣI = 6241 thickness = 31.30mm

Since ΣI = 6413 is larger and ΣI = 6241 is smaller than the calculated ΣI of 6360, only a test of sample springs made with each set of gradings will determine the selection for the spring. For this analysis we will assume that the grading of 1 leaf of 6.70, 3 leaves of 6.30 and 1 leaf of 6.00 has a test rate within the plus and minus 5% tolerance of the 17.5N/mm rate.

Stress Distribution Between The Leaves

For calculation of stress, consideration is given to the seat clamp. The inactive length in the seat for a spring without liners is estimated as the distance between the inside edges of the clamp bolts, which is 75mm. The active cantilever length for both front and rear will be 37.5mm less than the distance from center of eye to center of axle seat. The stress at 37.5mm from the center of axle seat can be calculated from

Front Cantilever $S = \dfrac{P_a \ell_a t}{2\Sigma I}$

Rear Cantilever $S = \dfrac{P_b \ell_b t}{2\Sigma I}$

Design Load = 3500N $P_a = 3500 \cdot \dfrac{830}{1400} = 2075N$

$P_b = 3500 \cdot \dfrac{570}{1400} = 1425N$

Maximum Load = 5425N

$P_a max = 3216N$

$P_b max = 2209N$

$\ell_a = 570 - 37.5 = 532.5mm$

$\ell_b = 830 - 37.5 = 792.5mm$

Front Cantilever $S_a = \dfrac{P_a \cdot 532.5t}{2 \cdot 6413} = 0.0415 P_a t$

Rear Cantilever $S_b = \dfrac{P_b \cdot 792.5t}{2 \cdot 6413} = 0.0618 P_b t$

The stress at design load (3500N) and at metal to metal position load (5425N) will be:

Leaf	Gage	Front Cantilever		Rear Cantilever	
		P=3500N	P=5425N	P=3500N	P=5425N
1	6.70	577MPa	894MPa	590MPa	915MPa
2, 3, 4	6.30	543MPa	841MPa	555MPa	860MPa
5	6.00	517MPa	801MPa	528MPa	819MPa

By using different free radii in each of the leaves, assembly stress can be added to or deducted from the load stresses of the assembled spring to obtain a more desirable stress pattern in the spring. The assembly stresses can be chosen more or less arbitrarily, except that they must be selected increasingly larger from main leaf to the short leaf, and the $\Sigma S_a t^2$ must be very nearly zero (S_a = Assembly Stress).

A negative assembly stress in the main leaf is required to reduce the maximum stress (metal to metal position) to about 830 MPa. The longitudinal and lateral forces imposed on the main leaf (with eyes) and its generally greater stress range are the reasons for reducing its bending stress.

In the thinner gage leaves the stress can be somewhat higher because of their lower stress range.

In view of the condition $\Sigma S_a t^2 = 0$, several selections of deducted and added stresses for the individual leaves may be analyzed before deciding on the best arrangement.

Leaf	Gage	Sa	Sat^2
1	6.70	−50MPa	−2245
2	6.30	−15MPa	− 595
3	6.30	+10MPa	+ 397
4	6.30	+24MPa	+ 953
5	6.00	+42MPa	+1512

$\Sigma Sat^2 = +22$ which is acceptable

The individual leaf stresses with these selected assembly stresses S_a will then be:

Leaf	Front Cantilever		Rear Cantilever	
	P=3500N	P=5425N	P=3500N	P=5425N
1	527 MPa	844 MPa	540 MPa	865 MPa
2	528 MPa	826 MPa	540 MPa	845 MPa
3	553 MPa	851 MPa	565 MPa	870 MPa
4	567 MPa	865 MPa	579 MPa	884 MPa
5	559 MPa	843 MPa	570 MPa	861 MPa

Stepping of Leaves

The overhangs shall be arranged so that at design load the chosen stresses are uniform along each leaf. This implies that loads between the leaf tips are all made equal at design load. The overhangs must then be made proportional to St^2.

Active cantilever length on front (short) end is 532.5mm and on the rear (long) end is 792.5mm.

$$\text{Overhang} = \text{Cantilever length} \cdot \frac{St^2 \text{ of the leaf}}{\Sigma St^2 \text{ of all the leaves}}$$

Leaf	t	Front Cantilever			Rear Cantilever		
		MPa (3500N)	St²	over hang	MPa (3500N)	St²	over hang
1	6.70	527	23657	115.4	540	24241	172.1
2	6.30	528	20956	102.2	540	21433	152.2
3	6.30	553	21949	107.0	565	22425	159.2
4	6.30	567	22504	109.7	579	22981	163.2
5	6.00	559	20124	98.1	570	20520	145.7
			$\Sigma St^2 = 109190$			$\Sigma St^2 = 111600$	

Leaf	Gage	Front Bar Length mm	Rear Bar Length mm	Total Bar Length mm	Mass kg/m	Mass of Bar kg
1	6.70	710	970	1680	3.270	5.494
2	6.30	472	675	1147	3.076	3.528
3	6.30	357	505	862	3.076	2.652
4	6.30	242	335	577	3.076	1.775
5	6.00	133	167	300	2.930	0.879

Total mass of steel bars = 14.328 kg

The leaf lengths are obtained by adding the successive overhangs and rounding off. For the shortest leaf 37.5mm per end must be added for the inactive length within the seat and at least 25mm for distance from bedding line to leaf end. If tip interliners are specified, the bedding line will be the ₵ of the tip interliner and the distance to leaf end will be measured from ₵ of the interliner. In this example 25mm will be used.

Leaf No.	Theoretical Lengths		Selected Lengths	
	Front	Rear	Front	Rear
1	570.0mm*	830.0mm*	570mm*	830mm*
2	479.6mm	682.9mm	505mm	725mm
3	377.4mm	530.7mm	390mm	555mm
4	270.4mm	371.5mm	275mm	385mm
5	160.7mm	208.3mm	165mm	215mm

*The Bedding Line of the Main Leaf is the ₵ of the Eye.

Calculate Mass of Steel in Spring

Spring manufacturers have established (based on their processing equipment) amounts in bar lengths required to be added for making eyes and to be deducted for tapering leaf ends.

For this Example, Eye Requirement = 3 · I.D. of Eye + 3 · Gage Thk. or 3 · 40 mm + 3 · 6.70 mm = 140 mm at each end of main leaf.

Assume tapered leaf end thickness $te = 2mm$ and a length of taper $TL = 100mm$ at front end and $TL = 150mm$ at rear end, extension of leaf end (Ext) due to tapering operation may be calculated by formula Ext. = $0.49 TL - 0.51 TL \cdot te/t$. The Ext. is deducted from the finished (specified) leaf length to obtain the bar length required before tapering.

For leaves of gage 6.30
Ext. = 0.49(100) − 0.51(100)·2/6.30 = 32.8 use 33mm
Ext. = 0.49(150) − 0.51(150)·2/6.30 = 49.2 use 50mm
For leaves of gage 6.00
Ext. = 0.49(100) − 0.51(100)·2/6.00 = 32.0 use 32mm
Ext. = 0.49(150) − 0.51(150)·2/6.00 = 48.0 use 48mm

Mass of each leaf will be calculated based on mass per meter length from Table 5.3 (see page 52).

Free Radii

Note: This section and the next section on, "Free Radii Including Shot Peening and Cold Setting," are based on the premise that the free arcs of the individual leaves are formed to a circular shape. However, it should be pointed out that in practical spring manufacturing this is rarely true, because the overriding importance lies in the desire to secure an outline for the assembled spring which conforms with the specified main leaf contour at "straight" main leaf (see "Seat Angle" paragraphs in Chapter 2, Section 1.)

The overhangs were developed so that the tip pressures at centerline of bedding are all equal at design load and the stresses in the leaves are the values calculated in the previous section on "Stress Distribution Between The Leaves." The free radii can now be determined from the condition that the change in curvature from free (unassembled) to design load must produce these stresses. Then the assumed conditions will be fulfilled.

The camber at design load is calculated from the given opening at design load, which is −20 mm., and the eye diameter, which is 40 mm).

$$\text{Camber} = \text{Opening} - \text{½ eye diameter}$$

$$= -20 - \frac{40}{2} = -40 \text{ mm}$$

Camber is converted to curvature according to the formula:

$$\text{Curvature} = -\frac{8 \times \text{camber}}{\text{length}^2}$$

Curvature is called positive in the direction of increasing load; camber is conventionally called positive in the opposite direction, therefore the minus sign.

The curvature at design load is (with a negative camber of −40 mm).

$$q = -\frac{8 \cdot (-40)}{1400^2} = +0.000163$$

The change of curvature corresponding to a stress "S" is 2S/Et and with E = 200 000 MPa gives the following free curvatures:

$$\text{Free curvature } q_a = q - \frac{2S}{Et}$$

1.50

Leaf No.	t mm	q	Front Cantilever				Rear Cantilever			
			MPa*	2S/Et	q_a	Free** Radius	MPa*	2S/Et	q_a	Free** Radius
1	6.70	0.000163	527	0.000787	−0.000624	1603	540	0.000806	−0.000643	1555
2	6.30	0.000163	528	0.000838	−0.000675	1481	540	0.000857	−0.000694	1441
3	6.30	0.000163	553	0.000878	−0.000715	1399	565	0.000897	−0.000734	1362
4	6.30	0.000163	567	0.000900	−0.000737	1357	579	0.000919	−0.000756	1323
5	6.00	0.000163	559	0.000932	−0.000769	1300	570	0.000950	−0.000787	1271

*Stress at design load 3500 N
**Without any shot peening or cold setting
Unsymmetrical springs require radii of different values for the front and rear cantilevers (see above). However, for cantilever length ratios of less than 1.30, a uniform radius may be used.

Free Radii Including Shot Peening and Cold Setting

The radii of the quenched leaves are obtained by including the changes produced by shot peening and cold setting.

Peening changes the curvature (in the direction of the change produced by load) equivalent to

$$q_p = \frac{0.003}{t^2}$$

Cold setting also changes the curvature (in the direction of the change produced by load):

$$q_s = q_c - q_a - \frac{0.014}{t}$$

q_c = average curvature at maximum set-down position (where "camber" is the arc height at this position)

$$= -\frac{8 \cdot camber}{length^2}$$

assuming the spring is to be set down 230 mm beyond metal-to-metal position,

$$q_c = -\frac{8(-230-110-40)}{1400^2} = +0.001551$$

q_a = (refer to the previous section on "Free Radii")

$\dfrac{0.014}{t}$ = curvature change due to 1400 MPa (stress at which cold setting begins)

q_b = quench curvature (or the inverse of quench radius), when spring is to be shot peened and cold set

$$q_b = q_a - q_p - q_s$$

Note: Both shot peening and cold setting add positive curvature, therefore the quench curvature (q_b) must be a larger negative value than the final free curvature (q_a).

Leaf	q_c	$\dfrac{0.014}{t}$	q_a	q_p	q_s	q_b	quench radius
Front Cantilever							
1	0.001551	0.002090	−0.000624	0.000067	0.000085	−0.000776	1289
2	0.001551	0.002222	−0.000675	0.000076	0.000004	−0.000755	1325
3	0.001551	0.002222	−0.000715	0.000076	0.000044	−0.000835	1198
4	0.001551	0.002222	−0.000737	0.000076	0.000066	−0.000879	1138
5	0.001551	0.002333	−0.000769	0.000083	none	−0.000852	1174
Rear Cantilever							
1	0.001551	0.002090	−0.000643	0.000067	0.000104	−0.000814	1229
2	0.001551	0.002222	−0.000694	0.000076	0.000023	−0.000793	1261
3	0.001551	0.002222	−0.000734	0.000076	0.000063	−0.000873	1145
4	0.001551	0.002222	−0.000756	0.000076	0.000085	−0.000917	1091
5	0.001551	0.002333	−0.000787	0.000083	0.000005	−0.000875	1143

When shot peening is not specified, replace q_b with q_r.
q_r = Quench curvature when spring is to be cold set only (that is, when shot peening is not specified) = $q_a - q_s$
The actual spring calculation, of course, takes much less time and paper than the sample on this page, which contains much text. This is particularly true if the calculation is carried out according to a definite plan.

TABLE 5.3—MASS OF LEAF SPRING BARS (kg per m length) OF MEAN DIMENSIONS PER TABLE 3.1 IN CHAPTER 3 (TABLE 1 IN SAE STANDARD J1123)

Thick-ness	Mass For Width Of									
	40.0	45.0	50.0	56.0	63.0	75.0	90.0	100.0	125.0	150.0
5.00	1.548	1.743	1.937	2.171	2.443					
5.30	1.640	1.847	2.053	2.301	2.589					
5.60	1.732	1.950	2.169	2.430	2.736					
6.00	1.855	2.089	2.322	2.603	2.930	3.497	4.198			
6.30	1.946	2.192	2.438	2.732	3.076	3.671	4.407			
6.70	2.068	2.330	2.591	2.904	3.270	3.904	4.687			
7.10	2.190	2.467	2.744	3.076	3.464	4.136	4.966	5.520		
7.50	2.311	2.604	2.897	3.248	3.658	4.368	5.245	5.830		
8.00	2.463	2.775	3.087	3.462	3.899	4.657	5.593	6.218		
8.50	2.613	2.945	3.277	3.675	4.140	4.946	5.941	6.605	8.270	
9.00	2.764	3.115	3.467	3.889	4.381	5.235	6.289	6.991	8.756	
9.50	2.913	3.285	3.656	4.101	4.621	5.523	6.635	7.377	9.241	
10.00	3.061	3.451	3.842	4.311	4.857	5.801	6.971	7.750	9.703	11.66
10.60	3.239	3.654	4.068	4.565	5.144	6.145	7.386	8.212	10.28	12.36
11.20	3.417	3.855	4.293	4.818	5.431	6.489	7.800	8.674	10.86	13.06
11.80	3.595	4.056	4.517	5.070	5.716	6.832	8.214	9.135	11.44	13.75
12.50	3.801	4.289	4.778	5.364	6.048	7.232	8.696	9.672	12.12	14.57
13.20			5.038	5.657	6.380	7.630	9.176	10.21	12.79	15.38
14.00			5.334	5.991	6.757	8.084	9.725	10.82	13.56	16.31
15.00			5.702	6.406	7.228	8.650	10.409	11.58	14.53	17.47
16.00			6.069	6.820	7.696	9.215	11.091	12.34	15.48	18.63
17.00				7.231	8.163	9.777	11.77	13.10	16.44	19.79
18.00				7.641	8.627	10.34	12.45	13.86	17.40	20.94
19.00				8.049	9.090	10.90	13.13	14.61	18.35	22.09
20.00				8.455	9.551	11.45	13.80	15.36	19.30	23.24
21.20						12.12	14.61	16.27	20.44	24.62
22.40						12.76	15.39	17.14	21.53	25.92
23.60						13.42	16.19	18.04	22.66	27.29
25.00						14.19	17.12	19.08	23.98	28.89
26.50								20.19	25.39	30.60
28.00								21.30	26.80	32.30
30.00								22.77	28.66	34.56
31.50									30.06	36.26
33.50									31.91	38.51
35.50										40.75
37.50										42.99

Note:—When using this table, divide by 1000 to reduce the length in meters to millimeters.

For derivation of formula for these tabulated values, see Appendix B.

6. Variable or Progressive Rate Springs

A popular type of spring, particularly in the light truck models, is the variable or progressive rate spring design. This is where a relatively low rate is desirable when the vehicle is in operation with only the driver and perhaps a light payload, but a higher rate is required when the vehicle is at design load.

Variable rate springs have the potential of improved ride quality over a wide load range, provided an adequate total deflection can be accommodated on the vehicle, and the design specifications assure manufacturing controls to obtain the desired spring rates.

In the light load range, the second stage leaf or leaves are inoperative. As the payload increases, the first stage, by rolling contact gradually engages the second stage. When complete contact is made between the first and second stages, the spring is then operating in the high rate range.

For a variable rate spring to produce the ride qualities desired, the specifications must clearly establish the type of load-deflection curve as shown in Fig. 2.9 with the initial contact between the two stages at a load greater than the curb load, and the complete contact at a load less than the design load. Where it is customary for constant rate springs to specify only a design load and height, it is necessary for variable rate springs to specify the curb load and height for the first stage and also the design load and height for the second stage. It is recommended that the

desired curve be drawn, showing a gradual transition from the first stage rate to the second stage rate, and that this curve be made an integral part of the specification.

Fig. 5.6 shows both the load-deflection diagram and the rate-deflection diagram for the Sample Calculation which follows. It should be understood that the rate-deflection diagram cannot be considered to be an exact depiction of the diagram obtained from an actual spring. It has been drawn without considering such details as the change in moment arm length when the spring is deflected from positive camber to flat to negative camber.

Sample Calculation

Design a variable or progressive rate spring for these conditions:

Width	w	= 63.0 mm
Length between eye centers	L	= 1320 mm
Load at curb height	P_1	= 2200 N
Load at design height	P_2	= 6000 N
Rate at curb or light loads	k_1	= 25 N/mm
Rate at design or heavy loads	k_2	= 43 N/mm
Metal to metal clearance	x_c	= 100 mm
Deflection between P_1 and P_2 loads		= 120 mm
Length of inactive spring seat		= 80 mm

Note that the deflection between curb and design loads is the difference in heights between curb and design load positions.

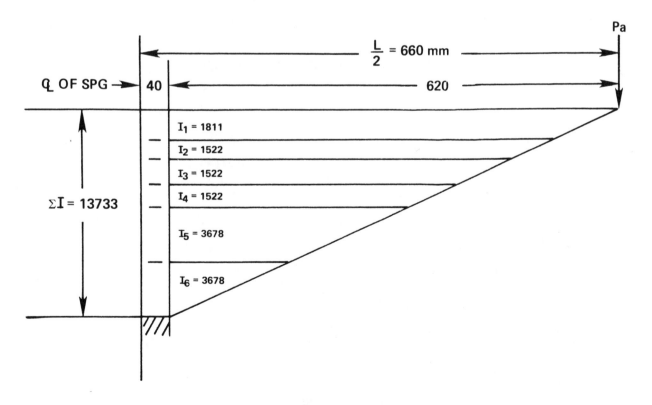

Fig. 5.5—Leaf lengths determined from moment of inertia diagram

1.53

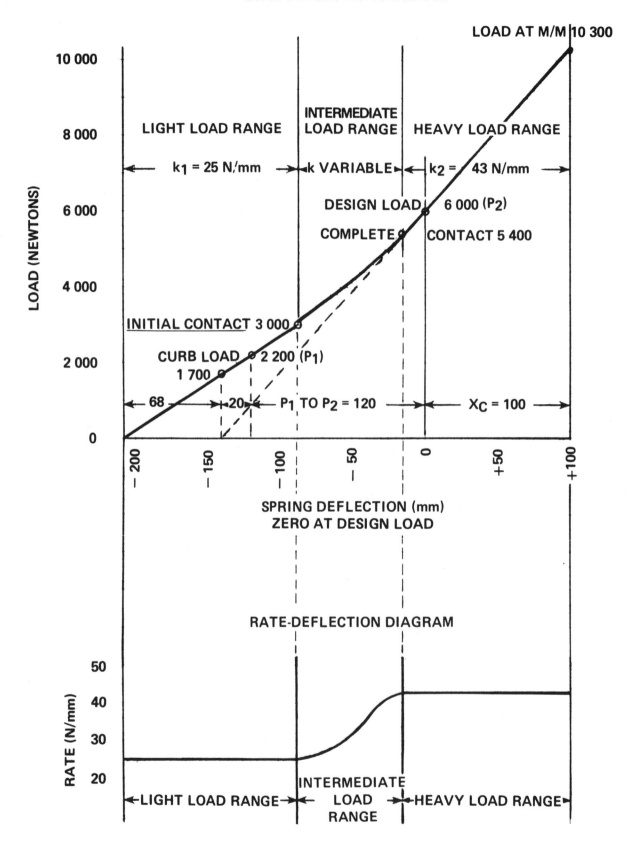

Fig. 5.6—Variable rate spring: Load-deflection and rate-deflection diagrams

All leaves except the main leaf are to have tapered leaf ends.

This spring will be designed assuming that the front and rear length ratio Y=b/a is 1.30 or less. With this condition the symmetrical formulae from Table 5.1 can be used.

Step One—For the first stage, consideration will be given to a three or a four leaf design. Because of the extended lengths of these leaves when operating without contact with the second stage, an SF value of 1.40 will be used in the design calculations, as explained in Section 2 of this chapter.

Calculate the moment of inertia requirement for the first stage leaves.

From Table 5.1 formula I

$$\Sigma I_1 = \frac{k_1 L^3}{32 E \cdot SF} = \frac{25 \cdot (1320)^3}{32 \cdot 200 \cdot 10^3 \cdot 1.40} = 6417 \text{ mm}^4$$

Selection Of Gages For The First Stage—Three leaves of 7.50 gage have $\Sigma I = 6402$ mm^4 or, considering a four leaf design, one leaf of 7.10 gage and three leaves of 6.70 gage have $\Sigma I = 6377$ mm^4. Either the three or the four leaf design should produce a first stage rate close to 25 N/mm. The three leaf design of course would be more economical.

Step Two—Calculate the moment of inertia requirement for the second stage leaves.

Consideration will be given to two leaves and when selecting the four leaf design for the first stage there will be a total of six leaves in the spring. Since all six leaves will be in active operation after the first stage makes complete contact with the second stage, the SF value will be 1.10. (see Section 2 in this chapter.)

From Table 5.1 formula I

$$\Sigma I_2 = \frac{k_2 L^3}{32 E \cdot SF} = \frac{43 \cdot (1320)^3}{32 \cdot 200 \cdot 10^3 \cdot 1.10} = 14048 \text{ mm}^4$$

The moment of inertia requirement for the two second stage leaves will be the difference between ΣI_2 and ΣI_1 or $14048 - 6377 = 7671$ mm^4.

Selection Of Gages For The Second Stage—Two leaves of 9.00 gage have a moment of inertia of 7356 mm^4. Adding this moment of inertia to the four leaf first stage as selected under the previous section, "Selection Of Gages For The First Stage," the total moment of inertia will be 13733 mm^4. This is within 2% of the calculated 14048 mm^4. It is recommended that sample springs are made and tested to verify the selected gages.

Length Of Leaves

With the gages of all the leaves established, the moment of inertia diagram can be drawn for the spring, see Fig. 5.5.

The length of leaf No. 6 will be determined by its mo-

ment of inertia relative to the total ΣI plus the inactive spring seat length ($\frac{1}{2} \cdot 80$ mm) plus the length at the ends for bedding (25 mm).

$$\text{Length of No. 6 leaf} = 2 \cdot \left[\frac{3678}{13733} \cdot 620 + 40 + 25 \right] =$$

$$2(166 + 65) = 2 \cdot 231 \text{ mm}$$

The length of leaf No. 5 will be determined in the same manner, where the moment of inertia I will equal 2(3678) = 7356 mm^4.

$$\text{Length of No. 5 leaf} = 2 \left[\frac{7356}{13733} \cdot 620 + 65 \right] =$$

$$2(332 + 65) = 2 \cdot 397 \text{ mm}$$

For eye support the length of the No. 2 leaf will extend to the center line of load application with the bedding line between leaf No. 1 and No. 2 at 25 mm from the end of the leaf.

The length between No. 2 leaf and No. 5 leaf is $660 - 397 = 263$mm.

The stepping between leaves No. 2, No. 3, and No. 4 will be established proportionate to their individual moment of inertia values. Since the moment of inertia of all three leaves is the same, the stepping will be equal or $263 \div 3 = 87.67$ mm.

The leaf lengths are as follows:

Leaf	Gage	I	Front End	Rear End	Total
1	7.10	1811	660	660	1320 mm
2	6.70	3333	660	660	1320 mm
3	6.70	4855	572	572	1144 mm
4	6.70	6377	484	484	968 mm
5	9.00	10055	397	397	794 mm
6	9.00	13733	231	231	462 mm

Stresses

Stress calculations for variable rate springs are more involved than they are for uniform rate springs where all the leaves are operative under all loadings.

Stresses in the main leaf will be calculated at the following sections from which a stress diagram can be drawn. See Fig. 5.8.

Section	Moment Arm ℓ	I mm⁴	Stress (MPa) in Main Leaf at Load of			
			2200 N CURB	3000 N CONTACT	6000 N DESIGN	10300 N M to M
A	660 − 572 + 25 = 113	3333	132	181	361	620
B	660 − 484 + 25 = 201	4855	162	220	441	757
C	660 − 397 + 25 = 288	6377	177	240	481	826
D*	660 − 40 = 620	6377	380	518	—	—
D	660 − 40 = 620	13733	—	—	637	982

For moment arm lengths see Fig. 5.8.
*At loads of 2200 N and 3000 N, only leaves 1, 2, 3, 4 are operative (I = 6377)

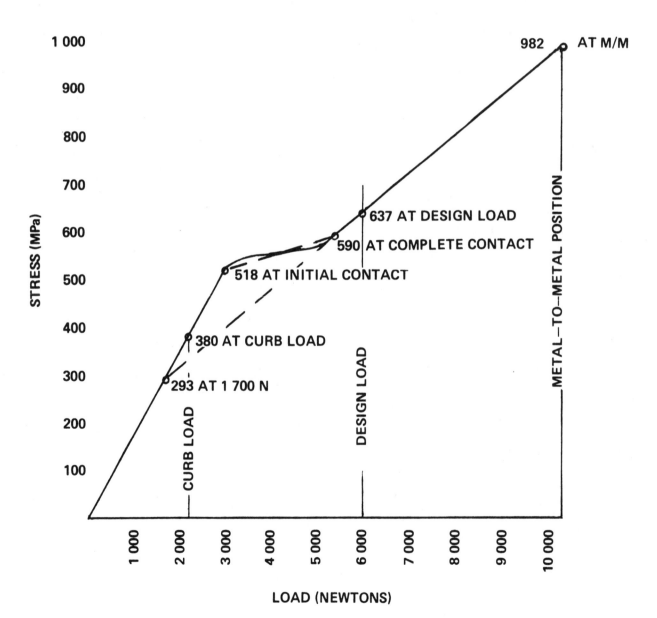

STRESS—LOAD DIAGRAM
FOR MAIN LEAF
AT LINE OF ENCASEMENT (SECTION D)

Fig. 5.7—Variable rate spring: Stress-load diagram for main leaf
at line of encasement

1.56

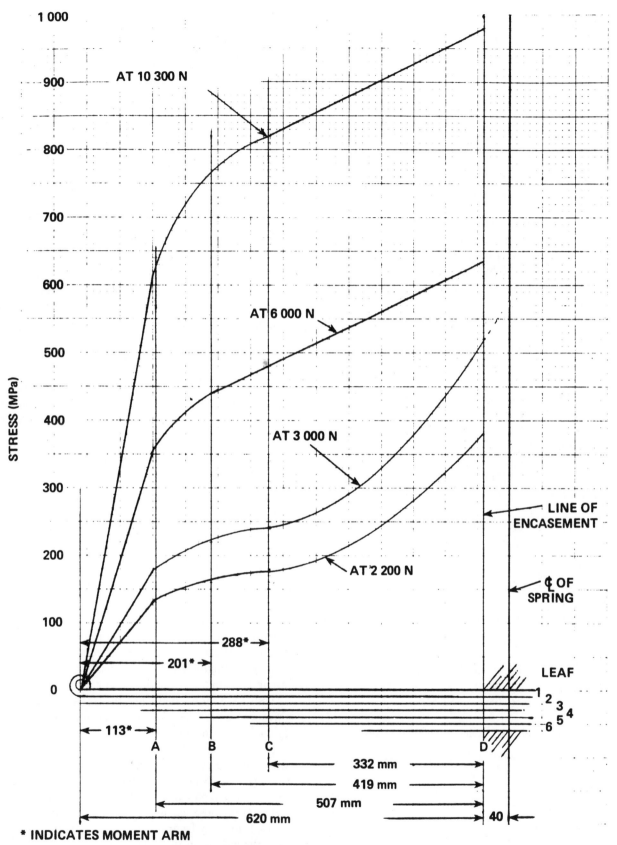

STRESS (MPa)

1 000

900   AT 10 300 N

800

700

600
                        AT 6 000 N

500

400
                AT 3 000 N

300
                                                    LINE OF
                                                    ENCASEMENT

200   AT 2 200 N
                                                    ¢ OF
                                                    SPRING
100
                    288*
            201*   LEAF
  0   1
                                                      2
                                                       3
                                                        4
        113*   5
                                                     6
      A   B   C   D

                                332 mm
                                419 mm
                            507 mm
                620 mm   40

* INDICATES MOMENT ARM

Fig. 5.8—Variable rate spring: Stress distribution along main
leaf under different loads

1.57

Sample stress calculations:

$$S_A(\text{at } 2200 \text{ N}) = \frac{P\ell t}{4I} = \frac{2200 \cdot 113 \cdot 7.10}{4 \cdot 3333} = 132 \text{ MPa}$$

$$S_B(\text{at } 3000 \text{ N}) = \frac{3000 \cdot 201 \cdot 7.10}{4 \cdot 4855} = 220 \text{ MPa}$$

$$S_C(\text{at } 6000 \text{ N}) = \frac{6000 \cdot 288 \cdot 7.10}{4 \cdot 6377} = 481 \text{ MPa}$$

$$S_D(\text{at } 3000 \text{ N}) = \frac{3000 \cdot 620 \cdot 7.10}{4 \cdot 6377} = 518 \text{ MPa}$$

The stresses at line of encasement (section D) for loads between design position and metal-to-metal position can be calculated by establishing the amount of load on the spring which is not supported by the second stage leaves after complete contact is made between the first and second stages.

The load on the spring at metal-to-metal position equals

$$6000 \text{ N (design load)} + 43 \text{ N/mm} \cdot 100 \text{ mm } (x_c) = 10300 \text{ N}$$

In Fig. 5.6, the dash line extension of the second stage rate line intersects the zero load at a position of minus 140 mm. At this position the first stage leaves have supported 1700 N (68 mm·25 N/mm); this load is not transmitted to the second stage leaves.

The stress for that 1700 N load unsupported by the second stage leaves is:

$$\frac{1700 \cdot 620 \cdot 7.10}{4 \cdot 6377} = 293 \text{ MPa}$$

At metal-to-metal position the stress for the load supported by all the leaves is:

$$\frac{(10300 - 1700) \cdot 620 \cdot 7.10}{4 \cdot 13733} = 689 \text{ MPa}$$

Thus the total stress at metal-to-metal position equals:

$$293 + 689 = 982 \text{ MPa}$$

Similarly, the total stress at design position equals:

$$293 + \frac{(6000 - 1700) \cdot 620 \cdot 7.10}{4 \cdot 13733}$$

$$= 293 + 344 = 637 \text{ MPa}$$

Figure 5.7 presents a detailed diagram of the stresses in the main leaf at line of encasement (section D).

The stresses in leaves No. 2, 3, 4 are 6% lower than the stresses in the main leaf since their gage thickness is 6.70 mm compared to 7.10 mm in the main leaf. The ratio of these gage thicknesses is 6.70/7.10 = 94%.

Since 1700 N is not supported by the second stage leaves, the stress in the second stage (leaves 5 and 6 with 9.00 mm gage thickness) can be calculated by using the transmitted load of 10300 − 1700 = 8600 N and a moment arm length from line of encasement to bearing line between leaves No. 4 and 5 (section C). This equals the length of No. 5 leaf (397 mm) less the inactive seat length (40 mm) and less the amount of bedding length (25 mm), thus equaling 332 mm.

The moment of inertia (I) for the two second stage leaves is 7356 mm^4. Then at metal-to-metal position, where the spring load is 10300 N, the stress in leaves No. 5 and 6 is:

$$\frac{(10300 - 1700) \cdot 332 \cdot 9.00}{4 \cdot 7356} = 873 \text{ MPa}$$

Note that these calculated stresses do not take into account any nip between the leaves which for the main leaf would slightly reduce the stress from the calculated value.

Optional Second Stage

The second stage can be made of a single leaf, provided the stress levels are acceptable and equipment is available for making the required taper contour of the leaf ends.

7. Strength of Spring Eyes

The bending stress in the main leaf adjacent to the spring eye or in the eye itself, due to longitudinal forces acting on the spring, can be calculated by the following formula:

$$S = \frac{3F(D + t)}{t^2 w}$$

The stress calculated by the above formula applies to upturned, downturned, and Berlin eyes, except that in the case of the Berlin eye the stress is zero under a compression type longitudinal force. Thus, the Berlin eye has a strength advantage over the other two types.

Tests have shown that when the stress, as calculated by this formula, reaches the yield point of the material, the eye will begin to open.

If leaf springs are used in Hotchkiss drive suspensions where they carry the longitudinal force due to braking and driving, a large factor of safety should be allowed. The longitudinal forces in such applications are often much larger than those calculated from static forces, because shock loads may be applied either by the driver, by hopping the wheels on the road, or by longitudinal shake of the axles. The maximum calculated stress should therefore not be allowed to exceed 350 MPa. Where vehicles such as buses are subject to frequent start and stop operations, the required eye strength should be determined by life tests which reflect these operating conditions.

The bending stress resulting from the press fit of metal bushings in spring eyes may be calculated by the following formula:

$$S = \frac{4}{\pi} \frac{\Delta E t}{(D + t)^2}$$

The maximum axial force and torque obtained without slippage between the bushing and spring eye will depend upon the finish of the engaging surfaces, their hardness, and the degree of lubrication present when the press fit is made.

Note that the stresses produced by bushing press fit and by horizontal forces on the main leaf eye may be algebraically additive.

Nomenclature

Δ = Difference between OD of bushing and ID of eye
S = Stress
D = ID of spring eye
t = Thickness of leaf at eye
w = Width of leaf at eye
E = Modulus of elasticity = 200 000 MPa
F = Longitudinal force

Chapter 6
Installation Effects

1. Characteristics of Shackles

The rate of deflection of a spring is defined as dP/df, the slope of a tangent to the load-deflection curve. A leaf spring tested on the rollers of a universal load scale has a rate which varies only slightly as the chord between the ends of the spring changes in length. However, due to the way the spring is installed, the rate on the vehicle may be different from that obtained on the load scale.

Knowledge of installation effects will reduce the amount of experimental testing required to obtain the desired installed rates, and it can also be useful in obtaining variable rates where desirable, through the use of ordinary springs with particular arrangement of shackles.

The installation may involve contact pads, bushed eyes with one spring end fixed and the other end shackled, or bushed eyes with both spring ends shackled.

If contact pads are used, the active length of the spring may decrease as it deflects under load, and the rate may accordingly be increased (see Fig. 3.17A).

When one eye is fixed and the other eye is shackled, two effects will result. As the spring deflects, the length of the chord changes, and the shackle will swing and change its angle. In swinging, the shackle may lift or lower the eye of the spring and with it the point of load application. This is the first shackle effect. When the shackle is not perpendicular to the datum line of the spring, the shackle load will have a longitudinal component either compressing or stretching the spring between the eyes. Compressing will decrease the rate of the spring, while stretching will increase the rate. This is the second shackle effect.

In the first shackle effect, the raising or lowering of the shackle eye changes direction when the shackle passes through the perpendicular position and also when the spring passes through the position where the distance between the eyes is a maximum. This is represented by "flat linkage," where the three-link equivalent (see Chapter 4, Sections 4 and 5), is stretched out in one straight line, except that in springs with eyes of unequal size or unequal offset, the "flat linkage" is only approximately flat. The terms "compression shackle" and "tension shackle" refer to the fact that load on the spring induces a compressive force in a compression shackle and a tensile force in a tension shackle.

In the second shackle effect, the compressing or stretching of the spring changes when the shackle passes through the perpendicular position. The amount of shackle effect depends on the load which the spring carries rather than on the rate of the spring. The rate of an installed spring with shackle may easily be 50% higher or lower than the nominal rate. To simplify calculations, the charts in Figs. 6.2–6.5 may be used for springs with one shackled eye and one fixed eye.

These charts give the installed rate in percent of the nominal rate. The nominal rate is the rate calculated as shown in Chapter 5, Section 1, or the rate obtained from tests with spring eyes on rollers and with the spring in the fully extended position represented by "flat linkage." For calculating these charts, the bending stiffness of the spring was replaced by torsional stiffness in the hinges of the equivalent linkage. Experimental checks confirmed the calculated results within the limits of accuracy of the measurements.

The rate of a spring with shackle depends on the nominal rate of the spring, the position of the shackle, the length of the shackle, the "camber" of the spring, and the load on the spring.

For the purpose of these charts, it is convenient to express the "camber" of the spring by a term which is characteristic of the linkage directly and includes the effect of eye offset and shackle. The *geometric deflection* is used for this purpose. It is measured along the line of load application. It is the distance by which the load application point is displaced from the position at which the equivalent linkage is flat (maximum distance between eyes). (See Fig. 6.1.) The geometric deflection is zero when the linkage is flat or stretched, positive for increased loads (with increasingly negative spring camber), and negative for decreased loads (with increasingly positive spring camber). It includes the effect of shackle displacement. In Figs. 6.2–6.5, the geometric deflection is given as a percentage of the (stretched) spring length (L), so that the curves can be used for all sizes of springs.

The "stretched" length is different from either the straight or the active length, but it is sufficiently accurate to use the straight length in this instance.

To express the effect of load on the rate, it is most convenient to define the load by the *geometric free camber* of the spring. When the linkage is flat, the geometric deflection is zero, and in that position the load does not depend on the shackle arrangement. This load is called P_o. The geometric free camber is defined as the quotient of the load P_o divided by the nominal rate of the spring (P_o/k). In Figs. 6.2–6.5, the geometric free camber is expressed as a fraction of the spring length, so that the curves can be used for all sizes of springs. Curves have been drawn for various values of geometric free camber, ranging from 0.05L–0.20L.

The curves for symmetrical springs are shown in Fig. 6.2 (with compression shackle) and in Fig. 6.3 (with ten-

A – SPRING WITH UPTURNED EYES AND COMPRESSION SHACKLE

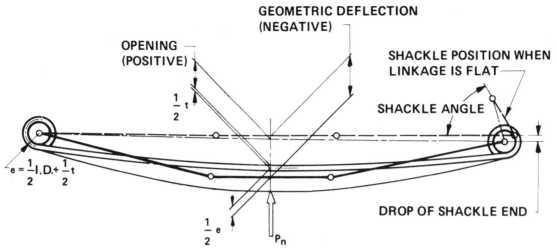

$$\text{GEOMETRIC DEFLECTION} = -\left(\text{OPENING} + \frac{1}{2}t + \frac{1}{2}e + \frac{1}{2}\text{ SHACKLE DROP}\right)$$

$$\text{GEOMETRIC DEFLECTION} = -\left(\text{OPENING} + \frac{1}{4}\text{ I.D.} + \frac{3}{4}t\right) \quad [\text{SHACKLE DROP NEGLECTED}]$$

B – SPRING WITH DOWNTURNED EYES AND TENSION SHACKLE

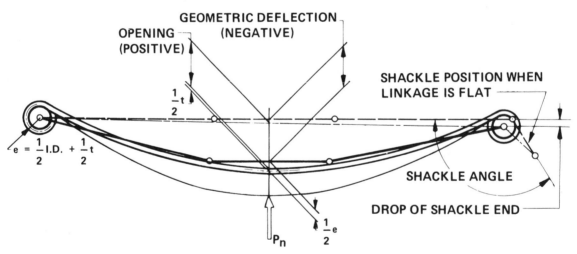

$$\text{GEOMETRIC DEFLECTION} = -\left(\text{OPENING} + \frac{1}{2}t - \frac{1}{2}e + \frac{1}{2}\text{ SHACKLE DROP}\right)$$

$$\text{GEOMETRIC DEFLECTION} = -\left(\text{OPENING} - \frac{1}{4}\text{ I.D.} + \frac{1}{4}t\right) \quad [\text{SHACKLE DROP NEGLECTED}]$$

Fig. 6.1—Explanation of geometric deflection

sion shackle). Refer to Fig. 6.1 for sketches of springs with compression shackle or with tension shackle. There are also curves for unsymmetrical springs in which the length ratio b/a or Y equals 2. These are shown in Fig. 6.4 (with compression shackle) and in Fig. 6.5 (with tension shackle). It is recommended that the curves in Figs. 6.2 and 6.3 be used for springs with length ratio up to 1.40, the curves in Figs. 6.4 and 6.5 for springs with length ratio exceeding 1.80. Where the length ratio is closer to 1.60, an interpolation between the curves may be used if they differ sufficiently to demand such a step.

In the interest of readability the number of curves on each chart has been kept to a minimum. Near zero geometric deflection, the curves for a given geometric free camber follow similar patterns regardless of shackle length. Curves for different geometric free camber values can be interpolated or extrapolated as follows: In a given chart (given shackle angle), with a given shackle length at a given abscissa (geometric deflection), equal differences in geometric free camber call for equal differences in ordinates (shackled rates).

Example: In the symmetrical spring with tension shackle of 0.100L length and with 110 deg shackle angle, the shackled rate at −0.087L geometric deflection is 0.80k for $P_o/kL = 0.100$ and 0.90k for $P_o/kL = 0.150$; therefore, it will be 0.95k for $P_o/kL = 0.175$.

Another example: In the unsymmetrical spring with compression shackle of 0.075L and with 90 deg shackle angle, the shackled rate at −0.079L geometric deflection is 0.70k for $P_o/kL = 0.200$ and 0.90k for $P_o/kL = 0.150$; therefore, it will be 1.00k for $P_o/kL = 0.125$.

The charts show that with compression shackles the load rate drops off to zero at high geometric deflections (or loads). To obtain the increase of rate with load which is generally desirable, springs with compression shackles should be used with negative geometric deflection (positive camber or upturned eyes) at light static load. In general, this type requires the use of stronger bumpers or helper springs.

With tension shackles the rate curves rise sharply with higher geometric deflections (or loads), which is desirable. The curves indicate that at low shackle angles the springs with relatively high static deflection should stand near zero geometric deflection under light static loads, while springs with relatively low static deflection should stand with positive geometric deflection (negative camber or downturned eyes) under light static loads.

The shackle effects are more pronounced with shorter shackles. Curves are given for three lengths of shackles. The relation of shackle length to spring length is what is important, and the shackle length is therefore expressed in percent of the spring length. Solid line curves show what happens if the shackle length equals 10% of spring length. Dot-and-dash line curves apply when the shackle length equals 7.5% of spring length, and broken line curves when it equals 5% of spring length.

The shackle setting is the remaining factor which must be considered. The different shackle settings are identified by the minimum shackle angle, that is, the shackle angle when the distance between spring eyes is longest (or the spring has zero geometric deflection, or the linkage is flat). It is measured as shown in Fig. 6.1. Curves are given on each of the charts for six different minimum shackle angles ranging from 60–110 deg.

Shackle angles increase when the spring is deflected from the flat position; therefore, the shackle may fall in line with the adjacent equivalent link. If this happens in rebound, the shackle may "toggle" or swing over from a compression position into a tension position. This danger exists often on compression shackles with large minimum shackle angles, and it is then necessary to provide either a rebound stop for the spring itself or for the shackle.

As an example of the application of these charts, the installed rate is determined for a spring with the following specifications:

Length center to center of eyes	L	= 1270 mm
Length fixed (front) cantilever	a	= 560 mm
Length shackled (rear) cantilever	b	= 710 mm
Nominal rate, clamped between rubber pads	k	= 21 N/mm
Design load	P	= 3340 N
Opening (positive) at design load		= 25 mm
Upturned eyes, inside diameter	ID	= 38 mm
Thickness of main leaf	t	= 6.70 mm
Clearance (metal-to-metal)	x_c	= 115 mm
Rebound	x_r	= 135 mm
Compression shackle, length		= 100 mm
Minimum shackle angle		= 75 deg

Since the cantilever length ratio $Y = b/a = 710/560 = 1.27$ is appreciably less than 1.40, the rate curves for symmetrical springs will give adequately accurate information. (See earlier discussion in this Section.)

The opening at zero geometric deflection is (see Figs. 4.1 and 4.2, and Figs. 6.1 and 6.6.)

$$-\frac{e + t}{2} = -\frac{0.5(ID + t) + t}{2}$$
$$= -\frac{0.5(38 + 6.70) + 6.70}{2}$$
$$= -14.53 \text{ mm}$$

The geometric deflection at design load is, therefore:

$$-25 - 14.53 = -39.53 \text{ mm}$$

This equals −3.1% of the spring length. Since it was obtained by neglecting shackle effects, a correction may have to be introduced later.

The load at zero geometric deflection (P_o) will be the same as the load on rollers at 39.53 mm beyond static deflection.

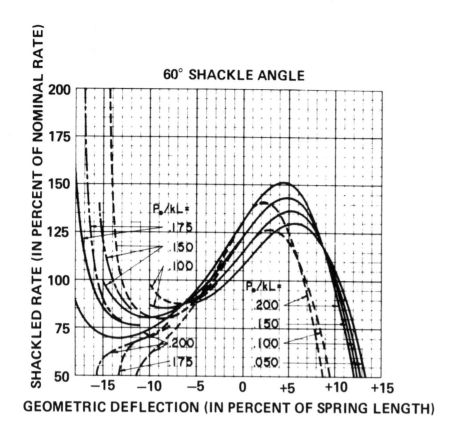

SHACKLE ANGLE AT

ZERO GEOMETRIC

CAMBER INDICATED

GEOMETRIC FREE

CAMBER AS FRACTION

OF SPRING LENGTH

INDICATED

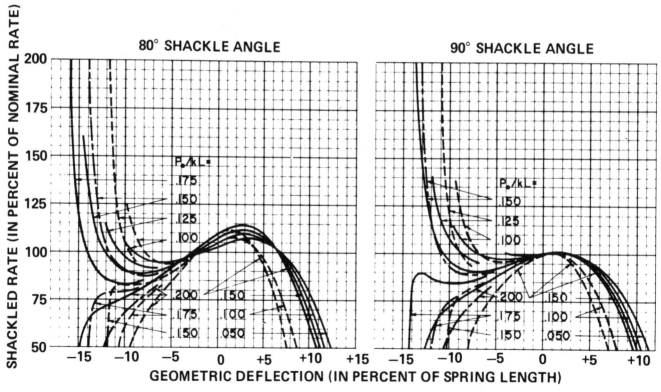

Fig. 6.2—Rate variations of springs with one compression
shackle and length ratio 1.00–1.40

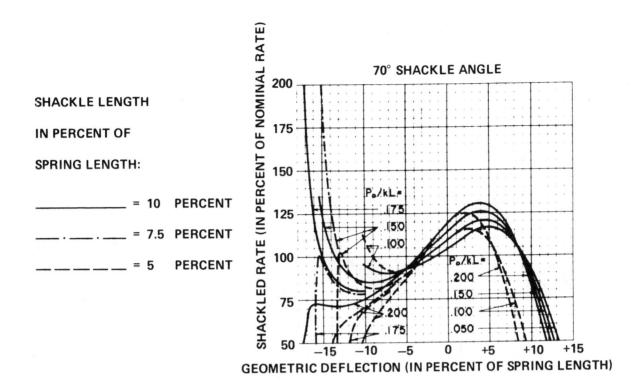

SHACKLE LENGTH

IN PERCENT OF

SPRING LENGTH:

_____ = 10 PERCENT

—— · —— · —— = 7.5 PERCENT

— — — — — = 5 PERCENT

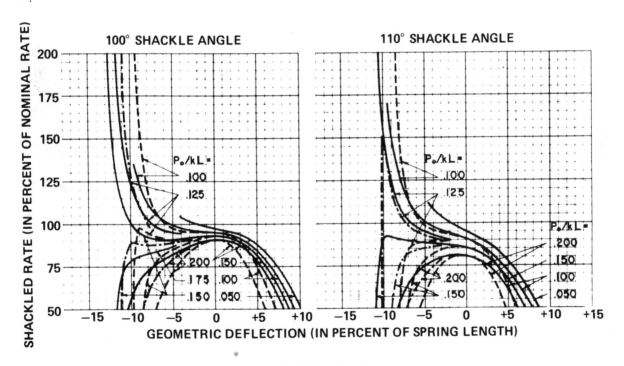

Fig. 6.2 continued

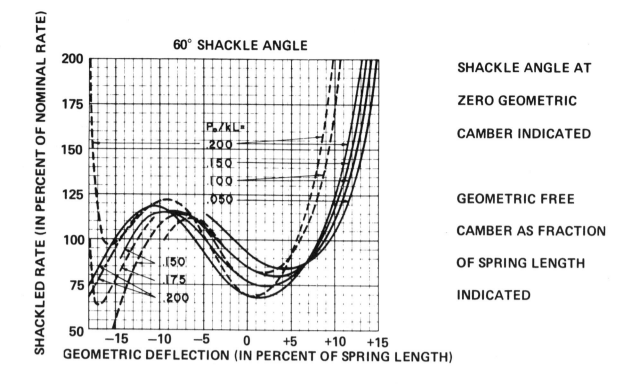

60° SHACKLE ANGLE

SHACKLE ANGLE AT
ZERO GEOMETRIC
CAMBER INDICATED

GEOMETRIC FREE
CAMBER AS FRACTION
OF SPRING LENGTH
INDICATED

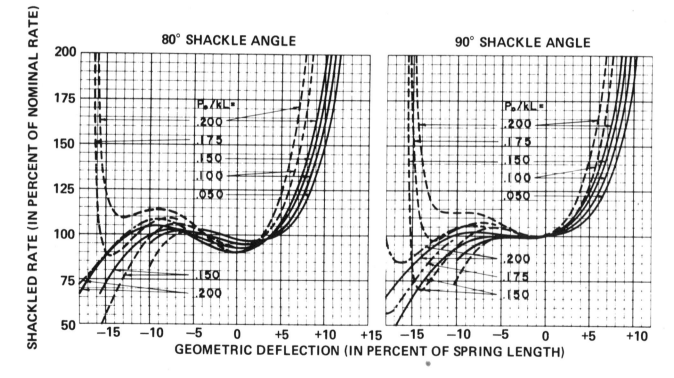

Fig. 6.3—Rate variations of springs with one tension shackle
and length ratio 1.00–1.40

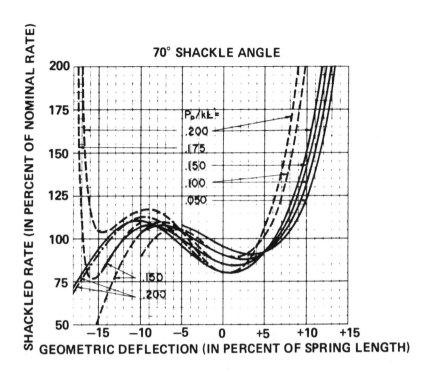

SHACKLE LENGTH

IN PERCENT OF

SPRING LENGTH:

_____ = 10   PERCENT

___ . ___ . ___ = 7.5 PERCENT

___ ___ ___ ___ = 5   PERCENT

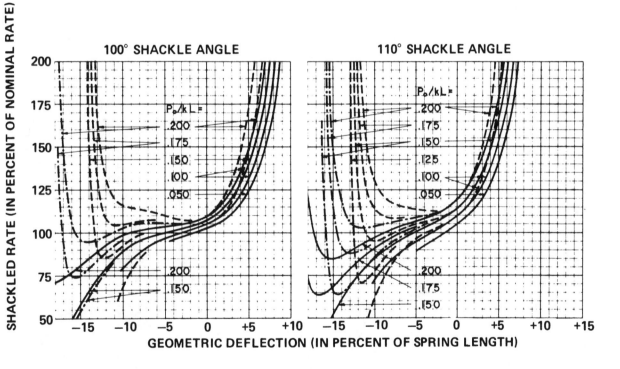

Fig. 6.3 continued

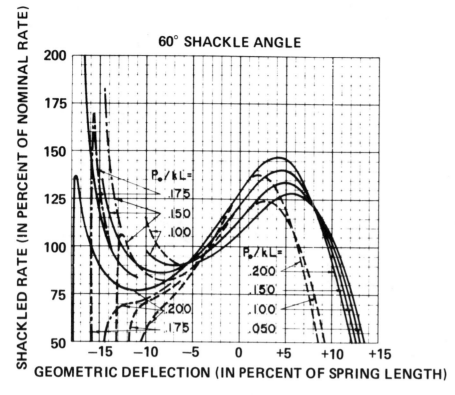

SHACKLE ANGLE AT
ZERO GEOMETRIC
CAMBER INDICATED

GEOMETRIC FREE
CAMBER AS FRACTION
OF SPRING LENGTH
INDICATED

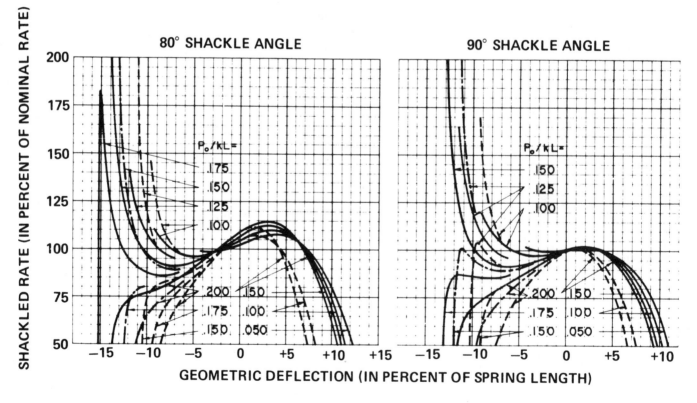

Fig. 6.4—Rate variations of unsymmetrical springs with one
compression shackle and length ratio exceeding 1.80

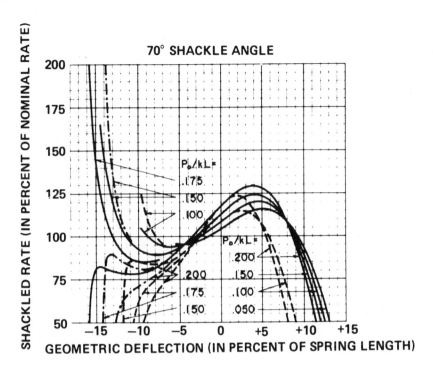

SHACKLE LENGTH

IN PERCENT OF

SPRING LENGTH:

─────────── = 10 PERCENT

─── · ─── · ─── = 7.5 PERCENT

─ ─ ─ ─ ─ ─ = 5 PERCENT

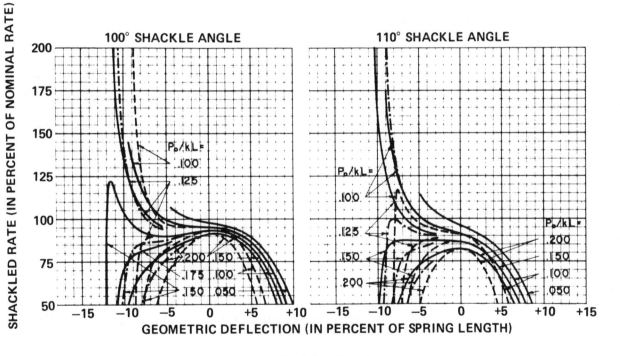

Fig. 6.4 continued

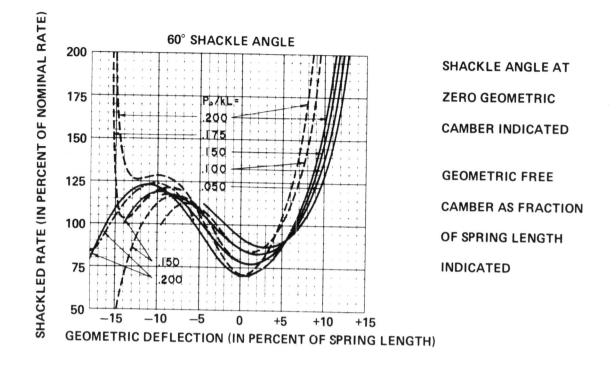

SHACKLE ANGLE AT
ZERO GEOMETRIC
CAMBER INDICATED

GEOMETRIC FREE
CAMBER AS FRACTION
OF SPRING LENGTH
INDICATED

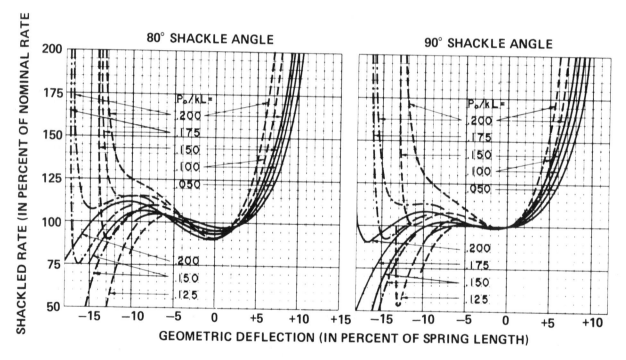

Fig. 6.5—Rate variations of unsymmetrical springs with one tension shackle and length ratio exceeding 1.80

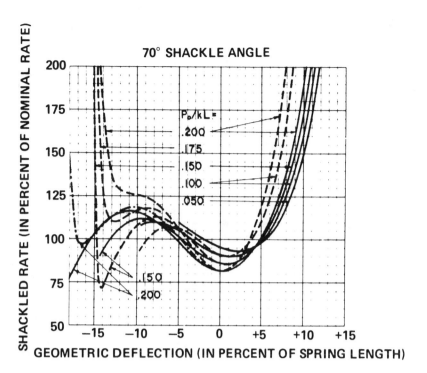

SHACKLE LENGTH

IN PERCENT OF

SPRING LENGTH:

———————— = 10 PERCENT

—— · —— · —— = 7.5 PERCENT

— — — — — = 5 PERCENT

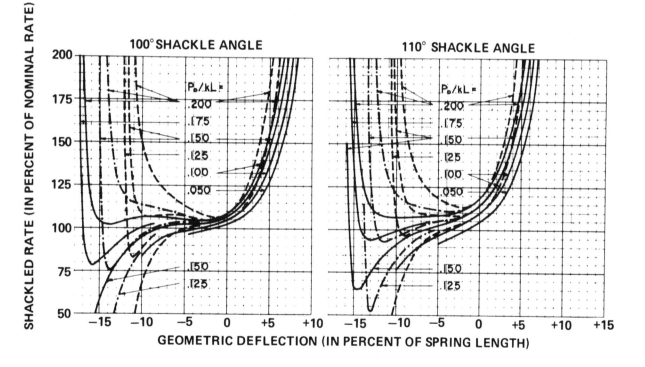

Fig. 6.5 continued

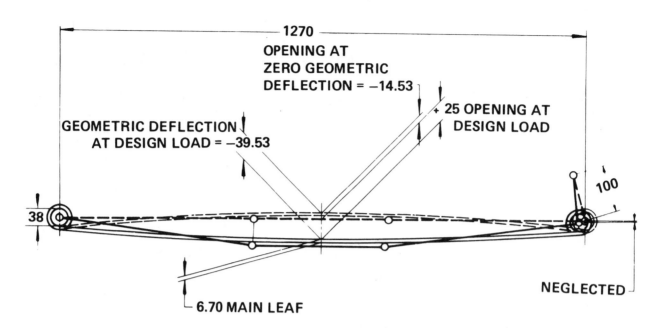

1270

OPENING AT
ZERO GEOMETRIC
DEFLECTION = −14.53

+ 25 OPENING AT
DESIGN LOAD

GEOMETRIC DEFLECTION
AT DESIGN LOAD = −39.53

38

100

NEGLECTED

6.70 MAIN LEAF

SOLID LINES: OPENING AS SPECIFIED FOR
DESIGN LOAD (ON ROLLERS)
BROKEN LINES: AT ZERO GEOMETRIC DEFLECTION

Fig. 6.6—Spring for sample use of charts

$$P_o = 3340 + 39.53(21) = 3340 + 830 = 4170 \text{ N}$$

The geometric free camber is:

$$P_o/k = 4170/21 = 198.58 \text{ mm}$$

This equals 15.6% of spring length, so that an interpolation about one-quarter of the way from curves for P_o/kL = 0.150 to curves for P_o/kL = 0.175 will be in order.

The shackle length is 100 mm which equals 7.9% of spring length. This makes the "dot-and-dash" curves nearly applicable, or an interpolation nearly halfway from broken line curves to solid line curves.

The geometric deflection at the metal-to-metal clearance position will be −39.53 + 115 = 75.47 mm. This equals 5.9% of spring length.

The geometric deflection at the rebound position will be −39.53 − 135 = − 174.53 mm. This equals −13.7% of spring length.

The specified minimum shackle angle is 75 deg, midway between the curves for 70 and 80 deg.

The following values for shackled rate (in percent of nominal rate) may now be established:

1. At design load position (where geometric deflection equals −3.1% of spring length):

Shackle Angle:	70 deg			80 deg			75 deg*
Shackle Length:	0.050L	0.100L	0.080L*	0.050L	0.100L	0.080L*	0.080L
P_o/kL = 0.150	98	100		96	98		
0.200	98	103		95	98		
0.156*	98	100	99	96	98	97	98

2. At metal-to-metal clearance position (where geometric deflection equals +5.9% of spring length):

Shackle Angle:	70 deg			80 deg			75 deg*
Shackle Length:	0.050L	0.100L	0.080L*	0.050L	0.100L	0.080L*	0.080L
P_o/kL = 0.150	98	122		85	105		
0.200	96	126		82	106		
0.156*	98	122	112	85	105	97	105

3. At rebound position (where geometric deflection equals − 13.7% of spring length):

Shackle Angle:	70 deg			80 deg			75 deg*
Shackle Length:	0.075L	0.100L	0.080L*	0.075L	0.100L	0.080L*	0.080L
P_o/kL = 0.150	117	102		160	126		
0.175	85	87		62	96		
0.156*	109	98	107	136	118	132	120

*Interpolated values.

To summarize:

	Shackled Rate	
	% of nominal rate	N/mm
At design load	98	20.6
At metal-to-metal clearance	105	22.0
Average in jounce range	102	21.4
At rebound	120	25.2
Average in rebound range	109	22.9

The data (and the curves) show that a reduction in shackle angle would produce a higher rate in the jounce

1.72

range and simultaneously a lower rate in the rebound range.

To obtain values for load at different deflections, the geometric deflection must be multiplied by the average rate between zero and that deflection. The resulting product is the load difference between that deflection and zero geometric deflection. The load at zero geometric deflection is not affected by the shackle and can therefore be calculated as shown above (4170 N).

For this example, the load at -39.53 mm geometric deflection would be:
$$4170 - 39.53(21.4) = 4170 - 846 = 3324 \text{ N},$$
and the geometric deflection under 3340 N design load would be:
$$-39.53 + 16/21 = -39.53 + 0.76 = 38.77 \text{ mm}.$$
This is the correction mentioned above. It is small in this case, but may be large in others.

Similar effects on rate may be produced by lengthwise tension or compression of the spring and can be caused by driving thrust, braking, or nonhorizontal position of the springs.

2. Windup of Springs

In many applications, leaf springs are loaded not only by vertical forces but also by horizontal forces and torques in the longitudinal vertical and transverse vertical planes.

Torque in the longitudinal vertical plane (windup) is usually produced by a longitudinal force applied above or below the spring seat. Where the spring does not carry the longitudinal force but only the vertical force and the windup torque, stresses and deflections can be calculated fairly simply with reasonable accuracy as shown in the section, "Springs Carrying Vertical Load and Windup Torque." However, when the longitudinal force is also carried by the spring, as in the Hotchkiss drive suspension, generally valid formulae become very complicated, and the problem must be treated as described in the section "Springs Carrying Vertical Load, Windup Torque, and Longitudinal Load."

In all formulae given in this section, the total length should be used as the flat length L. This will give correct values for deflections and sufficiently accurate ones for stresses.

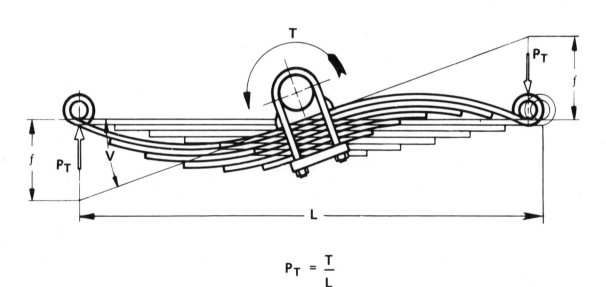

$$P_T = \frac{T}{L}$$

$$f = \frac{P_T}{.5k} = \frac{T}{.5kL}$$

$$V \approx \text{TAN } V = \frac{f}{.5L} = \frac{T}{.25kL^2}$$

$$\text{WINDUP STIFFNESS } \omega = \frac{T}{V} = \frac{kL^2}{4}$$

Fig. 6.7—Windup of a symmetrical spring under brake application

1.73

The stress formulae apply only to springs built as uniform strength beams. The deflection formulae, however, are not so restricted; they apply to all springs having certain rate and length ratios, but they are true only for small displacements because changes of ratio and length are neglected.

Springs Carrying Vertical Load and Windup Torque

The center or seat of a leaf spring is elastically restrained against deflection produced by a vertical force due to its rate k(N/mm); it is restrained against windup or rotation produced by a torque in the longitudinal vertical plane due to its windup stiffness or windup rate ω(N · mm/radian). This windup stiffness is often used to resist the driving and braking torque, as in the Hotchkiss drive suspension.

In "symmetrical" semi-elliptic springs, a vertical force produces no rotation of the seat, and a torque produces no vertical deflection of the seat. Such a spring with a torque applied to it is shown in Fig. 6.7.

The windup stiffness is $\omega = kL^2/4$ (N·mm/radian).

The stress is, expressed by the windup angle V

$$S_\omega = \frac{2Et}{L} \cdot V \cdot SF$$

or expressed by the torque T

$$S_\omega = \frac{8Et}{kL^3} \cdot T \cdot SF = \frac{2T}{kL} \cdot \frac{S}{f}$$

where SF is the stiffening factor (See Chapter 5, Section 2), S is the stress caused by a deflection f, and V is the windup angle in radians. The formulae show the great importance of spring length for resistance to windup. Windup stiffness increases as the square of the spring length for springs of equal load rate k. The stress produced by a given torque decreases inversely as the length for springs of equal load rate and equal stress rate S/f.

Semi-elliptic springs are frequently made unsymmetrical to obtain desirable geometry. This can be done by unequal division of length, or by adding leaves to one of the arms, or by combining both methods. Fig. 6.8 shows a spring with arms of unequal length under vertical and torque loading acting together. These loadings are produced under forward brake application in a Hotchkiss drive suspension.

In "unsymmetrical" springs, a vertical load will produce, beside vertical deflection, a tilting of the seat; and a torque will produce, besides tilting, a vertical deflection.

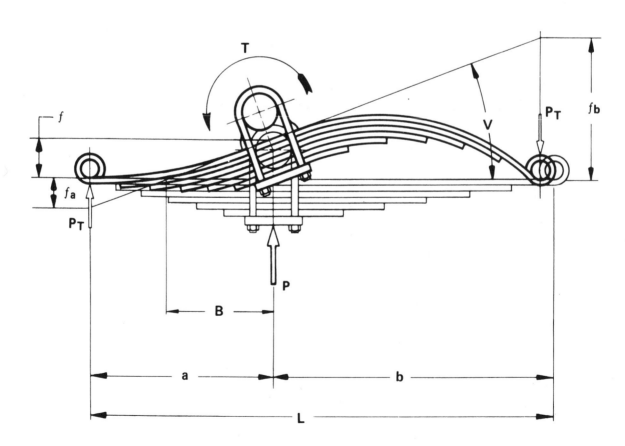

Fig. 6.8—Windup of an unsymmetrical spring under brake application

When the torque is applied with the spring seat prevented from deflecting vertically, the windup stiffness and the windup stress as expressed by the windup angle are increased in the ratio

$$\frac{k_a + k_b}{k} = \frac{(Z + 1)(Y^2 + Z)}{Z(Y + 1)^2}$$

The relationship between the vertical deflection f (mm) and the tilt angle V (radian), due to a vertical load, may be expressed as the equivalent torque arm Q (mm) which is shown in Fig. 4.2. There the deflection is called "x" and the tilt angle "θ."

$$Q = f/V$$

The length of the equivalent torque arm Q can be calculated if the lengths of the spring cantilevers and the ratio of their stiffnesses are known:

With length of fixed cantilever   = a
  length of shackled cantilever = b
  rate of fixed cantilever   = k_a
  rate of shackled cantilever = k_b

and with

$$L = a + b$$
$$Y = b/a$$
$$Z = k_a/k_b$$

Consideration of cantilever deflections will show that

$$Q = \frac{a^2 k_a + b^2 k_b}{a k_a - b k_b} = \frac{L(Z + Y^2)}{(Z - Y)(Y + 1)}$$

For springs designed to deflect in circular arcs (without reinforcing leaves), $Z = Y^3$ and the formula becomes

$$Q = \frac{LY}{Y^2 - 1} = \frac{ab}{b - a}$$

This has been shown in Table 4.1, and the relationship between L, Q, Y, and Z is graphically demonstrated in Fig. 4.5. The zero ordinate for L/Q (where Q becomes infinitely large and the seat angle change or tilt angle becomes infinitely small) is crossed by each Z line at the point of Y = Z. Thus, whenever Z equals Y, a vertical force will produce no tilting of the spring seat. When Z becomes smaller than Y, the equivalent torque arm Q will go in the direction of the longer spring arm, as indicated by the negative ordinate values.

The relationship between the vertical deflection f (mm) and the tilt angle V (radian), due to a torque while the spring is under a constant vertical load P, may be expressed as the effective swing radius B (mm), which is shown in Fig. 6.8.

$$B = f/V = \frac{\omega}{Qk} = \frac{ak_a - bk_b}{k_a + k_b} = \frac{L}{1 + Y} - \frac{L}{1 + Z}$$

The virtual center of rotation for this torque loading is a point at a distance B from the spring seat. B equals zero if Y equals Z, and the center of rotation is then at the spring seat. For larger or smaller values of Z, the center moves closer to the end of the spring arm which is stiffer than the value corresponding to Z = Y.

The windup stiffness of unsymmetrical springs can be calculated from the cantilever rates and lengths

$$\omega = (a^2 k_a + b^2 k_b) \cdot \frac{k}{k_a + k_b}$$
$$= \frac{k_a k_b L^2}{k_a + k_b} = \frac{(Y^2 + Z)kL^2}{(Z + 1)(Y + 1)^2}$$

For convenience, it is expressed as a fraction of the stiffness of a symmetrical spring of the same length L and of the same rate k. The stiffness ratio is

$$\frac{\omega}{kL^2/4} = \frac{4(a^2 k_a + b^2 k_b)}{(k_a + k_b)L^2} = \frac{4k_a k_b}{k(k_a + k_b)}$$
$$= \frac{4(Y^2 + Z)}{(Z + 1)(Y + 1)^2}$$

This is plotted on Fig. 6.9 over a "Y" abscissa. It shows that the value $\omega/(kL^2/4)$ can exceed the value 1 only in a few cases which have little practical application:

 1. When Z equals 1 while Y is either substantially more or substantially less than 1.
 2. When Z exceeds 1 while Y is less than 1.
 3. When Z is less than 1 while Y exceeds 1.

Lines have been drawn both in Figs. 4.5 and 6.9 for $W = Z/Y^3$ to indicate the extent of the distortion (compared with a conventional spring layout) which is required for a given relationship between Y and Z. (see Chapter 4.)

The stress due to windup in unsymmetrical springs is normally higher in the longer (usually the shackled) cantilever b; the windup stresses in the two cantilevers have the relationship

$$S_{\omega b} = S_{\omega a} \cdot \frac{Z}{Y^2} \cdot \frac{SF_b}{SF_a}$$

The stress in the longer cantilever expressed by the windup angle V

$$S_{\omega b} = \frac{2Et}{L} \cdot \frac{Z(Y + 1)^2}{2Y^2(Z + 1)} \cdot V \cdot SF_b$$

by the torque T

$$S_{\omega b} = \frac{8Et}{kL^3} \cdot \frac{Z(Y + 1)^4}{8Y^2(Y^2 + Z)} \cdot T \cdot SF_b$$

1.75

where SF_a and SF_b are stiffness factors that may be determined as stated in Chapter 5.

When Q, k, and ω are known, they can be used to express the relations between vertical load P, torque T, vertical deflection f, and tilt angle V:

$$f = \frac{P}{k} + \frac{T}{Qk} \qquad V = \frac{P}{Qk} + \frac{T}{\omega}$$

For vertical load only, the deflection $f = \dfrac{P}{k}$,

the tilt angle $V = \dfrac{P}{Qk}$, and

$\dfrac{f}{V} = Q$ as previously defined.

For torque only, the deflection $f = \dfrac{T}{Qk}$,

the tilt angle $V = \dfrac{T}{\omega}$, and

$\dfrac{f}{V} = B = \dfrac{\omega}{Qk}$ as previously defined.

Springs Carrying Vertical Load, Windup Torque, and Longitudinal Load

The three-link equivalent shown in Figs. 4.2 and 4.3 very closely reproduces the action of the spring under vertical force and windup torques, if the bending stiffness of the leaves is replaced by torsional stiffness in the two hinges of the three links. It also gives a reasonably accurate equivalent under longitudinal forces. It may therefore be used to calculate the effect of these forces.

Under longitudinal forces such a linkage evidently has a marked toggle effect which increases if the spring is made shorter. Under braking and accelerating forces, the results depend upon whether the spring cantilever which transmits these forces from spring seat to fixed eye is in tension or compression. If the fixed cantilever is in tension, there is a stiffening effect, and both windup and vertical deflection are decreased. If the fixed cantilever is in compression, the toggle action produces a decrease of rate, and both windup and vertical deflection are increased. In compression, this effect can be strong enough to make the

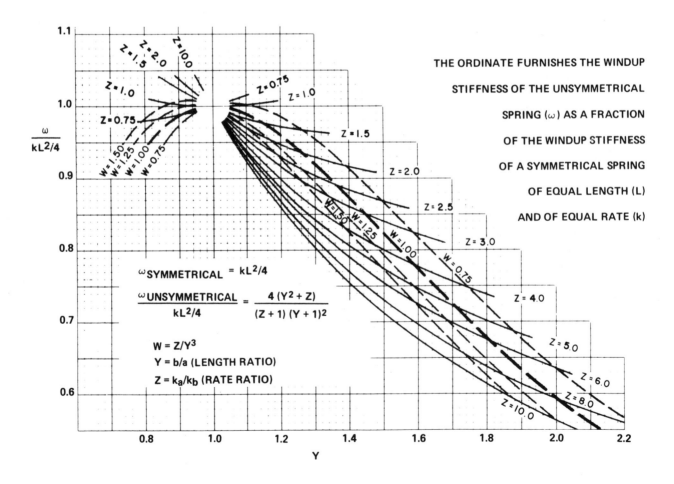

THE ORDINATE FURNISHES THE WINDUP STIFFNESS OF THE UNSYMMETRICAL SPRING (ω) AS A FRACTION OF THE WINDUP STIFFNESS OF A SYMMETRICAL SPRING OF EQUAL LENGTH (L) AND OF EQUAL RATE (k)

$$\omega_{SYMMETRICAL} = kL^2/4$$

$$\frac{\omega_{UNSYMMETRICAL}}{kL^2/4} = \frac{4(Y^2 + Z)}{(Z+1)(Y+1)^2}$$

$$W = Z/Y^3$$

$$Y = b/a \text{ (LENGTH RATIO)}$$

$$Z = k_a/k_b \text{ (RATE RATIO)}$$

Fig. 6.9—Windup stiffness of unsymmetrical springs

mechanism completely unstable and buckle the spring if it is relatively short or if the main leaf is too thin. The buckling tendency also increases in proportion to the static deflection.

An unequally divided (unsymmetrical) spring offers greater resistance to buckling when the shorter cantilever is between the seat and the fixed eye, as in the usual Hotchkiss drive suspension, due to its greater bending resistance and the additional vertical load produced by a horizontal accelerating force. In this case, the horizontal force not only produces the windup torque T, but also the additional vertical load P that opposes the windup lifting force T/Q. The vertical load is obtained from the following formula:

$$P = \frac{FH}{W_B}$$

where:

F = Tire friction force
H = Vehicle center of gravity height above ground
W_B = Vehicle wheelbase

The net spring deflection from the static position may be obtained by substituting the value for vertical load P in the deflection formula, which shows that vertical load and torque are related in the following manner to produce vertical deflection:

$$f = \frac{P}{k} + \frac{T}{Qk}$$

In acceleration, deflection (P/k) is always positive (spring in compression) and deflection (T/Qk) is always negative (spring in rebound). In braking, the signs of these deflections are always opposite because the horizontal (tire friction) force, vertical load, and windup torque are reversed. When the deflection f is positive, the vehicle drops, and when f is negative, the vehicle rises.

Rates of Unsymmetrical Springs Under Bounce and Roll

It should be noted that the vertical rate k equals the sum of the cantilever rates $(k_a + k_b)$ only if Z = Y, which happens in symmetrical springs where Z = Y = 1. In unsymmetrical springs, k is usually smaller than $(k_a + k_b)$. However, k may become equal to $(k_a + k_b)$ when the longer cantilever is made stiffer by additional leaves or longer leaves to balance the deflections. Fig. 6.10 shows the relationship of k_a, k_b, and k. Fig. 6.11 shows the ratio $(k_a + k_b)/k$ for various values of Y and Z.

The fact that in unsymmetrical springs $(k_a + k_b)$ is larger than k can be used to provide additional roll stiffness without using a roll stabilizer. The lower diagram of Fig. 6.10 shows the normal position and the deflected positions of two unsymmetrical springs which are rigidly

mounted on a stiff axle, one end of which is moved up and the other end down.

If both ends were moved down together, the spring seats would tilt one way, as shown in the upper diagram. If they were moved up together, the seats would tilt the other way. If one end is moved up and the other end down, tilting of the spring seats is prevented because they are forced by the axle to remain almost parallel. The axle imposes a windup torque on the springs and is itself stressed by a twisting torque. The roll stiffness of a pair of springs rigidly mounted on an axle is

$$k_{roll} = (k_a + k_b) \, C^2/2 \; (N \cdot mm \, / \, radian)$$

where C is the distance between spring seat centers. Any torsional deflection of the axle will reduce this roll stiffness.

For symmetrical springs or for springs whose seat is not forced to tilt with the axle, the roll stiffness is

$$k_{roll} = k \, C^2/2$$

Additional roll stiffness is produced by twisting of the spring (see Section 3 in this chapter). This raises the total roll stiffness by 20–40% above the values given in the last two equations.

Summary

Springs are made unsymmetrical either for reasons of geometry, to obtain a desirable equivalent torque arm or center of rotation, or to secure increased roll stiffness without increasing the ride rate by the ratio $(k_a + k_b)/k$.

Under torque loading only, windup stiffness is reduced if springs are unequally divided as shown in Fig. 6.8. However, in a Hotchkiss drive suspension, an unsymmetrical spring has less buckling tendency because of the stabilizing effect of additional (dynamic) vertical load and the increased bending resistance of the fixed cantilever.

The most economical springs are designed to bend in a (nearly) circular arc when loaded at the seat. In such springs, the rate ratio k_a/k_b equals the cube of the length ratio b/a.

For these useful springs, most of the formulae can be simplified by substituting Y^3 for Z. In Figs. 6.9 and 6.11, these springs are indicated by the heavy line for $W = Z/Y^3 = 1$.

3. Twist of Springs

In the usual suspension applications, leaf springs may be subjected to twisting, for example, by an obstacle under one wheel of an axle.

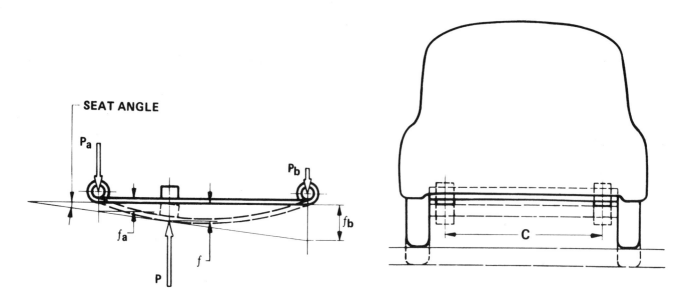

IN PITCH OR BOUNCE k IS LESS THAN $(k_a + k_b)$

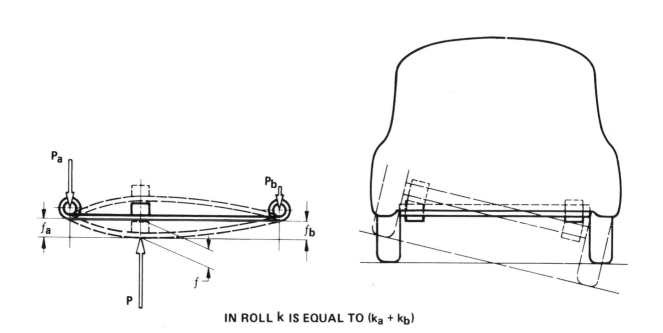

IN ROLL k IS EQUAL TO $(k_a + k_b)$

$$k = P / f \qquad k_a = P_a / f_a \qquad k_b = P_b / f_b$$

IN ROLL THE AXLE PREVENTS THE CHANGE OF SEAT ANGLE.

THIS ADDITIONAL RESTRAINT PRODUCES INCREASE OF RATE.

THE TWISTING OF THE SPRINGS RESULTS IN A FURTHER RATE INCREASE.

Fig. 6.10—Unsymmetrical leaf springs with rigidly connected axle

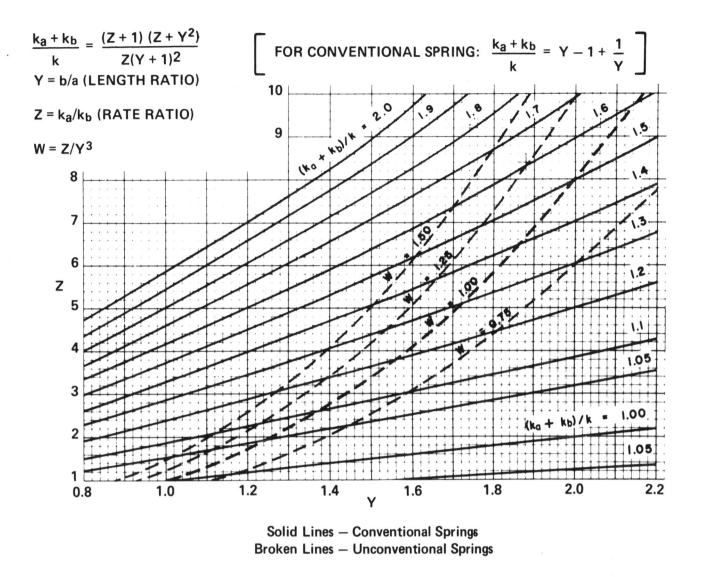

$$\frac{k_a + k_b}{k} = \frac{(Z + 1)(Z + Y^2)}{Z(Y + 1)^2}$$

$Y = b/a$ (LENGTH RATIO)

$Z = k_a/k_b$ (RATE RATIO)

$W = Z/Y^3$

FOR CONVENTIONAL SPRING: $\dfrac{k_a + k_b}{k} = Y - 1 + \dfrac{1}{Y}$

Solid Lines — Conventional Springs
Broken Lines — Unconventional Springs

Fig. 6.11—Relationship between spring rate (k) and sum of cantilever rates ($k_a + k_b$)

Twisting a spring leaf (having the substantially rectangular cross section of the SAE leaf spring steel with width w and thickness t) through α degrees in length ℓ (for example, between eye and seat) will produce a shear stress

$$S = 1400 \, t \, \alpha/\ell \text{ MPa (Approx.)}$$

and a torque

$$T = 420 \, w \, t^3 \, \alpha/\ell \text{ N} \cdot \text{mm (Approx.)}$$

To keep the twisting stress low, it is necessary to distribute the total twist angle over as long a length of the spring as possible, which means that the clips should not restrain the main leaf from twisting.

When military wrappers or reinforced eyes are used, both first and second leaf will be twisted and their torques will add.

The flexibility of brackets and shackles will reduce twisting of the spring.

The resistance of the spring against twisting increases the roll stiffness in the usual application.

Sometimes an axle is supported by two springs which are not parallel. Convenience of attachment is the usual reason for this practice. If the angle included between the springs is kept small (up to 10 deg), the action of the spring is changed only a little. Parallel motion of the axle (ride) will produce a small twist of the spring, and in unsymmetrical springs, a small windup torque, and the spring rate will be increased in this way.

Chapter 7
Interleaf Friction

1. Characteristics

Interleaf friction can be defined as the force which opposes the relative motion of adjacent leaves. The friction force provides damping in the suspension system. It also resists the initial deflection of the spring, making the suspension system less responsive to dynamic forces.

Many heavy trucks using multiple leaf springs rely solely on interleaf friction to provide damping and, therefore, shock absorbers are not required. In suspensions that use single leaf springs or springs where tip inserts have been installed, shock absorbers must be used to damp out the vibrations.

The magnitude of the friction force depends on the condition of the leaf surfaces (coefficient of friction), on the load carried by the leaf surfaces, and on the speed of sliding between leaves.

Fig. 7.1 shows a load-deflection diagram as the result of a test conducted at a slow rate of deflection without rapping the spring. It shows that at any given deflection, the load may be any value between an upper and lower limit, depending on the direction of motion and on the distance from the last reversal of motion. Tests conducted at faster rates of loading and unloading will show the same basic characteristics, but the width and the shape of the diagram will be different.

The procedure for testing springs to obtain the friction loop is given in Chapter 2, Section 1.

Springs with high frictional forces will have a higher effective (dynamic) rate when operated through small amplitudes. At larger amplitudes, the effective rate will approach the rate determined by a slow test.

2. Measurement

It has been found convenient to compare one spring with another by means of a friction factor:

$$\text{Friction Factor} = \frac{\text{Friction Force}}{\text{Average Load}}$$

Though not strictly correct, it is permissible to assume that the friction force equals one half the difference between the compression load and the release load. The average of the compression load and the release load at a given deflection is termed the average load. (See Chapter 2, Section 1).

3. Control

The friction factor, as measured by a slow test, may be as high as 0.10. To obtain lower friction factors, the contacting surfaces should be smooth, leaf ends should be flexible (tapered) and lubricated. Under these conditions, the friction factors may be between 0.02–0.05 on a new spring. However, when the spring becomes dry, dirty, and the contact surfaces are scored, the friction factor will increase.

In some applications, various types of interleaf liners or tip inserts are used to eliminate squeaks and obtain low friction factors. A typical tip insert installation is shown in Fig. 3.16E, Section A-A.

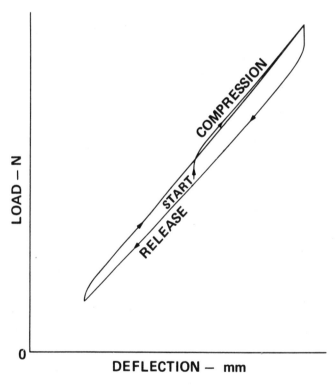

Fig. 7.1—Typical diagram of a leaf spring tested at low speed and without rapping to show interleaf friction

CHAPTER 8
Operating Stress and Fatigue Life

1. Operating Stress

In order to hold the mass of a spring to a minimum, it is necessary to use the highest stress that will give satisfactory operation. This stress is limited by three factors: settling under load, fatigue life, and quality of spring material and processing.

The settling of a spring of given hardness is a function of the maximum stress. Any spring which is repeatedly loaded to a high stress will settle somewhat during the first few load cycles. This settling can be reduced to a negligible amount by the proper presetting as outlined in Chapter 5, near the end of Section 6, and Chapter 9, Section 4.

For automotive suspensions, the design load stress is usually in the range of 600–750 MPa for passenger cars and 350–550 MPa for trucks. There is some possible overlap of design load stress between passenger cars and trucks for special vehicles (light trucks, station wagons, taxicabs, etc.). The maximum stress should not exceed the minimum yield stress of the spring material. For properly heat treated alloy steel, the minimum yield stress is generally accepted as 1200 MPa.

When a spring is subjected to windup under engine or brake torque or any other forced external means, stresses in addition to those due to vertical load may be present and should be considered in computing the stress.

2. Fatigue Life

Fatigue life is expressed by the number of deflection cycles a spring will withstand without failure and can be estimated by the use of Fig. 8.1. This has been constructed with the help of data obtained from laboratory fatigue tests which were conducted on various spring designs between 1950 and 1970. Improvements in the processing of the steel which may be expected in the future both at the steel mills and at the spring plants can be taken into account by shifting of the "cycles to failure" lines in Fig. 8.1. Such an adjustment would be based on the results obtained from extensive controlled fatigue tests. In order to establish the fatigue life cycles which are acceptable in any spring design, it is desirable to have road durability tests run over a prescribed course so that fatigue life test data and actual road durability results may be correlated.

It must be understood that the number of estimated life cycles is a statistical average and that fatigue test results will show scatter (or dispersion) even under closely controlled test conditions. The extent of the scatter will depend on the consistency of surface condition, fabrication, and the general quality of the springs which are tested.

The average life "cycles to failure" lines, as shown in Fig. 8.1, apply to present springs but do not include the effect of surface cold working as produced by shot peening (see Chapter 9, Section 4). Tests have proved that the fatigue life of a spring can be greatly improved by shot peening the tension surface. The increase in fatigue life will depend on a number of factors, such as peening intensity and coverage and the condition of the surface prior to shot peening.

Fatigue testing is an accelerated method of examining springs for design adequacy and for quality control purposes.

Reexamination of the design will be in order if the fatigue tests result in failures which are confined to one section of the spring. It is recommended that the fatigue setup produce at least an average of 50 000 and preferably an average of 100 000 cycles.

At higher stresses (shorter lives) the scatter is theoretically reduced, so that fewer test samples will produce a given degree of precision in the estimated life of the entire population. However, lower stresses (longer lives) produce more realistic results, since they duplicate more nearly the actual service conditions and the spring is less likely to settle during the test. Also, comparisons between different groups of springs will be more distinct at lower stresses, since different S-N curves tend to diverge the more they approach the fatigue limit (or limiting value of the stress at which 50% of the population would survive a very large number of cycles, usually 10 000 000.)

A leaf spring used in a suspension will undergo a large number of cycles of small amplitude near the design load position without failure. Under greater amplitudes the number of cycles without failure will be reduced, since the maximum stress as well as the stress range are increased, and both are determining factors in the fatigue life of a spring. See examples D versus C (both with 700 MPa stress at design load) in Fig. 8.1.

The metal-to-metal position (vertical load limit) is frequently used as the maximum deflection position of the spring in a fatigue test; but in heavy truck springs this deflection is often considered excessive for the test setup, as it is rarely reached in actual service.

The length of the test stroke is selected from experience. A frequently used method of establishing the

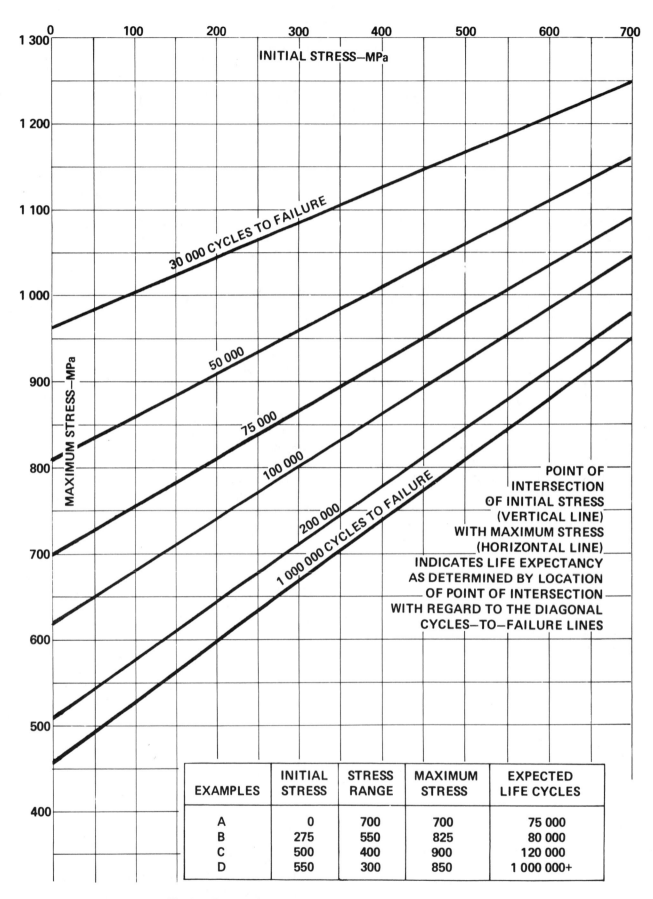

The diagram shows a graph with INITIAL STRESS—MPa on the top horizontal axis (0 to 700) and MAXIMUM STRESS—MPa on the vertical axis (400 to 1 300).

Diagonal lines labeled:
- 30 000 CYCLES TO FAILURE
- 50 000
- 75 000
- 100 000
- 200 000
- 1 000 000 CYCLES TO FAILURE

POINT OF
INTERSECTION
OF INITIAL STRESS
(VERTICAL LINE)
WITH MAXIMUM STRESS
(HORIZONTAL LINE)
INDICATES LIFE EXPECTANCY
AS DETERMINED BY LOCATION
OF POINT OF INTERSECTION
WITH REGARD TO THE DIAGONAL
CYCLES—TO—FAILURE LINES

EXAMPLES	INITIAL STRESS	STRESS RANGE	MAXIMUM STRESS	EXPECTED LIFE CYCLES
A	0	700	700	75 000
B	275	550	825	80 000
C	500	400	900	120 000
D	550	300	850	1 000 000+

Fig. 8.1—Diagram for estimating fatigue life cycles of steel leaf
springs (preset but not shot peened)

length of test stroke is to add to the compression stroke (from design load to maximum test load) one-half of this length for the release stroke (from design load to initial test load). This practice may require modification in those cases where it would produce less than 50 000 or much more than 100 000 cycles, according to Fig. 8.1.

Example: To estimate the expected fatigue life of the first example shown in Chapter 5, Section 3, compute the stresses at maximum and initial test loads as follows:

Deflection to design load = 3500/17.5 = 200 mm
Metal-to-metal clearance
   (compression stroke)   110 mm
Total deflection to maximum load   310 mm
Stress at metal-to-metal position
   (6.30 gage)   985 MPa
Stress rate = 985/310 =   3.18 MPa/mm

Release stroke = ½ · 110 =   55 mm
Fatigue test stroke = 110 + 55 =   165 mm
Initial stress = 985 − (165 · 3.18) =   460 MPa

From Fig. 8.1, the intersection of the horizontal line representing 984 MPa and the vertical line representing 460 MPa gives the expected average fatigue life cycles as 65 000.

A fatigue life of 100 000 cycles under amplitudes representing these stress conditions will generally assure a road life of more than 160 000 km of "on highway" operation. In this example, shot peening of the leaves of the spring is recommended if 160 000 km of highway operation is expected. Any "off-the-road" operation will reduce the service life of the spring. Springs for "off-the-road" duty should, therefore, have a lower operating stress than those designed for highway operation.

Another example: To estimate the expected fatigue life of the third example shown in Chapter 5, Section 3, (clearance of 90 mm to one-half bumper compression specified additionally), compute stresses at maximum and initial test loads as follows:

Deflection to design load = 14 700/110   134 mm
Clearance to one-half
   bumper compression   90 mm
Stress at design load   535 MPa
Stress rate = 535/134   3.99 MPa
                                   mm
Release stroke = ½ · 90   45 mm
Fatigue test stroke = (90+45)   135 mm
Initial stress = 535 − (45 · 3.99)   355 MPa
Maximum stress = 535 + (90 · 3.99)   894 MPa

From Fig. 8.1, the intersection of the horizontal line representing 894 MPa and the vertical line representing 355 MPa gives the expected average fatigue life cycles as 75 000.

3. Evaluation of Fatigue Test Results

It has long been recognized that considerable variation is present in the fatigue life of springs loaded with the same stress cycle, and that the average life of the tested springs is not sufficient by itself to establish a judgment either on the design, or on the material, or on the production method which they represent. The relationship between the number of applied cycles and the percentage of springs which failed at these cycles can best be analyzed with the help of statistical techniques which will systematically describe the "dispersion" or "spread" or "scatter" of the recorded test results.

Sampling

One of the main purposes of statistical analysis is to draw inferences about the properties of a large group (the "population") from the results of tests on a small group (the "sample"). If the entire population were tested, one would not have to infer anything; one would know how the population reacted to the test. This would be called 100% confidence. If 99% of the population were tested, one would be 100% confident about that 99%. Also, if the sample of 99% were considered representative of the remaining 1%, one would be close to 100% confident of predicting the results if that 1% were tested. If only 5% are tested, one would know about that 5%, but how much could one infer about the remaining 95%? Actually, if certain conditions of sampling are met, one can infer a great deal about the entire population from tests on small samples. What is required is a good, honest sample.

The primary condition for a good sample is that it be taken at random under conditions which ensure that all springs of the population have an equal chance of being chosen. This is obviously impossible in the case where the sample consists of a few handmade springs of a design which has not yet gone into production. Only experience can tell the engineer whether the various properties of the sample which can affect the test result (in regard to material as well as to production methods and controls) will also be present in the production springs (the "population"). This determination is outside the realm of statistics. However, statistical mathematics are based on the inherent assumption that the sample is a true representative of the population.

Distribution

If the entire population were tested under identical test conditions, the results could be shown in graphical form by arranging them in ascending numerical order and plotting the cumulative fraction (or percent) of failures over an abscissa of "life cycles." A sample selected at random from this population can be expected to exhibit a similar distribution of fatigue life; the larger the number of springs in the sample, the closer will be the similarity.

It is possible to calculate the likelihood of similarity for samples of any given size. Tables are available based on such likelihoods; Table 8.1 is an example. It presents the "median rank" of each test result for a sample size between 1 and 30. A rank is assigned to each individual test result corresponding to that portion of the population which it is most likely to represent. The median rank is used as an estimate of the true rank because it is just as likely to be high as low. Table 8.1 lists percent figures.

A good approximation formula for the median rank (which may be used for larger sample sizes than those in Table 8.1) is:

$$100 \cdot \frac{J - 0.3}{N + 0.4}$$

where:

J = Position (in ascending order) for each test result in the sample

N = Total quantity of springs in the sample

The median rank line constructed from such data predicts that certain percentages of the population will survive specific cycles-to-failure. But any such estimate may err substantially on either the high or low side. The question is: How confident can the engineer be of such an estimate? When he has accumulated a great deal of experience in comparing the results of small samples which represent springs of different materials or different designs or different production methods, he may judge, after contemplating two such median rank lines, that he should give preference to one set of springs over the other because of its apparent superiority in fatigue life. However, when judgment based on experience is not considered adequate for a final decision, then it will be necessary to construct lines of higher confidence.

In many cases a confidence level of 90 or 95 or even 99% will be required, so that there will remain only a 10 or 5 or 1% risk of the estimate being either too high or too low. For a chosen confidence level, the life cycles of a given percentage of the population will be found within a certain "tolerance interval". On the median rank graph this may be represented by a "tolerance band" to either side of the median rank line. Wider bands indicate increased doubt about the line truly representing the population. With a given sample size, the bands will be wider for higher confidence levels. With a given confidence level, the bands will be wider for a smaller sample size.

TABLE 8.1—MEDIAN RANKS (PERCENT) FOR SAMPLE SIZES 1 TO 30

Sample Size

Rank Order	1	2	3	4	5	6	7	8	9	10	Rank Order
1	50.000	29.289	20.630	15.910	12.945	10.910	9.428	8.300	7.412	6.697	1
2		70.711	50.000	38.573	31.381	26.445	22.849	20.113	17.962	16.226	2
3			79.370	61.427	50.000	42.141	36.412	32.052	28.624	25.857	3
4				84.090	68.619	57.859	50.000	44.015	39.308	35.510	4
5					87.055	73.555	63.588	55.984	50.000	45.169	5
6						89.090	77.151	67.948	60.691	54.831	6
7							90.572	79.887	71.376	64.490	7
8								91.700	82.038	74.142	8
9									92.587	83.774	9
10										93.303	10

Sample Size

Rank Order	11	12	13	14	15	16	17	18	19	20	Rank Order
1	6.107	5.613	5.192	4.830	4.516	4.240	3.995	3.778	3.582	3.406	1
2	14.796	13.598	12.579	11.702	10.940	10.270	9.678	9.151	8.677	8.251	2
3	23.578	21.669	20.045	18.647	17.432	16.365	15.422	14.581	13.827	13.147	3
4	32.380	29.758	27.528	25.608	23.939	22.474	21.178	20.024	18.988	18.055	4
5	41.189	37.853	35.016	32.575	30.452	28.589	26.940	25.471	24.154	22.967	5
6	50.000	45.951	42.508	39.544	36.967	34.705	32.704	30.921	29.322	27.880	6
7	58.811	54.049	50.000	46.515	43.483	40.823	38.469	36.371	34.491	32.795	7
8	67.620	62.147	57.492	53.485	50.000	46.941	44.234	41.823	39.660	37.710	8
9	76.421	70.242	64.984	60.456	56.517	53.059	50.000	47.274	44.830	42.626	9
10	85.204	78.331	72.472	67.425	63.033	59.177	55.766	52.726	50.000	47.542	10
11	93.893	86.402	79.955	74.392	69.548	65.295	61.531	58.177	55.170	52.458	11
12		94.387	87.421	81.353	76.061	71.411	67.296	63.629	60.340	57.374	12
13			94.808	88.298	82.568	77.525	73.060	69.079	65.509	62.289	13
14				95.169	89.060	83.635	78.821	74.529	70.678	67.205	14
15					95.484	89.730	84.578	79.976	75.846	72.119	15
16						95.760	90.322	85.419	81.011	77.033	16
17							96.005	90.849	86.173	81.945	17
18								96.222	91.322	86.853	18
19									96.418	91.749	19
20										96.594	20

If a 90% confidence level has been chosen, the lower and upper limits to the estimated fatigue life distribution of the population can be shown by constructing "5% rank" and "95% rank" lines (see "Theory and Technique of Variation Research" by Leonard G. Johnson. Elsevier Publishing Co., 1964). The numbers 5% and 95% represent the chance of being either too high or too low in assigning the given ranks to the individual springs in the sample, and they are used to establish the limits within which the true population is expected to lie.

For a concise summary of distribution mathematics see "Engineering Considerations of Stress, Strain, and Strength," by Robert C. Juvinall, McGraw-Hill Book Co., 1967.

Weibull Plot

Several systems of mathematically organizing the test result data have been established. In the past, the normal (or Gaussian) distribution has been most widely used. It is graphically represented by the familiar symmetrical bell-shaped distribution curve which is completely defined by two statistical parameters:

1. The *mean life*—In the test sample it is the sum of all the recorded test result values, divided by the sample size. It then becomes an estimate of the population mean life. A population with normal distribution has the *mean* coinciding with the *median* (which is the middle result when all individual results are arranged in order of magnitude), and also coinciding with the *mode* (which is the cycle value at which the greatest number of failures occurs).

2. The *standard deviation*—It describes the scatter on either side of the mean. For the test sample it is mathematically defined as the square root of the sample variance (which in turn is the sum of the squares of the difference between each recorded test result and the mean, divided by the sample size minus 1). It then becomes an estimate of the population standard deviation.

When the test results are arranged in ascending numerical order, and the cumulative percent of failures is plotted (using median ranks) over an abscissa of life cycles on *normal probability graph paper*, it will be found that a straight line can be fitted to the results as long as the distribution is normal. This becomes an estimate of the population distribution.

While the normal distribution has a number of attractive attributes and has been the subject of many publications, it must be recognized that in spring fatigue testing the results are usually not normal in that they cannot be

Sample Size

Rank Order	21	22	23	24	25	26	27	28	29	30	Rank Order
1	3.247	3.101	2.969	2.847	2.734	2.631	2.534	2.445	2.362	2.284	1
2	7.864	7.512	7.191	6.895	6.623	6.372	6.139	5.922	5.720	5.532	2
3	12.531	11.970	11.458	10.987	10.553	10.153	9.781	9.436	9.114	8.814	3
4	17.209	15.734	15.734	15.088	14.492	13.942	13.432	12.958	12.517	12.104	4
5	21.890	20.015	20.015	19.192	18.435	17.735	17.086	16.483	15.922	15.397	5
6	26.574	25.384	24.297	23.299	22.379	21.529	20.742	20.010	19.328	18.691	6
7	31.258	29.859	28.580	27.406	26.324	25.325	24.398	23.537	22.735	21.986	7
8	35.943	34.334	32.863	31.513	30.269	29.120	28.055	27.065	26.143	25.281	8
9	40.629	38.810	37.147	35.621	34.215	32.916	31.712	30.593	29.550	28.576	9
10	45.314	43.286	41.431	39.729	38.161	36.712	35.370	34.121	32.958	31.872	10
11	50.000	47.762	45.716	43.837	42.107	40.509	39.027	37.650	36.367	35.168	11
12	54.686	52.238	50.000	47.946	46.054	44.305	42.685	41.178	39.775	38.464	12
13	59.371	56.714	54.284	52.054	50.000	48.102	46.342	44.707	43.183	41.760	13
14	64.057	61.190	58.568	56.162	53.946	51.898	50.000	48.236	46.592	45.056	14
15	68.742	65.665	62.853	60.271	57.892	55.695	53.658	51.764	50.000	48.352	15
16	73.426	70.141	67.137	64.379	61.839	59.491	57.315	55.293	53.408	51.648	16
17	78.109	74.616	71.420	68.487	65.785	63.287	60.973	58.821	56.817	54.944	17
18	82.791	79.089	75.703	72.594	69.730	67.084	64.630	62.350	60.225	58.240	18
19	87.469	83.561	79.985	76.701	73.676	70.880	68.288	65.878	63.633	61.536	19
20	92.136	88.030	84.266	80.808	77.621	74.675	71.945	69.407	67.041	64.832	20
21	96.753	92.488	88.542	84.912	81.565	78.471	75.602	72.935	70.450	68.128	21
22		96.898	92.809	89.013	85.507	82.265	79.258	76.463	73.857	71.424	22
23			97.031	93.105	89.447	86.058	82.914	79.990	77.265	74.719	23
24				97.153	93.377	89.847	86.568	83.517	80.672	78.014	24
25					97.265	93.628	90.219	87.042	84.078	81.309	25
26						97.369	93.861	90.564	87.483	84.603	26
27							97.465	94.078	90.885	87.896	27
28								97.555	94.280	91.186	28
29									97.638	94.468	29
30										97.716	30

plotted in the symmetrical bell-shaped distribution curve but in a skewed curve. This has led to other mathematical formulations. In the automotive industry the *Weibull plot* is used because it permits straight-line plotting of the cumulative failure probability versus life cycles on *Weibull probability graph paper*, even when the distribution is skewed.

In the Weibull distribution the relationship between the number of applied cycles and the cumulative percent of failures at these cycles is expressed by a formula which uses three parameters:

1. The *minimum life,* which may or may not be zero. It is denoted by the letter "a."

It is generally assumed that a = zero because that is the condition for which the Weibull formula assures straight-line plotting on Weibull paper. See Section 3 in this chapter, for those cases where minimum life is greater than zero.

2. The *Weibull slope,* which is an indicator of the skewness of the distribution. It is called the "shape parameter" and is denoted by the letter "b". It also is a measure of the scatter of the distribution; a low slope value indicates a high degree of scatter, and vice versa.

The slope is the tangent of the angle formed by the distribution line with the abscissa on Weibull probability paper, when the scales are such that the distance representing the factor 100 on the (logarithmic) abscissa scale for life cycles equals the distance from 2.3 to 90.0% on the (log-log) ordinate scale for percentages of failure.

In the Weibull distribution, the mean, the median, and the mode never coincide exactly. But when the Weibull slope is within the range of 3.2–3.5, the differences are small enough to give the Weibull distribution an appearance of symmetry. The Weibull comes nearest to the normal distribution when the Weibull slope equals 3.44 (thus representing an angle of 73.8 deg), because there the mean and the median have identical values.

The more the slope value increases above 3.44, the more the distribution curve will be skewed to the left (with a long tail to the left), where the mean is to the left of (or less than) the median. The more the slope value decreases below 3.44, the more the distribution curve will be skewed to the right (with a long tail to the right), where the mean is to the right of (or more than) the median. Fig. 8.2 shows a graph (for Weibull slopes 1 through 12) which locates the percent of failed springs at the life cycles representing the mean of the population.

3. The *characteristic life,* which is the 63.2% failure point for the population. It is called the "scale parameter" and is denoted by the Greek letter theta (θ).

$$63.2 = 100(1-1/e)$$

where:

$$e = 2.7183 \text{ (the Napierian base)}$$

Example 1 (Fig. 8.3)-Eight springs have been fatigue tested under identical conditions. The results are arranged in ascending order of failure cycles and are given rank order numbers accordingly. In this order they are assigned median ranks from Table 8.1 as follows:

Spring	Order No.	Cycles to Failure	Median Rank
E	1	61 000	8.30
A	2	91 000	20.11
F	3	114 000	32.05
H	4	135 000	44.02
C	5	155 000	55.98
B	6	177 000	67.95
G	7	205 000	79.89
D	8	245 000	91.70

These points are plotted on Weibull graph paper in Fig. 8.3. Drawing a straight line of best fit through the median rank points produces an estimate for the failure rate of the entire population with the parameters b = 2.4 and $\theta = 170\,000$.

From Fig. 8.2 it will be seen that for b = 2.4, the percent of failed springs at the mean is 52.7. At that failure level the median rank line in Fig. 8.3 shows 150 000 cycles. At the B-10 life level (that is at the number of cycles where 10% of the population are estimated to fail), the median rank line shows 66 000 cycles.

Significant Differences

In most cases the fatigue testing of springs will be undertaken for the purpose of comparing different samples, and the probability graph will be expected to convey information on the relative life distribution of the populations represented by these samples. For example, the comparison may involve a sample representing a first design, and another sample representing a second design.

When the two median rank lines for the test data of the two samples are plotted on the same graph, they will readily show if the second design promises some improvement in fatigue life. However, in order to establish if there is a "significant difference" between the two designs, it will be necessary to find quantitative values for the degree of improvement (or degradation) between one design and the other. The question is this: How confidently can one say the limited test results indicate that the second design assures an improvement in fatigue life for the entire spring population? The answer depends not only on the amount of separation between the two plotted slopes, but also on the size of the two test samples.

Furthermore, the degree of confidence in the superiority of one design over the other need not be constant from one quantile level to another. For example, it is possible to have a significant improvement at the B-50 life (50% failure level) without any improvement at the B-10 life (10%

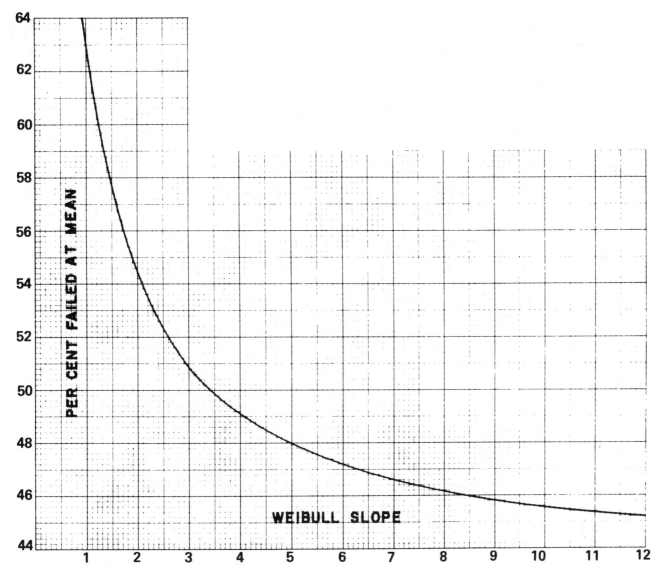

Fig. 8.2—Location of the mean for Weibull slopes 1–12

failure level), or vice versa. This is partly due to differences in the Weibull slope, and partly due to the greater width of the tolerance bands at the lower quantile levels.

Example 2 (Fig. 8.3)- It has been proposed that the spring design represented by the sample of eight in Example 1 be replaced by a new design. Seven springs of the new design have been fatigue tested with results which are shown arranged in ascending order and are assigned median ranks from Table 8.1 as follows:

Spring	Order No.	Cycles to Failure	Median Rank
M	1	132 000	9.43
O	2	195 000	22.85
K	3	233 000	36.41
L	4	275 000	50.00
P	5	315 000	63.59
J	6	365 000	77.15
N	7	440 000	90.57

For this second plot it will be seen from Fig. 8.3 that the Weibull slope b = 2.7 and the characteristic life $\theta = 310\,000$; the mean life level is at 51.8% (see Fig. 8.2), therefore the estimated mean life is 280 000 cycles. The estimated B-10 life is 138 000 cycles.

Since the estimated mean life in Example 1 was 150 000 cycles, the mean life ratio on the median rank lines is 280 000/150 000 = 1.87. This represents an estimated improvement of 87%. The "confidence number" corresponding to this mean life ratio (that is, the probability that the true mean life ratio of the population is greater than 1) is found by reference to the mean life nomograph (Fig. 8.4).

In order to work with the nomograph it is necessary to establish the "degree of freedom" (= freedom of movement of the individual test results about a fixed mean) for the samples representing the two designs. The degree of freedom equals (N − 1) where N is the size of the sample.

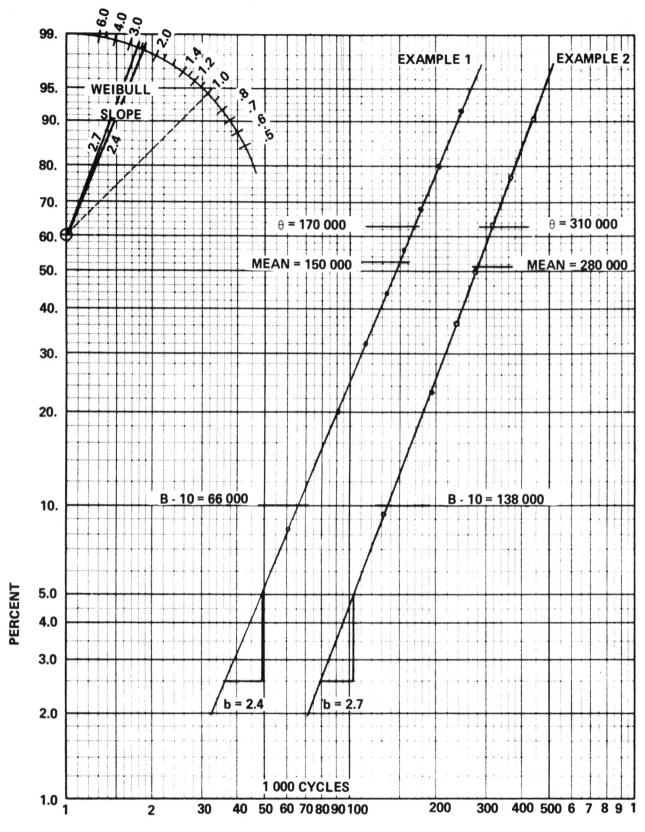

Fig. 8.3—Median rank lines

1.90

Thus $N_1 - 1 = 7$ and $N_2 - 1 = 6$, and the product of the two, known as "total degrees of freedom," is 42.

The nomograph furnishes the confidence number 98.8 for b = 2.4 (first design) and 99.4 for b = 2.7 (second design). The average is 99.1, and this means that 99.1 times out of 100 the second design is superior to the first design at the mean life level.

The confidence number corresponding to the B-10 life ratio will be found by reference to the B-10 level nomograph (Fig. 8.5). Since the estimated B-10 life in Example 1 was 66 000 cycles, the B-10 life ratio on the median rank line is 138 000/ 66 000 = 2.10. The nomograph furnishes the confidence number 91.0 for the first design and 93.0 for the second design. The average is 92.0, so the second design is superior at the B-10 level 92.0 times out of 100.

Thus, the confidence numbers obtained from the foregoing "significant difference" study indicate a certain superiority of the second design over the first. Quantitative values for the degree of this superiority are obtained by using the information from the nomographs on so-called "confidence interpolation graph paper" or simply "ratio paper" which has a log-log ordinate for percent confidence and an arithmetical abscissa for life ratio (Fig. 8.6).

To obtain quantitative values at the mean life level, the mean life ratio on the median rank lines (in this case 1.87) is plotted on the ratio paper at the 50% confidence level and is connected by a straight line with the confidence number (in this case 99.1) at abscissa 1. The life ratio values at other confidence levels are then found on this line. It will be seen that at the 60% confidence level the ratio is 1.75, so there are 6 out of 10 chances that 75% improvement will occur. At 90% confidence (which is frequently used as a standard) the ratio is 1.32, so there are 9 out of 10 chances that 32% improvement will be realized.

The same procedure will establish quantitative values at the B-10 level. There the percent of improvement will always be comparatively lower for the same degree of confidence. This is due to the greater width of the tolerance bands at the lower quantile levels. In this example, the life ratio at 90% confidence is 1.08, so there are 9 out of 10 chances that 8% improvement will be realized.

Minimum Life Greater than Zero

As stated earlier, Weibull plots are generally constructed with the assumption that the minimum life is zero. The sample data will then plot a straight line on Weibull probability graph paper.

When the minimum life is greater than zero, the sample points can usually be fitted with a fairly smooth curved line. However, this would mean foregoing one of the major advantages of the Weibull process, which is to analyze the data with the help of a straight line even when the number of test results is small.

A relatively simple technique permits the test results, which indicate a minimum life greater than zero, to be converted to straight-line plotting on the Weibull graph. This requires that the curve drawn through the sample points be extended downward until the abscissa value which it approaches asymptotically can be approximately established as an estimate of the finite minimum life "a." When this is subtracted from each of the plotted data points, it may be possible to fit the new points thus obtained with a straight line.

If it develops that the new points can still be better fitted by a curve than by a straight line, then the estimate of the minimum life was incorrect. If the new line is still concave downward, the minimum life was estimated too small; if the new line is concave upward, the minimum life was estimated too large. A second (and possibly a third) estimate will then be required until the plotted points can be successfully fitted with a straight line. It is well to remember that before any one of the life cycle values found on this straight line is used for comparison with a corresponding value on any other median rank line, it must be increased by the "a" value which was subtracted from the curved median rank line to obtain the straight line.

Example 3 (Fig. 8.7). Seven springs have been fatigue tested with results listed below, arranged in ascending order and shown with their assigned median ranks from Table 8.1. The plotting of these values produces a curve which is concave downward. A tentative extension indicates "a" to approximate 50 000 cycles. When the points are replotted with each result reduced by 50 000, they can only be fitted with a curve which is concave upward; therefore, a second attempt is made with a = 45 000, and this brings about a successful straight line fit.

Order No.	Cycles to Failure	Median Rank	Cycles less 50 000	Cycles less 45 000
1	85 000	9.43	35 000	40 000
2	110 000	22.85	60 000	65 000
3	135 000	36.41	85 000	90 000
4	155 000	50.00	105 000	110 000
5	180 000	63.59	130 000	135 000
6	210 000	77.15	160 000	165 000
7	250 000	90.57	200 000	205 000

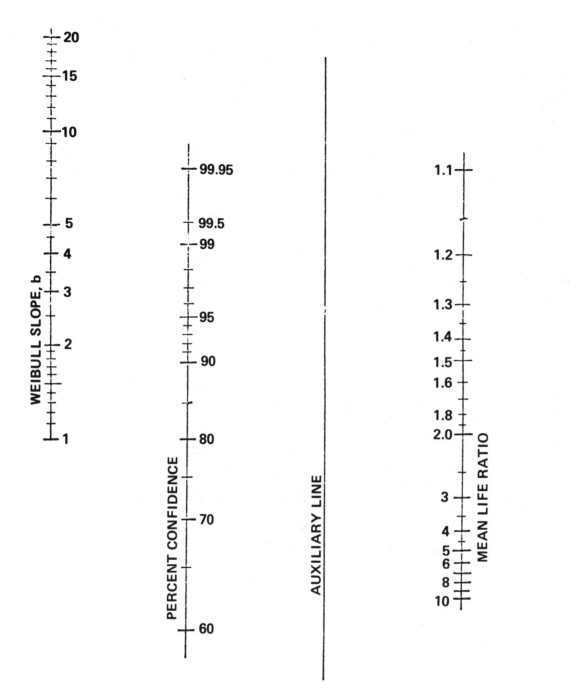

1. **Connect total degrees of freedom with Weibull slope and locate intersection point on auxiliary line.**

2. **Connect life ratio with intersection point and continue to intercept on confidence number.**

3. **For unequal Weibull slopes perform operation for each slope and average the confidence numbers so obtained.**

Fig. 8.4—Confidence nomograph at mean life level

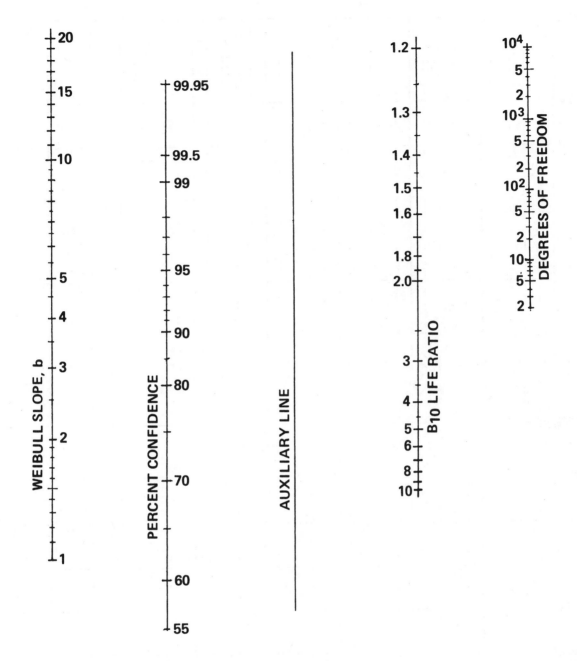

1. Connect total degrees of freedom with Weibull slope and locate intersection point on auxiliary line.

2. Connect life ratio with intersection point and continue to intercept on confidence number.

3. For unequal Weibull slopes perform operation for each slope and average the confidence numbers so obtained.

Fig. 8.5—Confidence nomograph at B_{10} life level

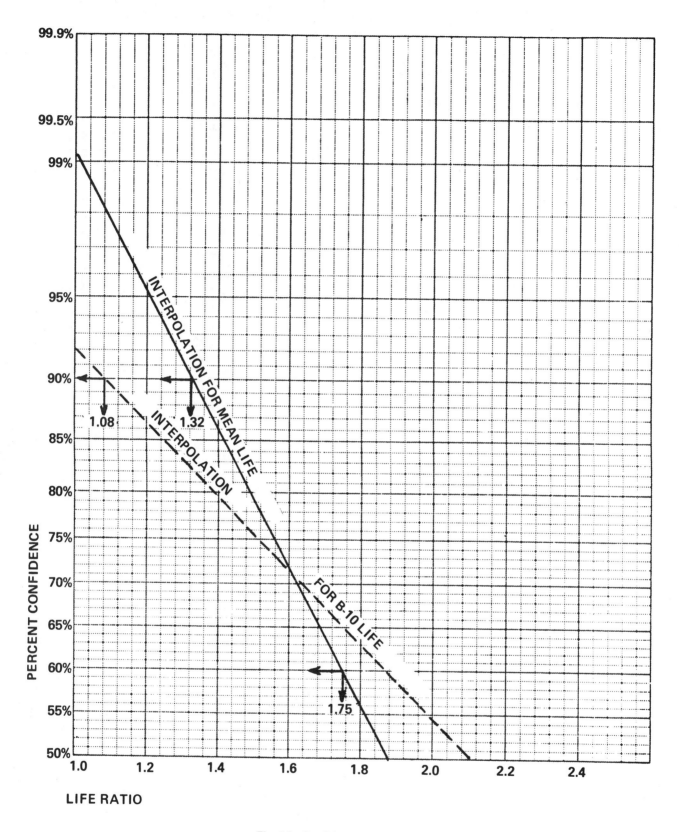

Fig. 8.6—Confidence interpolation

1.94

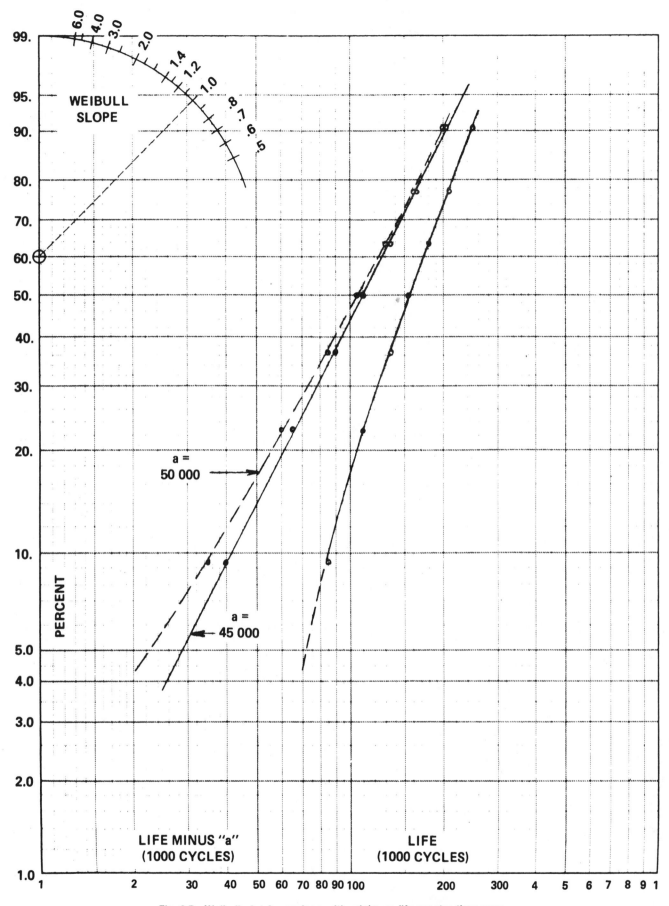

Fig. 8.7—Weibull plot for springs with minimum life greater than zero

1.95

Chapter 9
Material and Processing

1. Steel

The basic requirement of a leaf spring steel is that the selected grade of steel must have sufficient hardenability for the size involved to insure a fully martensitic structure throughout the entire section. Nonmartensitic transformation products detract from the fatigue properties.

Automotive chassis leaf springs have been made from various fine grained alloy steels such as:

G92600 (SAE 9260)	G86600(SAE 8660)
G40680 (SAE 4068)	G51600(SAE 5160)
G41610 (SAE 4161)	G51601(SAE 51B60)
G61500 (SAE 6150)	H51600(SAE 5160H)
	G50601(SAE 50B60)

In the United States almost all leaf springs are currently made of chromium steels such as G51600, G51601 or their H equivalents. For example, with G51600, the chemistry is specified as an independent variable (while the hardenability of the steel is a dependent variable which will vary with the chemistry). H51600 is essentially the same steel except that the hardenability is specified as an independent variable (while the chemistry is a dependent variable which may be adjusted to meet the hardenability band requirement).

In general terms, higher alloy content is necessary to insure adequate hardenability when the thicker leaf sections are used. When considering the grade of steel to be used, it is recommended either that the hardenability be calculated from the chemistry (for example, G51600), or that the hardenability band charts (for example, H51600) for the various H steels, as published in the SAE Handbook, be consulted.

The following "rule of thumb" may be useful for correlating section size and steel grade:

Thickness-mm	Steel
8.0 max	G51600
16.0 max	H51600
37.5 max	G51601

2. Mechanical Properties

Steels of the same hardness in the tempered martensitic condition have approximately the same yield and tensile strengths. The ductility, as measured by elongation and reduction of area, is inversely proportional to the hardness. Based upon experience, the optimum mechanical properties for leaf spring applications are obtained within the range of Brinell hardness numbers 388-461. This range contains the six standard Brinell hardness numbers 388, 401, 415, 429, 444, and 461 (corresponding to the ball indentation diameters 3.10, 3.05, 3.00, 2.95, 2.90, 2.85 mm obtained with an applied mass of 3000 kg). A specification for leaf springs usually consists of a range covered by four of these Bhn's, such as 415-461 for thin section sizes.

Measurements of typical mechanical properties of leaf spring steel are given below:

Hardness:	Bhn 388-461 (3000 kg mass)
	Brinell indentation diameter 3.10–
	2.85 mm
	Rockwell C 42-49
Tensile strength:	1300–1700 MPa
Yield strength	
(0.2% offset):	1170–1550 MPa
Reduction of area:	25% min
Elongation:	7% min

3. Surface Decarburization

Surface decarburization may reduce the fatigue durability of the springs; therefore, it is important that surface decarburization be at a minimum.

Hot rolled steel bars, as received from the mills, have some decarb, at least of the minimum Type 3 (see SAE J419 and SAE J1123 in the SAE Handbook), where more than 50% of the base carbon content remains at the surface (i.e., some partial but not more than 50% loss of carbon).

If decarb is of Type 2, where 50% or less of the base carbon content remains at the surface (i.e., appreciable partial but not total loss of carbon), the decarb normally does not exceed a depth of 0.25 mm for steels of thicknesses 5.00 through 12.50 mm, nor a depth of 0.50 mm for steels of thicknesses over 12.50 through 37.50 mm.

With sections over 25.00 mm in thickness, some of the hot rolled steel bars may have decarb of Type 1, in which virtually carbon free ferrite (i.e., total loss of carbon) exists for a measurable distance below the surface.

The depth of decarb varies from mill to mill, from rolling to rolling, and from bar to bar. The extent to which the depth and type of the decarb can be acceptable will be subject to agreement between the steel producer and the spring manufacturer.

The edges of the bars are somewhat higher in decarb than the flat surfaces; decarb on both the edges and the flat surfaces usually has greater depth with increased bar thickness.

After forging and non-atmospheric controlled heat treating, the spring leaves will have greater decarb. Scaling of the steel in this processing reduces the thickness of the leaf. While some of the surface decarb is removed with the scale, the final depth of decarb is usually greater than it was in the steel bars as received from the mills.

4. Mechanical Prestressing

Presetting, shot peening, and/or stress peening at ambient temperatures produces large increases in fatigue durability without increasing the size of the spring. These prestressing methods are more effective in increasing the fatigue properties of a spring than are changes in material.

When a load is applied to a leaf spring, the surface layers are subject to the maximum bending stress. One surface of each leaf will be in tension and the opposite surface will be in compression. The surfaces which are concave in the free position will generally be in tension under load, while the convex surfaces will generally be in compression. Fatigue failures of the leaves usually start at or near the surface on the tension side. Since residual stresses are algebraically additive to load stresses, the introduction of residual compressive stresses in the tension surface by prestressing reduces the operating stress level, thereby increasing the fatigue life.

Presetting (synonymous terms are: cold setting, bulldozing, setting-down, scragging) produces residual compressive stresses in the tension surface and residual tensile stresses in the compression surface by forcing the leaves to yield or take a permanent set in the direction of subsequent service loading. While this operation is beneficial to fatigue life, its primary effect is the reduction of "settling" (load loss) in service. Presetting is usually done on the spring assembly.

Shot Peening introduces compressive residual stresses by subjecting the tension side of the individual leaves to a stream of high-velocity shot. The SAE, "Manual on Shot Peening." HS-84, deals with the control of process variables, while techniques for control of peening effectiveness and quality are explained in, "Procedures for Using Standard Shot Peening Test Strip," SAE J443. Cut wire shot, size CW-23 to CW-41, and cast steel shot, size S-230 to S-390, are generally used for this purpose. Shot peening intensity is expressed as a dimensionless Almen number. The intensity of shot peening applied to light and medium type springs is usually in the range of 10A-20A as read on the Almen gage. For heavy springs, the intensity is usually 6C-14C as read on the Almen gage. Coverage in both cases should be at least 90%.

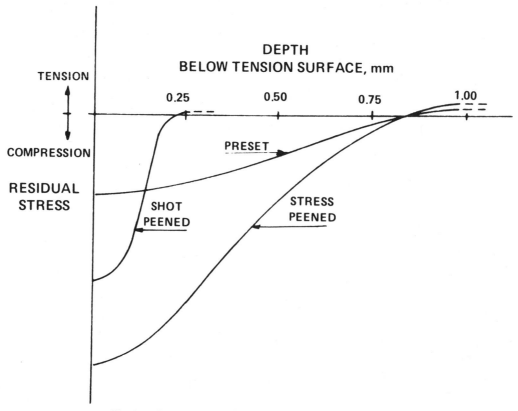

Fig. 9.1—Beneficial stress patterns induced by presetting and peening in the absence of surface decarburization

Stress Peening (strain peening) is a means of introducing higher residual compressive stresses than is possible with shot peening with the leaf in free (unloaded) position. Stress peening is done by shot peening the leaf while it is loaded (under stress) in the direction of subsequent service loading.

Curvature of a leaf spring will be changed by mechanical prestressing. The magnitude of the changes due to shot peening and presetting can be calculated using the formulae given in Chapter 5 Section 5.

The stress patterns induced by these processes are compared schematically in Fig. 9.1. The effects of shot peening and presetting are cumulative to some extent, but the results are influenced by the sequence of operations.

5. Surface Finishes and Protective Coatings

Surface Finish is defined as the surface condition of the spring leaves after the steel has been formed and heat treated, and prior to any subsequent coating treatment. Normally, automotive leaves are utilized "as heat treated" or in the "shot peened" condition. "As heat treated" finish will be a tight oxide produced by the quenching and tempering operations and will exhibit a blue or blue-black appearance. "Shot peened" finish is characterized by a matte luster appearance as a result of the removal by the peening operation of the blue or blue-black surface oxide.

Protective Coating refers to a material added to the surface of individual leaves or the exposed areas of leaf spring assemblies. Its primary purpose is to prevent corrosion both in storage and in operational environments. All exposed surfaces to be coated must be free from loose scale and dirt. Peened surfaces for which a coating is specified should be coated as soon as possible to prevent the formation of any corrosion. It is important that an enveloping coat is applied. An unprotected area or a break in the coating may contribute to localized corrosion and a reduction in fatigue life. Before a coating is specified, whether grease, oil, paint, or plastic, it should be evaluated for effects that its application might have on the fatigue life of the spring steel. The thickness and adhesion characteristics of the coating must be within the tolerances which have been established for the type of material being used in order to provide adequate corrosion protection and assure satisfactory performance.

Chapter 10
Design Data for
Single Leaf Springs

1. Single Leaf Types

The term "single leaf spring" has become popular terminology for a leaf spring which is made by tapering either in width or in thickness, or by a combination of both. Tables 10.1 and 10.2 deal with various types of single leaf springs. Since the semi-elliptic spring can be considered as the combination of two cantilever springs, the tables have been set up for cantilever springs. The combining of the two cantilever springs (front and rear) is discussed in Section 2 of this chapter and Chapter 5, Section 3.

Table 10.1 contains design descriptions for 12 types of single leaf springs, and Table 10.2 lists the applicable formulae and values for the computation of the volume, the vertical and lateral load rates, the peak stress under vertical load, and the volume efficiency for the 10 most important types.

The single leaf designs have been divided into three basic types:

Type F have uniform thickness.
Type P have parabolic contour thickness.
Type T have tapered thickness.

Types F-1, F-2, and F-4 are applicable to multi-leaf springs with the width split into stepped leaves. As multi-leaf designs, they are covered in the preceding chapters of this Manual.

Types P-2, T-1, and T-2 are used for design calculations where the width can be maintained constant in the tapering operation.

Types P-4, T-3, and T-4 are used for design calculations where the width is not restrained in the tapering operation so that a variable width is produced.

The formula for volume efficiency gives the amount of energy which each type of spring is capable of storing at a specified peak stress, compared to the active volume of material in the spring.

For a steel spring the energy (mJ or N·mm) per kg of active spring mass is obtained when the volume efficiency figure is divided by $7.85 \cdot 10^{-6}$ (which is the mass in kg of 1 mm^3 of spring steel). (See Chapter 1, Section 2).

The volume efficiency can also be converted into a dimensionless figure called "specific volume efficiency" or η by multiplication with the factor

$$\frac{2E}{S^2_{peak}}$$

This furnishes a comparison with a rod under tension which has $\eta = 1$. For cantilever beams in bending, the η values range between a low of 1/9 for type F-1 and a high of 1/3 for types F-3 and P-1.

The most efficient leaf spring has a constant stress pattern from line of encasement to line or point of load application. The constant stress pattern designs are types F-3, P-1, and P-3, but they are not functional as single leaf springs because of additional material requirements for support of load at the line or point of application. The three types are included in the tables for their basic formula values.

2. Rate Calculations

The formulae and data given in Table 10.2 are for cantilever springs. For springs of the semi-elliptic type, the load rates for front and rear cantilevers must be combined as shown in Chapter 5, Section 3. Most of the terms are defined in Chapter 4, Section 6, but ℓ, c, i, w, C_V, C_L are defined in Chapter 10.

In an *unconventional* semi-elliptic spring design, where Z does not equal Y^3, and the control of the center link (Φ) is specified, the vertical load rate can be calculated from

$$k_a = k \cdot \frac{(Z + Y^2)}{(Y + 1)^2}$$

and

$$k_b = \frac{k_a}{Z}$$

In a *conventional* semi-elliptic spring design, where Z equals Y^3, these formulae are reduced to

$$k_a = k \cdot \frac{b^2}{aL} \quad \text{and} \quad k_b = k \cdot \frac{a^2}{bL}$$

When the spring is a symmetrical semi-elliptic design, both cantilevers will have the same rate which is equal to one half of the combined rate, or

$$k_a = k_b = \frac{k}{2}$$

Note that when combining the cantilever formulae for a semi-elliptic spring design, the spring length which is called ℓ in Tables 10.1 and 10.2 becomes "a" for the fixed end cantilever and "b" for the shackled or free end cantilever.

The vertical rate formula is the usual starting point for the spring calculation. When the spring type and the val-

TABLE 10.1—DESCRIPTION OF VARIOUS TYPES OF SINGLE LEAF SPRINGS

With Flat Profile: "F" Types	With Parabolic Profile:

F-1 RECTANGULAR CANTILEVER

Constant in thickness (t_o) and in width (w_o). Under load (P) the stress is greatest at line of encasement and decreases at a constant rate to zero at line of load application. The elastic curve in bending (from an initially flat spring) has its smallest radius at line of encasement, that is, the rate of change of curvature is greatest at this line.

This design is inefficient.

P-1 PARABOLIC CANTILEVER

Constant in width (w_o). Thickness decreases from (t_o) at line of encasement in a parabolic profile which terminates in zero thickness at line of load application. Under load (P) the stress is constant throughout the length. The elastic curve in bending (from an initially flat spring) has its smallest radius at line of load application, that is, the rate of change of curvature is greatest at this line.

Although this design is highly efficient, it is impractical as no material is provided for load application.

F-2 TRAPEZOIDAL CANTILEVER

Constant in thickness (t_o). Width decreases at a constant rate from (w_o) at line of encasement to a specified dimension (w_e) at line of load application. Under load the stress is greatest at line of encasement.

This design is more efficient than Type F-1.

P-2 MODIFIED PARABOLIC CANTILEVER

Same as Type P-1 except for an end portion of length (c) with constant cross section ($t_c \times w_o$) to facilitate load application.

This design is slightly less efficient than Type P-1.

F-3 TRIANGULAR CANTILEVER

Constant in thickness (t_o). Width decreases at a constant rate from (w_o) at line of encasement to zero at point of load application. Under load the stress is constant throughout the length, the elastic curve in bending is circular, that is, the rate of change of curvature is constant throughout the length.

Although this design is highly efficient, it is impractical as no material is provided for load application.

P-3 PARABOLIC—TRAPEZOIDAL CANTILEVER

Thickness decreases from (t_o) at line of encasement in an approximately parabolic profile which terminates in zero thickness at line of load application. Width increases at a substantially constant rate from (w_o) at line of encasement to (w_e) at line of load application. Under load the stress is constant throughout the length.

Although this design is highly efficient, it is impractical as no material is provided for load application.

F-4 MODIFIED TRIANGULAR CANTILEVER

Same as Type F-3 except for an end portion of length (c) with constant cross section ($t_o \times w_c$) to facilitate load application.

This design is slightly less efficient than Type F-3.

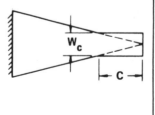

P-4 MODIFIED PARABOLIC—TRAPEZOIDAL CANTILEVER

Same as Type P-3 except for an end portion of length (c) with constant cross section ($t_c \times w_c$) to facilitate load application.

This design is slightly less efficient than Type P-3.

"P" Types	With Tapered Profile: "T" Types

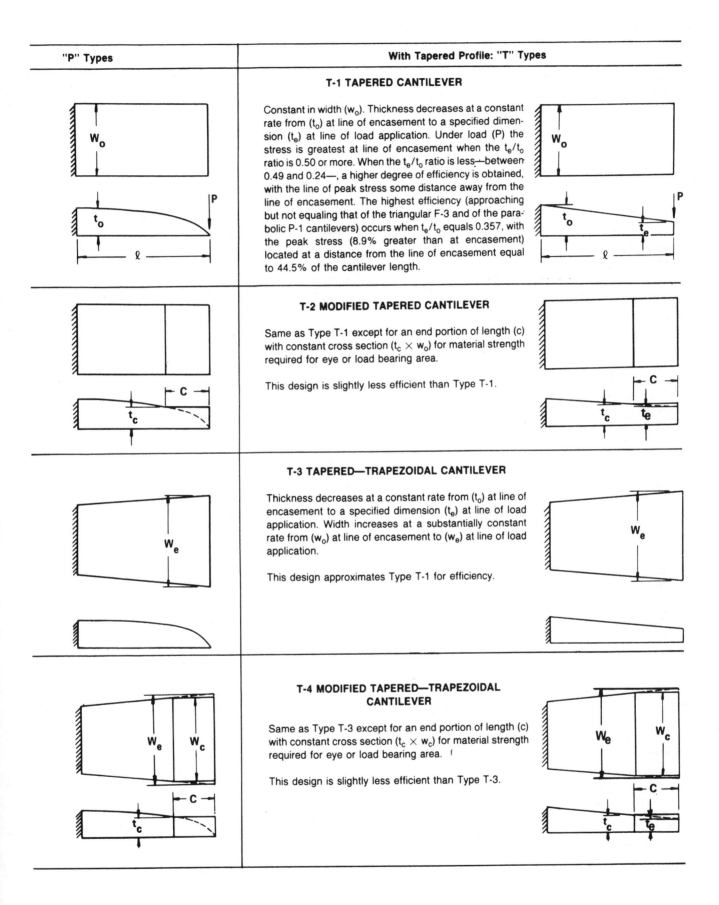

T-1 TAPERED CANTILEVER

Constant in width (w_o). Thickness decreases at a constant rate from (t_o) at line of encasement to a specified dimension (t_e) at line of load application. Under load (P) the stress is greatest at line of encasement when the t_e/t_o ratio is 0.50 or more. When the t_e/t_o ratio is less—between 0.49 and 0.24—, a higher degree of efficiency is obtained, with the line of peak stress some distance away from the line of encasement. The highest efficiency (approaching but not equaling that of the triangular F-3 and of the parabolic P-1 cantilevers) occurs when t_e/t_o equals 0.357, with the peak stress (8.9% greater than at encasement) located at a distance from the line of encasement equal to 44.5% of the cantilever length.

T-2 MODIFIED TAPERED CANTILEVER

Same as Type T-1 except for an end portion of length (c) with constant cross section ($t_c \times w_o$) for material strength required for eye or load bearing area.

This design is slightly less efficient than Type T-1.

T-3 TAPERED—TRAPEZOIDAL CANTILEVER

Thickness decreases at a constant rate from (t_o) at line of encasement to a specified dimension (t_e) at line of load application. Width increases at a substantially constant rate from (w_o) at line of encasement to (w_e) at line of load application.

This design approximates Type T-1 for efficiency.

T-4 MODIFIED TAPERED—TRAPEZOIDAL CANTILEVER

Same as Type T-3 except for an end portion of length (c) with constant cross section ($t_c \times w_c$) for material strength required for eye or load bearing area.

This design is slightly less efficient than Type T-3.

TABLE 10.2—VOLUME, RATES, EFFICIENCY OF VARIOUS TYPES OF SINGLE LEAF SPRINGS

	Flat Types				
	F-1	F-2	F-3	F-4	P-1
Length (mm): Cantilever Length is ℓ Length of constant cross section at end is c $c/\ell = H =$	1	0	0	B_e	0
Width (mm): At Encasement it is w_o At Load Application it is w_e $w_e/w_o = B_e =$	1	< 1	0	(0)	1
except for Types F-4, P-4, T-4 where it is w_c $w_c/w_o = B_c =$				H	
Thickness (mm): At Encasement it is t_o At Load Application it is t_e $t_e/t_o = J_e =$	1	1	1	(1)	0
except for Types P-2, P-4, T-2, T-4 where it is t_c $t_c/t_o = J_c =$				1	
Volume (mm³): Volume Factor =	1	$\frac{1}{2}\cdot(1 + B_e)$	1/2	$\frac{1}{2}\cdot(1 + H^2)$	2/3
Vertical Rate (N/mm): Vertical Rate Factor (C_v) =	1	$\frac{2}{3}\cdot\dfrac{(1-B_e)^3}{(1-B_e)(1-3B_e)-2B_e^2\ln B_e}$ (see Fig. 10.3)	2/3	$\dfrac{2}{3-H^2}$ (see Fig. 10.4)	1/2
Lateral Rate (N/mm): Lateral Rate Factor (C_L) =	1	$\frac{2}{3}\cdot\dfrac{(1-B_e)^3}{(B_e-3)(1-B_e)-2\ln B_e}$ (see Fig. 10.3)	Undefined	$\dfrac{1}{1-3\ln H}$ (see Fig. 10.4)	5/6

$$V = w_o\, t_o\, \ell \cdot \text{Volume Factor}$$

$$k = \frac{E\, w_o\, t_o^3}{4\,\ell^3} \cdot C_v$$

with i = length of constant cross section adjacent to encasement:

$$k = \frac{E\, w_o\, t_o^3}{4[\ell^3 - (\ell-i)^3 + (\ell-i)^3/C_v]}$$

$$k_L = \frac{E\, t_o\, w_o^3}{4\,\ell^3} \cdot C_L$$

(TABLE CONTINUED ON FOLLOWING PAGES)

	Parabolic Types			Tapered Types	
	P-2	**P-3**	**P-4**	**T-1**	**T-2**
	J_e^2	0	$J_c^2 B_c = J_c^2 \dfrac{B_e}{1 + J_c^2(B_e-1)}$	0	$(J_c - J_e)/(1-J_e)$
	(1)	> 1	(> 1)	1	(1)
	1		$H + B_e(1-H)$		1
	(0)	0	(0)	< 1	(<1)
	$\sqrt{H}$		$\sqrt{H/B_c} = \sqrt{\dfrac{H}{H+B_e(1-H)}}$		$J_e + H(1-J_e)$
	$\dfrac{1}{3}(2 + \sqrt{H^3})$ (see Fig. 10.1)	$\dfrac{B_e^2 \tan^{-1}\sqrt{B_e-1}}{4\sqrt{(B_e-1)^3}}$ $- \dfrac{2-B_e}{4(B_e-1)}$ (see Fig. 10.1)	see Fig. 10.1	$1/2 \cdot (1 + J_e)$ (see Fig. 10.2 with H = 0)	$1/2[1 + J_e + H^2(1-J_e)]$ (see Fig. 10.2 with H=0.1, 0.2, 0.3, 0.4, 0.5)
	$\dfrac{1}{2 - \sqrt{H^3}}$ (see Fig. 10.5)	see Fig. 10.5	see Fig. 10.5	$\dfrac{2(1-J_e)^3}{3(1-J_e)(J_e-3) - 6\ln J_e}$ (see Fig. 10.7 with H = 0)	see Fig. 10.7 with H=0.1, 0.2, 0.3, 0.4, 0.5
	$\dfrac{1}{6 - \sqrt{H^5}}$ (see Fig. 10.6)	see Fig. 10.6	see Fig. 10.6	$\dfrac{2(1-J_e)^3}{3(3J_e-1)(J_e-1) - 6J_e^2 \ln J_e}$ (see Fig. 10.8 with H = 0)	see Fig. 10.8 with H=0.1, 0.2, 0.3, 0.4, 0.5

	Flat Types				P-1
	F-1	F-2	F-3	F-4	P-1
Peak Stress (MPa) Under Vertical Load P (N): $S_{peak} =$ S_O = Stress at Encasement $= \dfrac{6 \ell P}{w_O t_O^2}$ [for location of S_{peak} see Chapter 10, Section 4]	S_O	S_O	S_O Uniform Over Entire Length	S_O From Encasement to Distance $(\ell - c)$	S_O Uniform Over Entire Length
Elastic Energy (mJ = N·mm): Energy Factor = $U = \dfrac{S_{peak}^2}{18 E} w_O t_O \ell \cdot \dfrac{\text{Energy Factor}}{C_V}$	1	1	1	1	1
Volume Efficiency (mJ/mm³): $U/V =$	$\dfrac{S_{peak}^2}{18 E}$	$\dfrac{S_{peak}^2}{9 E (1 + B_e)} \cdot \dfrac{1}{C_V}$	$\dfrac{S_{peak}^2}{6 E}$	$\dfrac{S_{peak}^2}{18 E} \cdot \dfrac{3 - H^2}{1 + H^2}$	$\dfrac{S_{peak}^2}{6 E}$

ues for length, width, and rate are specified, the thickness remains to be established.

The vertical rate for the cantilever spring is

$$k = \frac{E w_o t_o^3}{4 \ell^3} \cdot C_V$$

where C_v is the vertical rate factor for whatever single leaf type is under consideration (see Section 3 in this chapter). The subscript "o" refers to the values at the line of encasement for the cantilever spring or at the start of the active length for each half of the semi-elliptic spring (located within the axle seat area as discussed in Chapter 5, Section 2).

In the *conventional* semi-elliptic spring, the thickness t_o can be established by combining the rate formulae either for the fixed end or for the shackled (or free) end. The rate formulae at the fixed end are combined to

$$k_a = \frac{kb^2}{aL} = \frac{Ew_o t_o^3}{4a^3} \cdot C_V$$

then

$$k = \frac{Ew_o t_o^3 L}{4a^2 b^2} \cdot C_V$$

and

$$t_o^3 = k \cdot \frac{4a^2 b^2}{Ew_o L} \cdot \frac{1}{C_V}$$

In the same manner, the lateral rate for the semi-elliptic spring can be determined from

$$k_{La} = \frac{k_L b^2}{aL} = \frac{Et_o w_o^3}{4a^3} \cdot C_L$$

then

$$k_L = \frac{Et_o w_o^3 L}{4a^2 b^2} \cdot C_L$$

where: C_L = lateral rate factor.

In the *unconventional* spring, the value of "Z" will be established by means of the formula shown in Table 4.1

Parabolic Types			Tapered Types	
P-2	P-3	P-4	T-1	T-2
S_O From Encasement to Distance $(\ell - c)$	S_O Uniform Over Entire Length	S_O From Encasement to Distance $(\ell - c)$	For $J_e \geqslant 0.5$: S_O For $J_e < 0.5$: $\dfrac{S_O}{4J_e(1-J_e)}$	For $J_e \geqslant 0.5$: S_O For $J_e < 0.5$: $\dfrac{S_O}{4J_e(1-J_e)}$
1	1	1	For $J_e \geqslant 0.5$: 1 For $J_e < 0.5$: $16J_e^2(1-J_e)^2$	For $J_e \geqslant 0.5$: 1 For $J_e < 0.5$: $16J_e^2(1-J_e)^2$
$\dfrac{S_{peak}^2}{6E} \cdot \dfrac{2-\sqrt{H^3}}{2+\sqrt{H^3}}$	$\dfrac{S_{peak}^2}{18E} \cdot \dfrac{1}{\text{Volume Factor}} \cdot \dfrac{1}{C_V}$	$\dfrac{S_{peak}^2}{18E} \cdot \dfrac{1}{\text{Volume Factor}} \cdot \dfrac{1}{C_V}$	For $J_e \geqslant 0.5$: $\dfrac{S_{peak}^2}{9E(1+J_e)} \cdot \dfrac{1}{C_V}$ For $J_e < 0.5$: $\dfrac{S_{peak}^2}{9E(1+J_e)} \cdot \dfrac{16J_e^2(1-J_e)^2}{C_V}$	For $J_e \geqslant 0.5$: $\dfrac{S_{peak}^2}{9E[1+J_e+H^2(1-J_e)]} \cdot \dfrac{1}{C_V}$ For $J_e < 0.5$: $\dfrac{S_{peak}^2}{9E[1+J_e+H^2(1-J_e)]} \cdot \dfrac{16J_e^2(1-J_e)^2}{C_V}$

for specific values of "a", "b", and "Φ". Then at the fixed end of the semi-elliptic spring, the rate formulae are combined to

$$k_a = k \cdot \frac{(Z + Y^2)}{(Y + 1)^2} = \frac{Ew_o t_{oa}^3}{4a^3} \cdot C_{Va}$$

then

$$t_{oa}^3 = k \cdot \frac{(Z + Y^2)}{(Y + 1)^2} \cdot \frac{4a^3}{Ew_o} \cdot \frac{1}{C_{Va}}$$

At the shackled or free end the formulae are combined to

$$k_b = \frac{k}{Z} \cdot \frac{(Z + Y^2)}{(Y + 1)^2} = \frac{Ew_o t_{ob}^3}{4b^3} \cdot C_{Vb}$$

then

$$t_{ob}^3 = \frac{k}{Z} \cdot \frac{(Z + Y^2)}{(Y + 1)^2} \cdot \frac{4b^3}{Ew_o} \cdot \frac{1}{C_{Vb}}$$

Note that in the unconventional single leaf design the rate factor may have a different value for the fixed and for the shackled end. It must be realized that in the unconventional design the stress level may not be the same for both cantilevers, and that some sacrifice in design efficiency for vertical loads must then be accepted (see Chapter 4, Section 3).

From the basic rate formula in Chapter 5, Section 3

$$k = \frac{k_a k_b L^2}{k_a a^2 + k_b b^2}$$

it follows that after establishing the rate factors C_{Va} and C_{Vb} for the two ends, a control check on the calculated rate can be made with the formula

$$k = \frac{E \cdot w_o \cdot t_{oa}^3 \cdot t_{ob}^3 \cdot L^2}{4a^2 b^2 \left[\dfrac{b \cdot t_{oa}^3}{C_{Vb}} + \dfrac{a \cdot t_{ob}^3}{C_{Va}} \right]}$$

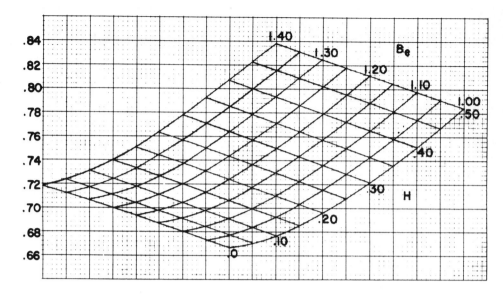

Fig. 10.1—Volume factor for parabolic ("P") types

3. Rate Factors

Both the vertical and the lateral rate formulae contain "rate factors." The values for the vertical rate factor are to be selected from the graphs in Figs. 10.3 and 10.4 when dealing with one of the F types; from the carpet plot in Fig. 10.5 when dealing with one of the P types; and from the carpet plots in Fig. 10.7 when dealing with one of the T types, depending upon the thickness ratio J_e which is specified or chosen.

Corresponding values for the lateral rate factor are available from the graphs in Figs. 10.3 and 10.4 for F types; from the carpet plot in Fig. 10.6 for P types; and from the carpet plots in Fig. 10.8 for T types.

The "carpet plots" are characterized by the absence of a numerical abscissa scale. They show the interaction of three variables by plotting a family of curves on a horizontally staggered scale. In Fig. 10.5 the value of the vertical rate factor for the P types is related to both the length ratio "H" (which equals zero for types P-1 and P-3) and the width ratio "B_e" (which equals 1.0 for types P-1 and P-2). The value of the factor can be read on the ordinate scale opposite the intersection of the applicable constant value curves for "H" and "B_e."

Interpolation is accomplished by drawing new constant value curves at the appropriate horizontal distances between the existing curves and establishing their intersection points. It should be noted that all curves of values which are multiples of 0.10 intersect on common vertical lines.

When dealing with one of the T type designs, the choice of the thickness ratio J_e and therefore the value of the rate factor may have to be determined by stress and manufacturing considerations. These considerations, together with the wide range of the rate factor values, often make it

necessary to carry out a series of calculations. In these cases the availability of a computer for such iterations will prove a major convenience for the designer.

4. Stress Calculations

It will be seen from the descriptions in Table 10.1 and from the stress formulae in Table 10.2 that for a given vertical load the stress at the line of encasement (S_o) is also the peak stress for any of the F type and P type cantilevers.

For the T type cantilevers S_o is the peak stress only as long as the thickness ratio J_e equals or exceeds 0.5. Thus the designer has an opportunity to place the peak stress at some distance from the encasement or spring seat by choosing a value for J_e of less than 0.5. This is desirable for highly stressed springs because any fretting corrosion which may occur due to the clamp effect will be less harmful to fatigue life if it does not occur in the area of peak stress. Table 10.2 shows how to calculate the peak stress and its location for the T-1 and T-2 types.

The formula for the peak stress in the T type cantilevers with J_e values of less than 0.5 (which is also shown in Table 10.2) is

$$S_{peak} = \frac{S_o}{4J_e(1 - J_e)}$$

This peak stress is located at a distance from the line of encasement which may be expressed in two different ways: either in terms of the thickness ratio J_e ($= t_e/t_o$); or, for type T-2, in terms of the thickness ratio J_c ($= t_c/t_o$) and of the length ratio H ($= c/\ell$). The distance is

$$\frac{\ell(1 - 2J_e)}{1 - J_e} \text{ or } \frac{\ell(1 - 2J_c + H)}{1 - J_c}$$

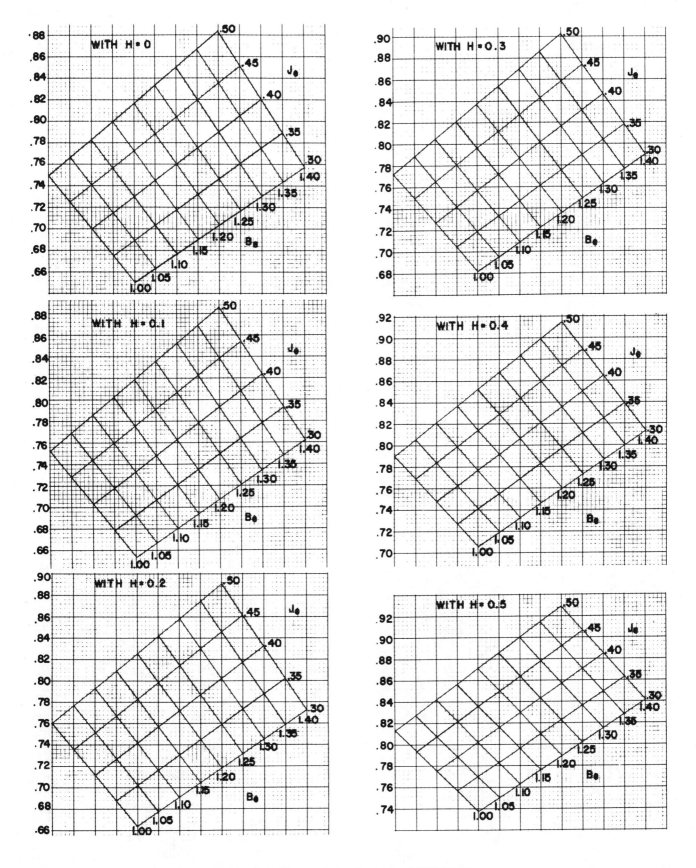

Fig. 10.2—Volume factor for tapered ("T") types

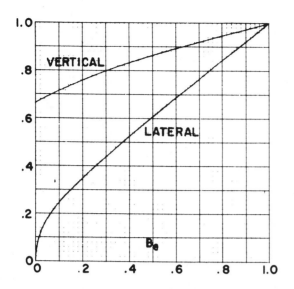

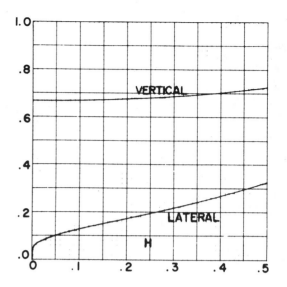

B_e	C_V	C_L	B_e	C_V	C_L	B_e	C_V	C_L
0	.667	0	.35	.815	.486	.70	.921	.771
.05	.695	.180	.40	.832	.528	.75	.937	.810
.10	.719	.244	.45	.848	.570	.80	.949	.849
.15	.741	.299	.50	.863	.612	.85	.962	.887
.20	.761	.349	.55	.879	.652	.90	.975	.925
.25	.780	.396	.60	.893	.692	.95	.988	.963
.30	.798	.442	.65	.908	.732	1.00	1.000	1.000

Note: B_e = 0 represents type F-3;
B_e = 1 represents type F-1

Fig. 10.3—Rate factors for type F-2

H	C_V	C_L	H	C_V	C_L
0	.667	0	.25	.681	.194
.02	.667	.079	.30	.687	.217
.05	.667	.100	.35	.695	.241
.10	.669	.126	.40	.704	.267
.15	.672	.149	.45	.715	.295
.20	.676	.172	.50	.727	.325

Note: H = 0 represents type F-3

Fig. 10.4—Rate factors for type F-4

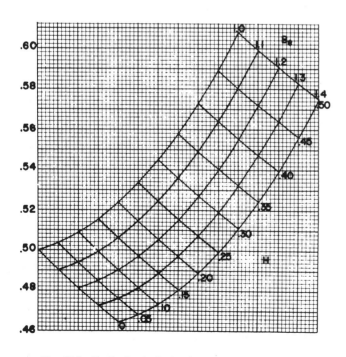

Fig. 10.5—Vertical rate factor for parabolic ("P") types

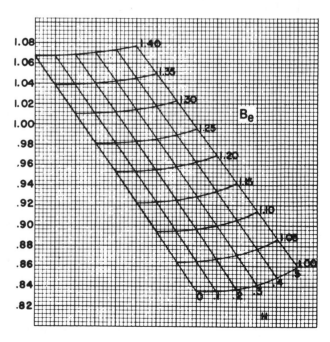

Fig. 10.6—Lateral rate factor for parabolic ("P") types

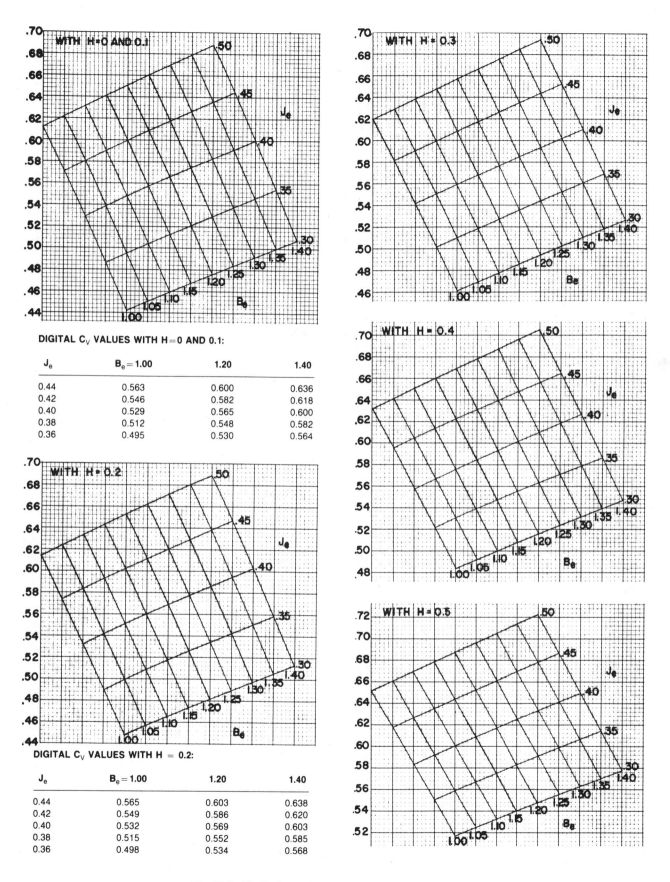

DIGITAL C$_V$ VALUES WITH H=0 AND 0.1:

J$_e$	B$_e$=1.00	1.20	1.40
0.44	0.563	0.600	0.636
0.42	0.546	0.582	0.618
0.40	0.529	0.565	0.600
0.38	0.512	0.548	0.582
0.36	0.495	0.530	0.564

DIGITAL C$_V$ VALUES WITH H = 0.2:

J$_e$	B$_e$ = 1.00	1.20	1.40
0.44	0.565	0.603	0.638
0.42	0.549	0.586	0.620
0.40	0.532	0.569	0.603
0.38	0.515	0.552	0.585
0.36	0.498	0.534	0.568

Fig. 10.7—Vertical rate factor for tapered ("T") types

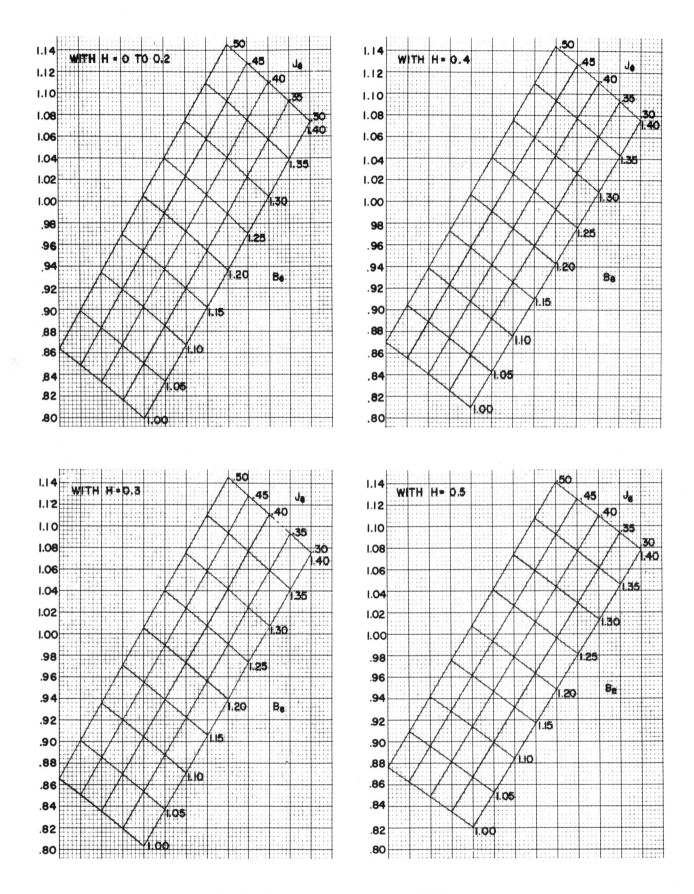

Fig. 10.8—Lateral rate factor for tapered ("T") types

When there is a length "i" of constant cross section adjacent to the line of encasement, the distance is

$$i + \frac{(\ell - i)(1 - 2J_e)}{1 - J_e} \quad \text{or} \quad i + \frac{(\ell - i)(1 - 2J_c + H)}{1 - J_c}$$

The thickness of the leaf at the line of this peak stress is

$$2J_e t_o \quad \text{or} \quad 2t_e \quad \text{or} \quad \frac{2(J_c - H)}{1 - H} \cdot t_o$$

5. Practical Details

The calculations for single leaf springs are generally based on the active length of the spring. This requires that the load rate in the formulae of Table 10.2 must be the rate of the spring as clamped in the vehicle position.

If the active length is not specified, it may be established by reducing the total length between centerlines of supports by the distance between the inner edges of the U or clamping bolts. This applies even when the spring is clamped in the seat area with rubber liners which actually extend the active spring length beyond the inner edges of the U or clamping bolts.

For the conventional semi-elliptic spring design, the relationship between the unclamped rate (k) and the clamped rate (k_c) is expressed by the formula

$$k_c = k \cdot \frac{L_c a^2 b^2}{La_c^2 b_c^2}$$

where L_c, a_c, and b_c are the active lengths.

When clamping is required for checking the rate in a test machine, as in cantilever springs, but also in some semi-elliptic springs, obviously the drawing must specify the clamped rate and the clamp details.

In designs where the load cannot be supported by one single leaf within the width, length, and stress limitations, several single leaves may be stacked to make the spring (Fig. 10.9B). In such a spring of stacked single leaves having identical configuration, each leaf will support its proportional share of the load and rate. Thus, where there are two single leaves stacked, each leaf will support one-half of the load and rate; where there are three single leaves stacked, each leaf will support one-third of the load and rate, etc.

In order to alleviate fretting corrosion, it is recommended that liners of plastic material or rubber isolate the single leaf from the steel pressure or U-bolt plate and from the axle seat.

The tension surface of the single leaf spring at the seat clamp is particularly critical in requiring this isolation. The need for the liner between the compression surface of the spring and the attaching part will depend on the spring application and will frequently be dictated by the designer's experience. Where single leaves are stacked to make a spring assembly, a liner should separate the leaves in the axle seat area (See Fig. 10.9B).

Rubber liners are generally used with passenger car single leaf springs. Liners of high compressive strength are required for trucks because they are subject to greater loads. Since the proper liner material is critical in the

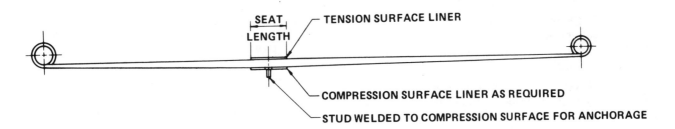

A. SINGLE LEAF SPRING

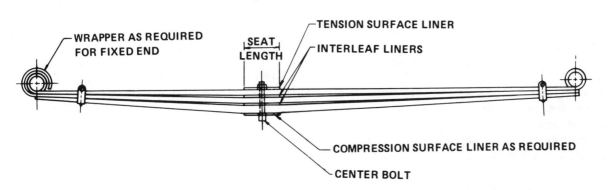

B. STACKED SINGLE LEAF SPRING

Fig. 10.9

fatigue life of single leaf springs, its selection must be based on test results with springs.

6. Camber of Single Leaf Springs

Leaf springs are generally designed to operate on the vehicle in a more or less flat (zero camber) position at design load.

Multi-leaf springs or single leaf springs of types F-2 and F-4 have more or less circular bending under change in load. (See Chapter 4, Section 1). Circular bending occurs when the camber of a spring changes due to increase or decrease in load from an arc of a circle to another arc of a circle.

Theoretically, single leaf springs of types P and T will have a more or less parabolic contour in the free camber. The actual contour including allowances for shot peening, presetting, and "spring back" from the camber forms is usually established by fitting a sample so that the desired shape and height is obtained under load. While exact calculations for the shape of the camber forms can be worked out from the elastic equation for deflection, they are too difficult for most practical purposes.

Under deflection, the eye center of a single leaf spring of type F-1 will move in a path which closely approximates an arc of a circle of 0.83ℓ radius, while the eye center of types P-1 and P-3 will move in a path of 0.67ℓ radius.

The eyes of other types of springs will move in a path with a radius which is between these extremes of 0.83ℓ and 0.67ℓ.

As shown in Chapter 4, Section 2, and in Fig. 4.1, the center of the eye of types F-2 and F-4 (which have circular bending) will move in a path which closely approximates an arc of a circle of 0.75ℓ radius.

For P-2, P-4, T-1, and T-2 type springs, an 0.70ℓ radius is generally used for the layout path of the eye.

7. Sample Calculations

First Example—Design a semi-elliptic passenger car spring to meet the following specifications:

Rate, as tested without center clamp	k	= 19.00	N/mm
Design load	P	= 3300	N
Clearance (metal-to-metal)	x_c	= 115	mm
Length, center to center of eyes	L	= 1500	mm
Length of front cantilever	a	= 625	mm
Length of rear cantilever	b	= 875	mm
Minimum leaf thickness at eyes	t	= 6.70	mm
Width (constant)	w	= 63.0	mm
Stress at M to M not to exceed	S_{max}	= 1175	MPa
Conventional design where	Z	= Y^3	

Single leaf spring calculations are based on the active length of the spring. Therefore, the rate to be used in the calculations must be the clamped rate.

Since the example is a passenger car spring and mounted to the axle seat with rubber insulators, the active length will extend well within the seat area. For this example, the inactive length is assumed to be 50% of the seat length or 50 mm.

Therefore: front end active length a_c = 600 mm
rear end active length b_c = 850 mm
total active length L_c = 1450 mm

Step One—Calculate clamped rate:

$$k_c = k \cdot \frac{L_c \, a^2 \, b^2}{L \, a_c^2 \, b_c^2}$$
$$= 19.00 \cdot \frac{1450 \cdot 625^2 \cdot 875^2}{1500 \cdot 600^2 \cdot 850^2}$$
$$= 21.12 \text{ N/mm}$$

Step Two—First trial: Design as Type T-1 single leaf spring (where H = 0 and width ratio B_e = 1.0) with thickness ratio J_e = 0.40.
Vertical rate factor (from Fig. 10.7). C_v = 0.528

$$t_o^3 = k_c \cdot \frac{4 \, a_c^2 \, b_c^2}{E \, w_o \, L_c} \cdot \frac{1}{C_V}$$
$$= 21.12 \cdot \frac{4 \cdot 600^2 \cdot 850^2}{200 \cdot 10^3 \cdot 63 \cdot 1450} \cdot \frac{1}{0.528}$$
$$= \frac{1202.7}{0.528} = 2278$$

$$t_o = 13.16 \text{ mm}$$

Load at M to M:

$$P_{max} = 3300 + 115 \cdot 21.12 = 5729 \text{ N}$$

Stress at M to M:

$$S_o = \frac{6 \cdot P_{max} \cdot a_c \cdot b_c}{w \cdot t_o^2 \cdot L_c}$$
$$= \frac{6 \cdot 5729 \cdot 600 \cdot 850}{63 \cdot 13.16^2 \cdot 1450}$$
$$= \frac{191907}{13.16^2} = 1108 \text{ MPa}$$

Peak stress is greater than the stress at t_o and is located in the front and rear ends at some distance from t_o.
Peak Stress:

$$S_p = \frac{S_o}{4 \cdot J_e(1 - J_e)}$$
$$= \frac{1108}{4 \cdot 0.4(1 - 0.4)} = 1154 \text{ MPa}$$

Design is within stress limitations.

Eye Thickness:
$$t_e = J_e \cdot t_o$$
$$= 0.4 \cdot 13.16 = 5.26 \text{ mm}$$

Eye thickness does not meet the design requirements.

Step Three—Second trial: Design as type T-2 single leaf spring with $t_c = 6.70$ mm. Compared to the first trial, this means added material in the vicinity of the eyes which will tend to increase the spring rate. By way of compensation t_o will have to be slightly less than the 13.16 mm thickness which was computed in the first trial. Try $t_o = 13.00$ mm and keep $J_e = 0.40$ as in the first trial. Then

$$J_c = \frac{t_c}{t_o} = \frac{6.70}{13.00} = 0.515$$

and

$$H = \frac{J_c - J_e}{1 - J_e}$$
$$= \frac{0.515 - 0.400}{1 - 0.400} = 0.192$$

The vertical rate factor for type T-2 with $H = 0.192$ (see Fig. 10.7) is close enough to the factor shown for $H = 0.20$ where $C_V = 0.532$ to use this value

$$t_o^3 = \frac{1202.7}{0.532} = 2261$$

$$t_o = 13.13 \text{ mm}$$

Since $t_o = 13.16$ mm in the first trial where $H = 0$, and $t_o = 13.13$ mm with $H = 0.192$, it is obvious that there will be only minor differences between spring rates as long as H does not exceed 0.20. Stress at M to M:

$$S_o = \frac{191907}{13.13^2} = 1113 \text{ MPa}$$

Peak Stress:

$$S_p = \frac{1113}{4 \cdot 0.4(1 - 0.4)} = 1159 \text{ MPa}$$

Location of peak stress:
Distance "X" from beginning of active length:

At front end: $X_f = \dfrac{a_c \cdot (1 - 2J_e)}{1 - J_e}$
$$= \frac{600 \cdot (1 - 0.8)}{1 - 0.4}$$
$$= 200 \text{ mm}$$

Measured from centerline of spring seat:

$$200 + 25 = 225 \text{ mm}$$

At rear end: $X_r = \dfrac{b_c \cdot (1 - 2J_e)}{1 - J_e} = \dfrac{850 \cdot (1 - 0.8)}{1 - 0.4}$

$$= 283 \text{ mm}$$

Measured from centerline of spring seat:

$$283 + 25 = 308 \text{ mm}$$

The second trial provides leaf thickness contour within all design limitations.

Step Four—Lateral rate may be calculated from

$$k_L = \frac{Et_o w_o^3 L_c}{4a_c^2 b_c^2} \cdot C_L$$

$C_L = 0.833$ from Fig. 10.8.

$$= \frac{200 \cdot 10^3 \cdot 63^3 \cdot 1450 \cdot 13.13 \cdot 0.833}{4 \cdot 600^2 \cdot 850^2}$$

$$= 69.70 \cdot 13.13 \cdot 0.833$$

$$= 762 \text{ N/mm}$$

Step Five—Calculate mass of spring:

Volume for active length $= w_o t_o \ell \cdot$ Volume factor

The volume factor for type T-2 spring with $J_e = 0.4$, $B_e = 1.0$, and $H = 0.192$, is between those shown in Fig. 10.2 for $H = 0.10$, (VF = 0.703), and $H = 0.20$ (VF = 0.712). By interpolation we obtain VF = 0.711.

Front end active volume:
$63 \cdot 13.13 \cdot 600 \cdot 0.711$ $\qquad = \qquad 352880$
Rear end active volume:
$63 \cdot 13.13 \cdot 850 \cdot 0.711$ $\qquad = \qquad 499910$
Inactive volume at seat:
$63 \cdot 13.13 \cdot (25 + 25)$ $\qquad = \qquad \underline{41360}$
Total volume exclusive of extensions
necessary for making eyes $\qquad\qquad 894150$

$$\text{Mass} = 894150 \text{ (mm}^3) \cdot 7.85 \cdot 10^{-6} \text{ (kg/mm}^3)$$
$$= 7.02 \text{ kg}$$

Step Six—Calculate volume efficiency:

$$\frac{U}{V} = \frac{S_p^2}{9E[1 + J_e + H^2(1 - J_e)]} \cdot \frac{16J_e^2(1 - J_e)^2}{C_V}$$

$$= \frac{1159^2 \cdot 16 \cdot 0.4^2 \cdot 0.6^2}{9 \cdot 200 \cdot 10^3 \cdot 1.4221 \cdot 0.532}$$

$$= 0.909 \text{ N·mm/mm}^3$$

Step Seven—Calculate specific volume efficiency:

$$\eta = \frac{U}{V} \cdot \frac{2E}{S_p^2} = 0.909 \cdot \frac{2 \cdot 200 \cdot 10^3}{1159^2}$$

$$= 0.271$$

Second Example—Design a semi-elliptic passenger car spring with the same parameters as those for the first example, but now design it as a type P-2 spring.

Step One—Same as in first example: $k_c = 21.12$ N/mm

Step Two—For first trial assume length ratio H = 0.25

$$C_V \text{ (from Fig. 10.5)} = 0.534$$

$$t_o^3 = k_c \cdot \frac{4a_c^2 b_c^2}{Ew_oL_c} \cdot \frac{1}{C_V} = \frac{1202.7}{0.534} = 2252$$

$$t_o = 13.11 \text{ mm}$$

Stress at M to M, $S_o = \dfrac{191907}{13.11^2} = 1117$ MPa

This is the peak stress which extends for the length of spring with parabolic contour thickness. The stress is within design limitations.

Eye thickness $= t_c = \sqrt{H} \cdot t_o = \sqrt{0.25} \cdot 13.11$

$$= 6.56 \text{ mm}$$

Step Three—Since this does not meet the minimum eye section requirement, a second trial becomes necessary with a new length ratio greater than 0.25. (Conversely, if t_c had been sufficiently greater than 6.70 mm to warrant further design work, a second trial would have been calculated with H less than 0.25.)
Using H = 0.27 with $C_V = 0.538$:
$t_o^3 = 2236$ $t_o = 13.08$ mm
$S_o = S_p = 191907/13.08^2 = 1122$ MPa
$t_c = \sqrt{0.27} \cdot 13.08 = 6.80$ mm
This meets all requirements.
Step Four—Lateral rate: See rate formula in step four of first example:

$$C_L \text{ (from Fig. 10.6)} = 0.840$$
$$k_L = 69.70 \cdot 13.08 \cdot 0.840$$
$$= 766 \text{ N/mm}$$

Step Five—Mass of spring: Volume factor for type P-2 with H = 0.27 (from Fig. 10.1):

$$VF = 0.712$$

Front end active volume:
$63 \cdot 13.08 \cdot 600 \cdot 0.712$ $\qquad = \qquad 352030$
Rear end active volume:
$63 \cdot 13.08 \cdot 850 \cdot 0.712$ $\qquad = \qquad 498710$
Inactive volume at seat:
$63 \cdot 13.08 \cdot (25 + 25)$ $\qquad = \qquad \underline{41200}$
Total volume exclusive of extensions
necessary for making eyes $\qquad$ 891940

$$\text{Mass} = 891940 \text{ (mm}^3\text{)} \cdot 7.85 \cdot 10^{-6} \text{ (kg/mm}^3\text{)}$$
$$= 7.00 \text{ kg}$$

Step Six—Volume efficiency:

$$\frac{U}{V} = \frac{S_p^2}{6E} \cdot \frac{2 - \sqrt{H^3}}{2 + \sqrt{H^3}}$$

$$= \frac{1122^2}{6 \cdot 200 \cdot 10^3} \cdot \frac{2 - \sqrt{0.27^3}}{2 + \sqrt{0.27^3}}$$

$$= 0.912 \text{ N·mm/mm}^3$$

Step Seven—Specific volume efficiency:

$$\eta = \frac{U}{V} \cdot \frac{2E}{S_p^2}$$
$$= 0.912 \cdot \frac{2 \cdot 200 \cdot 10^3}{1122^2}$$
$$= 0.290$$

Third Example—Design a semi-elliptic passenger car spring with the same parameters as those for the first example, but now design it as a type P-4 spring with $w_o = 56.0$ mm and $w_e = 70.0$ mm, and with a minimum eye section thickness of 6.50 mm.

Step One—Same as in first example: $k_c = 21.12$ N/mm

Step Two—With width ratio $B_e = 70.0/56.0 = 1.25$ and assuming a length ratio H = 0.27 (as in second example): C_V (from Fig. 10.5) = 0.516

$$t_o^3 = 21.12 \cdot \frac{4 \cdot 600^2 \cdot 850^2}{200 \cdot 10^3 \cdot 56 \cdot 1450} \cdot \frac{1}{0.516}$$

$$= \frac{1353.0}{0.516} = 2622$$

$$t_o = 13.79 \text{ mm}$$

Stress at M to M:

$$S_o = \frac{6 \cdot 5729 \cdot 600 \cdot 850}{56 \cdot 13.79^2 \cdot 1450}$$

$$= \frac{215896}{13.79^2} = 1135 \text{ MPa}$$

This is the peak stress, and it is within design limitations.

$$\text{Eye thickness} = t_c = t_o \sqrt{\frac{H}{B_c}}$$

where

$$B_c = H + B_e (1 - H)$$

$$= 0.27 + 1.25 (1 - 0.27)$$

$$= 0.27 + 0.91 = 1.18$$

then

$$t_c = 13.79 \sqrt{\frac{0.27}{1.18}} = 6.60 \text{ mm}$$

This meets the minimum eye section requirement. The width at the eyes is

$$w_c = w_o \cdot B_c$$

$$= 56 \cdot 1.18 = 66.1 \text{ mm}$$

The front and rear lengths at this uniform width are

$$c_f = H \cdot a_c$$
$$= 0.27 \cdot 600 = 162 \text{ mm}$$

$$c_r = H \cdot b_c$$
$$= 0.27 \cdot 850 = 230 \text{ mm}$$

Step Three—Since both stress and eye section are within design limitations, no second trial is required.

Step Four—Lateral rate:
$$C_L = 0.984 \text{ from Fig. 10.6}$$
$$k_L = \frac{200 \cdot 10^3 \cdot 56^3 \cdot 1450}{4 \cdot 600^2 \cdot 850^2} \cdot 13.79 \cdot 0.984$$

$$= 48.95 \cdot 13.79 \cdot 0.984$$

$$= 664 \text{ N/mm}$$

Step Five—Mass of spring:

$$VF = 0.748 \text{ from Fig. 10.1}$$

Front end active volume:
$56 \cdot 13.79 \cdot 600 \cdot 0.748$	$= \quad 346580$

Rear end active volume:
$56 \cdot 13.79 \cdot 850 \cdot 0.748$	$= \quad 490990$

Inactive volume at seat:
$56 \cdot 13.79 \cdot (25 + 25)$	$= \quad \underline{38610}$

Total volume exclusive of extensions necessary for making eyes $\qquad$ 876180

$$\text{Mass} = 876180 \ (\text{mm}^3) \cdot 7.85 \cdot 10^{-6} \ (\text{kg/mm}^3)$$
$$= 6.88 \text{ kg}$$

Step Six—Volume efficiency:
$$\frac{U}{V} = \frac{S_p^2}{18 \ E} \cdot \frac{1}{C_V} \cdot \frac{1}{VF}$$

$$= \frac{1135^2}{18 \cdot 200 \cdot 10^3 \cdot 0.516 \cdot 0.748}$$

$$= 0.927 \text{ N} \cdot \text{mm/mm}^3$$

Step Seven—Specific volume efficiency:
$$\eta = \frac{U}{V} \cdot \frac{2E}{S_p^2}$$

$$= 0.927 \cdot \frac{2 \cdot 200 \cdot 10^3}{1135^2} = 0.288$$

Fourth Example—Design a semi-elliptic passenger car spring with the same parameters as those for the first example, but with an additional requirement for an axle control (Φ-see Table 4.1) of 0.030 deg/mm of vertical travel. This requires an unconventional design of type T-2.

Step One—In the unconventional design Z does not equal Y^3, therefore it will be necessary to calculate t_o for both the front and the rear cantilevers, and this requires calculating the clamped rates for both the front and the rear cantilevers. The formulae for the following developments are taken from Chapters 4 and 5.

$$Z = \frac{b_c (57.3 + \Phi \cdot b_c)}{a_c (57.3 - \Phi \cdot a_c)}$$

$$= \frac{850(57.3 + 25.5)}{600(57.3 - 18.0)} = 2.985$$

$$Y = \frac{b_c}{a_c} = \frac{850}{600} = 1.417$$

$$k_a = k_c \cdot \frac{(Z + Y^2)}{(Y + 1)^2}$$

$$= 21.12 \cdot \frac{2.985 + 1.417^2}{2.417^2}$$

$$= 18.05 \text{ N/mm}$$

$$k_b = \frac{k_a}{Z} = \frac{18.05}{2.985} = 6.05 \text{ N/mm}$$

As a control check on these values compute

$$k_c = \frac{L_c^2}{\dfrac{a_c^2}{k_b} + \dfrac{b_c^2}{k_a}}$$

$$= \frac{1450^2}{\dfrac{600^2}{6.05} + \dfrac{850^2}{18.05}} = 21.12 \text{ N/mm}$$

Step Two—For first trial, assume length ratio H = 0.15. With thickness ratio $J_e = 0.40$ (same as in first example) C_V (from Fig. 10.7) = 0.530 by interpolation between H = 0.10 and H = 0.20.

For front cantilever:

$$t_{oa}{}^3 = k_a \cdot \frac{4 a_c{}^3}{E w_o} \cdot \frac{1}{C_V}$$

$$= 18.05 \cdot \frac{4 \cdot 600^3}{200 \cdot 10^3 \cdot 63} \cdot \frac{1}{0.530}$$

$$= 2335$$

$$t_{oa} = 13.27 \text{ mm}$$

$$P_a \text{ at M to M} = 5729 \cdot \frac{850}{1450}$$

$$= 3358 \text{ N}$$

$$S_{oa} \text{ at M to M} = \frac{6 a_c P_a}{w_o t_{oa}{}^2}$$

$$= \frac{6 \cdot 600 \cdot 3358}{63 \cdot 13.27^2}$$

$$= 1090 \text{ MPa}$$

$$S_{pa} = \frac{1090}{4 J_e (1 - J_e)} = 1135 \text{ MPa}$$

$$t_{ca} = J_c \cdot t_{oa}$$
$$= [J_e + H(1 - J_e)] \cdot t_{oa}$$
$$= [0.4 + 0.15(1 - 0.4)] \cdot 13.27$$
$$= 0.490 \cdot 13.27 = 6.50 \text{ mm}$$

Step Three—Since $t_{ca} = 6.50$ mm is not acceptable, a second trial becomes necessary with a length ratio greater than 0.15.

Using H = 0.18 and $C_V = 0.531$:
$t_{oa}{}^3 = 2331$
$t_{oa} = 13.26$
$S_{pa} = 1091/0.96 = 1136$ MPa,
$t_{ca} = [0.4 + 0.108] \cdot 13.26 = 6.74$ mm

Therefore front end cantilever is within design limitations.

For rear cantilever:

$$t_{ob}{}^3 = k_b \cdot \frac{4 b_c{}^3}{E w_o} \cdot \frac{1}{C_V}$$

$$= 6.05 \cdot \frac{4 \cdot 850^3}{200 \cdot 10^3 \cdot 63} \cdot \frac{1}{0.531}$$

$$= 2221$$

$$t_{ob} = 13.05 \text{ mm}$$

$$P_b \text{ at M to M} = 5729 \cdot \frac{600}{1450}$$

$$= 2371 \text{ N}$$

$$S_{ob} \text{ at M to M} = \frac{6 b_c P_b}{w_o t_{ob}{}^2}$$

$$= \frac{6 \cdot 850 \cdot 2371}{63 \cdot 13.05^2}$$

$$= 1127 \text{ MPa}$$

$$S_{pb} = \frac{1127}{0.96} = 1174 \text{ MPa}$$

$$t_{cb} = J_c \cdot t_{ob}$$

$$= 0.508 \cdot 13.05 = 6.63 \text{ mm}$$

Rear end cantilever is within stress limitations, but eye section thickness is below the specified 6.70 mm minimum.

If 6.63 mm is not acceptable for the rear eye section thickness, a third trial becomes necessary for the rear cantilever with H larger than 0.18 (use 0.20).

For this example, 6.63 mm will be acceptable because the rear eye is shackled and does not take the braking and driving forces. Also, the rear cantilever is longer than the front cantilever, therefore the static and dynamic loads are less than on the front eye.

Since the leaf thickness required for the front cantilever ($t_{oa} = 13.26$) is greater than the leaf thickness required for the rear cantilever ($t_{ob} = 13.05$), a transition for a distance (T) must be created between the two. For clamping purposes, the front thickness (13.26) is extended for the full inactive length (which is 50 mm in this example).

At the start of the active length for the rear cantilever, the thickness transition must be made in as short a distance as possible in order to prevent an undue increase in the rate of the rear cantilever. On the other hand, the distance (T) must not be so short as to create too abrupt a transition. Depending on the capability of the tapering equipment, it may be postulated that the tapering rate (reduction in thickness per mm of length) within the distance (T), to be called TR_T, should not be greater than twice the tapering rate of the rear cantilever which is $TR_b = t_{ob} \cdot (1 - J_e) / b_c$

In this example

$$TR_b = \frac{13.05 \cdot 0.6}{850} = 0.00921$$

Therefore maximum $TR_T = 2 \cdot 0.00921$

The leaf thickness at the junction of the transition taper with the rear cantilever (to be called t_j) can be expressed in two ways, namely $t_j = t_{ob} - T \cdot TR_b = t_{oa} - T \cdot TR_T$

$$\text{Then } T = \frac{t_{oa} - t_{ob}}{TR_T - TR_b}$$

If TR_T is selected as the maximum, namely $2 \cdot 0.00921$, then

$$T = \frac{13.26 - 13.05}{2 \cdot 0.00921 - 0.00921} = 22.8 \text{ mm}$$

By using $T = 25.0$ mm,

$$t_j = 13.05 - 25 \cdot 0.00921 = 12.82 \text{ mm}$$

Step Four—Lateral rate: Since t_o is different for the front (13.26) and rear (13.05) cantilevers, it is suggested to use the average thickness (13.16 mm) for calculating the lateral rate.

The value of C_L will be found in the first carpet plot of Fig. 10.8 since the factor remains practically unchanged between $H = 0$ and $H = 0.20$. With $B_e = 1.00$ and $J_e = 0.40$ the lateral rate factor $C_L = 0.832$

$$k_L = 69.70 \cdot 13.16 \cdot 0.832$$

$$= 763 \text{ N/mm}$$

Step Five—Mass of spring: $VF = 0.711$ from Fig. 10.2 by interpolation between carpet plots for $H = 0.10$ and $H = 0.20$

Front end active volume:
$63 \cdot 13.26 \cdot 600 \cdot 0.711$	$=$	356370

Rear end active volume:
$63 \cdot 13.05 \cdot 850 \cdot 0.711$	$=$	496870

Transition 13.26 to 13.05:
$63 \cdot (0.21/2) \cdot 25$	$=$	170

Inactive volume at seat:
$63 \cdot 13.26 \cdot (25 + 25)$	$=$	41770

Total volume exclusive of extensions necessary for making eyes 895180

$$\text{Mass} = 895180 \ (\text{mm}^3) \cdot 7.85 \cdot 10^{-6} \ (\text{kg/mm}^3)$$
$$= 7.03 \text{ kg}$$

Step Six—Volume efficiency: In the unconventional design, the front and rear cantilevers have different volume efficiencies because of their different peak stresses.

$$\frac{U}{V} = \frac{S_p^2}{9 E[1 + J_e + H^2 (1 - J_e)]} \cdot \frac{16 J_e^2 (1 - J_e)^2}{C_V}$$

For front cantilever:

$$\frac{U}{V} = \frac{1135^2 \cdot 16 \cdot 0.4^2 \cdot 0.6^2}{9 \cdot 200 \cdot 10^3 [1 + 0.4 + 0.0324 \cdot 0.6]0.531}$$
$$= \frac{1135^2 \cdot 0.9216}{1.8 \cdot 10^6 \cdot 1.41944 \cdot 0.531} = 0.875$$

For rear cantilever:

$$\frac{U}{V} = \frac{1174^2 \cdot 0.9216}{1.8 \cdot 10^6 \cdot 1.41944 \cdot 0.531} = 0.936$$

Step Seven—Specific volume efficiency:
For front cantilever:

$$= 0.875 \cdot \frac{2 \cdot 200 \cdot 10^3}{1136^2} = 0.271$$

For rear cantilever:

$$= 0.936 \cdot \frac{2 \cdot 200 \cdot 10^3}{1174^2} = 0.272$$

SUMMARY OF THE FOUR EXAMPLES

Example	First	Second	Third	Fourth
Single leaf type	T-2	P-2	P-4	T-2
Control (deg/mm)	0.028	0.028	0.028	0.030
Thickness				
t_o (at seat)	13.13	13.11	13.79	13.26/13.05
t_c (at eye)	6.70	6.80	6.60	6.74/ 6.63
Width				
w_o (at seat)	63.0	63.0	56.0	63.0
w_c (at eye)	63.0	63.0	66.1	63.0
Lengths of constant sections				
front	115	162	162	108
rear	163	230	230	153
Stress S_p, at M to M (MPa)	1159	1122	1135	1136/1174
Distance of S_p from seat center				
front	225	25	25	225
rear	308	25	25	308
Lateral rate (N/mm)	762	766	664	763
Mass (kg)	7.02	7.00	6.88	7.03
Volume efficiency (J/mm³)	0.909	0.912	0.927	0.875/0.936
Specific volume efficiency	0.271	0.290	0.288	0.271/0.272

Points which may be considered to favor a choice between the four examples:
First and fourth examples have peak stresses at considerable distances from the axle seat.
Second example has the lowest peak stress and the highest specific volume efficiency.
Third example has the lowest mass, lowest lateral rate, narrowest width at axle seat.
Fourth example furnishes specified (higher) axle control.

Appendix A

Conversion Table

To Convert from SI Unit to U.S. Customary Unit, Divide by the Factor
To Convert from U.S. Customary Unit to SI Unit, Multiply by the Factor

Quantity	SI Unit		Factor (* = Exact)	U.S. Customary Unit	
Length	meter	m	0.304 8*	foot	ft
Length	millimeter	mm	25.4*	inch	in
Area Moment of Inertia		mm^4	416 231.425 6*	inch to the fourth power	in^4
Mass	kilogram	kg	0.453 592 4	pound-mass	lb$_m$
Force (or Load)	newton	N	4.448 222^a	pound-force	lb$_f$
Elastic Energy, Work	joule	J (= N·m = kN·mm)	0.112 984 8	inch pound	lb$_f$·in
Bending Moment, Torque	newton millimeter	N·mm	112.984 8	pound inch	lb$_f$·in
Spring Rate (k)	newton per mm	N/mm	0.175 126 8	pound per inch	lb$_f$/in
Stress, Modulus of Elasticity	pascal / megapascal	Pa (N/m^2) / MPa	6 894.757^b / 0.006 894 757	pound per square inch	lb$_f$/in^2 (= psi)
Modulus of Elasticity for Steel (E)	200 × 10^3 MPa		Equals (Nearly)	29 × 10^6 psi	
Density of Material e.g. for steel	kilogram per cubic meter	kg/m^3 7850 kg/m^3	27 679.90^c Equals (Nearly)	pound per cubic inch 0.283 6 lb$_m$/in^3	lb$_m$/in^3
Acceleration of Gravity "g" — adopted 1901 by International Committee on Weights and Measures		9.806 650 m/s^2 9.806 650 m/s^2	0.3048* 0.0254*	32.174 05 386.089	ft/s^2 in/s^2
Natural Frequency "f" $f = \dfrac{1}{2\pi}\sqrt{\dfrac{g}{F_s}}$ where F_s = static deflection = load/spring rate	cycles per second $f = \sqrt{0.2484/F_s(m)}$ $f = \sqrt{248.4/F_s(mm)}$	Hz	Equals (Nearly)	$f = \sqrt{9.780/F_s(in)}$	

^{a}4.448 222 = 0.453 592 37 · 9.806 650

^{b}6894.757 = $\dfrac{4.448\ 221\ 6}{0.000\ 645\ 16}$

^{c}27 679.90 = $\dfrac{0.453\ 592\ 37}{0.000\ 016\ 387\ 064}$

Appendix B

Derivation of Formulae for the Tabulated Values in Tables 5.2 and 5.3

Table 5.3 shows the mass of each of the leaf spring bars as listed in Table 3.1 (also listed in Table 1 in SAE Standard J1123). The mass values are expressed in terms of kg per m (or g per mm) length for bars of mean dimensions.

The calculation in this appendix shows the formula for cross section area of the bar. The mass values in Table 5.3 apply to steel bars having a mass of 7850 kg per m^3.

Table 5.2 shows the moment of inertia of each of the leaf spring bar cross sections listed in Table 3.1. The moments

of inertia are expressed in terms of mm^4 for bars of mean dimensions.

The calculation shows the formula for the moment of inertia.

The description in Chapter 3 of the cross section refers to a flatness tolerance which permits the two flat surfaces to be slightly concave. The calculations treat these surfaces as parabolic rather than circular. The difference amounts to considerably less than the accuracy to which the values in the Tables have been extended.

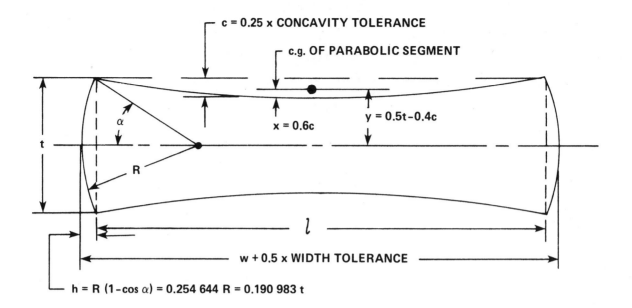

h = R (1−cos α) = 0.254 644 R = 0.190 983 t

Nomenclature

w = NOMINAL WIDTH
t = NOMINAL THICKNESS
R = 0.75 t
sin α = 0.50 t/0.75 t = 0.666 667
α = 0.729 722 (radian)
$\cos^2 α = 1 − (⅔)^2 = ⅝$
$\cos α = \sqrt{⅝} = 0.745\,356$
$l = w + 0.5 × \text{WIDTH TOLERANCE} − 2h$

Cross Section Area

AREA = ▭   + ◖◗   − ⟨⟩
AREA = lt + $R^2 (2α − \sin 2α)$ $− 2(⅔\,cl)$
AREA = lt + $0.5625t^2(1.459\,444 − 0.993\,807) − 1.333\,333cl$
AREA = lt + $0.261\,921t^2$ $− 1.333\,333cl$ (mm²)

Mass per Unit Length (See Table 5.3)

MASS = $(lt$ $+ 0.261\,921\, t^2$ $− 1.333\,333cl)·0.007\,850$ (kg/m)

Moment of Inertia (See Table 5.2)

I = ▭   + ◖◗   − ⟨⟩

▭ = $lt^3/12 = 0.083\,333\,lt^3$

$$◖◗ = \frac{2R^2}{4} \cdot \frac{A◖◗}{2} \left[1 − \frac{2\sin^3α·\cosα}{3(α − \sinα·\cosα)} \right]$$

$$'' = 0.140\,625\,t^2 \cdot 0.261\,921\,t^2 \left[1 − \frac{2 · 0.296\,296 · 0.745\,356}{3(0.729\,722 − 0.666\,667 · 0.745\,356)} \right]$$

$$'' = 0.013\,540\,t^4$$

$$⟨⟩ = 2\left[\frac{2c^3 l}{7} − \frac{2cl}{3}·0.36c^2 + \frac{2cl}{3}(0.5t − 0.4c)^2 \right]$$

$$'' = 1.333\,333\,cl\, [0.068\,571\,c^2 + (0.5\,t − 0.4\,c)^2]$$

I = $0.083\,333\,lt^3 + 0.013\,540\,t^4$ $− 1.333\,333\,cl\,[0.068\,571\,c^2 + (0.5\,t − 0.4\,c)^2]$ (mm⁴)

1.122

Part 2

Design and Application of Helical and Spiral Springs

SAE HS 795

SPRING COMMITTEE

E. C. Oldfield (Chairman), Hendrickson Spring
K. Campbell
D. Curtin, Chevrolet Motor Division, General Motors Corporation
M. Glass, Detroit Steel Prod. Co., Marmon Group Inc.
R. S. Graham, Suspension Systems Co., Rockwell Int'l.
R. A. Gray
D. J. Hayes
P. W. Hegwood, Jr., GMC Delco Prods.
E. H. Judd, Barnes Grp. Inc., Associated Spring Co.
J. F. Kelly
S. Landsman, Associated Spring, Barnes Group Inc.
M. Lea, GKN Composites
D. J. Leonard, Airspring Dev. Svcs., Firestone Tire & Rubber Co.
J. Marsland, Jeep Truck Engrg., Chrysler Corp.
W. T. Mayers, Peterson Amer. Corp.
J. E. Mutzner, Inland Div., GMC
J. P. Orlando, CPC, Engrg., General Motors Corp.
R. L. Orndorff, Jr., Dynamic Polymer Products, B. F. Goodrich Co.
W. Platko, Chevrolet Motor Division, General Motors Corporation
G. R. Schmidt, Jr., Suspension Engineering, Moog Automotive Inc.
A. Schremmer, Barnes Group Inc., Associated Spring Saline
G. A. Schremmer, Schnorr Corp.
K. E. Siler, Ford Motor Co.
R. W. Siorek, AMSTA-TMV, US Army Tank Auto Command
A. R. Solomon, Analytical Engineering & Res. Inc.
M. C. Turkish, Valve Gear Design Associates
F. J. Waksmundzki, Suspension Div. Prod. Engrg., Eaton Corp.
D. A. F. Whidden, Hendrickson Canada Ltd.
J. Dimmock, GKN Composites
J. A. Fader, Suspension System Co., Rockwell Int'l.
G. W. Folland
J. R. Hughlett, Quality Assurance QSO, Ford of Europe Inc.
D. Merriman, Engrg. Center, Navistar Int'l.
D. R. Olberts, Deere & Co.
J. Schindler

COIL SPRING SUBCOMMITTEE

W. T. Mayers (Chairman), Peterson Amer. Corp.
A. M. Peach (Vice Chairman)
W. J. Behnke, Chrysler Motors
D. E. Gartner, Union Corp. Eng., Union Spring Mfg. Co.
R. A. Gray
E. H. Judd, Barnes Grp. Inc., Associated Spring Co.
C. J. Meyer, John Deere Product Engrg. Center, Deere & Co.
I. Neimanis, Madison Heights Plant, Peterson American Corp.
J. Schindler
G. R. Schmidt, Jr., Suspension Engineering, Moog Automotive Inc.
K. E. Siler, Ford Motor Co.
A. R. Solomon, Analytical Engineering & Res. Inc.
M. C. Turkish, Valve Gear Design Associates
R. L. Van Eerden, Suspension Systems, Rockwell International

TABLE OF CONTENTS

LETTER SYMBOLS

Symbol	Definition	SI Unit
a	side of square wire	millimeter (mm)
b	width (long side) of rectangular wire	millimeter (mm)
C	spring index (D/d)	—
c	surge velocity of transient wave in dynamic loading	meter per second (m/s)
D	mean coil diameter	millimeter (mm)
D_a	diameter of arbor	millimeter (mm)
D_i	inside diameter	millimeter (mm)
D_o	outside diameter	millimeter (mm)
d	wire (and bar) diameter	millimeter (mm)
E	modulus of elasticity in tension	megapascal (MPa) ($= N/mm^2 = MN/m^2$)
F	linear deflection	millimeter (mm)
f	frequency (cycles per second)	hertz (Hz)
G	modulus of elasticity in shear	megapascal (MPa) ($= N/mm^2 = MN/m^2$)
g	standard acceleration of gravity	9.809 965 m/s²
γ	density (mass per unit volume)	kilogram per cubic meter (kg/m³)
K	correction factor	—
L	spring length	millimeter (mm)
M	moment	newton millimeter (N $\cdot$ mm)
N_a	number of active coils	—
N_t	total number of coils	—
n	number of cycles	—
P	load or force	newton (N) ($= kg \cdot m/s^2$)
P $\cdot$ F/2	energy [linear rate]	joule (J) ($= N \cdot m = kN \cdot mm$)
p	pitch (axial distance center to center of coils)	millimeter (mm)
R	linear spring rate	newton per millimeter (N/mm)
	torsional spring rate	newton millimeter per degree (N $\cdot$ mm/deg)
r	radius	millimeter (mm)
S	stress	megapascal (MPa) ($= N/mm^2 = MN/m^2$)
T_h	time in hours	hour (h)
T_s	time in seconds	second (s)
t	thickness (short side) of rectangular wire	millimeter (mm)
θ	angular deflection	degree (deg)
V	volume of active material	cubic millimeter (mm³)
v	velocity of spring under dynamic loading	meter per second (m/s)
W	mass	kilogram (kg)

CONVERSION TABLE

To Convert from SI Unit to U.S. Customary Unit, Divide by the <u>Factor</u>
To Convert from U.S. Customary Unit to SI Unit, Multiply by the <u>Factor</u>

Quantity	SI Unit		<u>Factor</u> (§ = Exact)	U.S. Customary Unit	
Length	kilometer	km	1.609 344§	mile	
	meter	m	0.304 8§	foot	ft
	millimeter	mm	25.4§	inch	in
Area	square millimeter	mm²	645.16§	square inch	in²
Volume	cubic millimeter	mm³	16 387.064§	cubic inch	in³
	cubic millimeter	mm³	3 785 412.0	gallon	gal (U.S.)
	liter	L	3.785 412	gallon	gal (U.S.)
Area Moment of Inertia	millimeter to the fourth power	mm⁴	416 231.425 6§	inch to the fourth power	in⁴
Mass	kilogram	kg	0.453 592 37	pound-mass	lbₘ
Force (or Load)	newton	N	4.448 221 6[a]	pound-force	lb_f
Elastic Energy, Work	joule $\quad$ J (N · m) $(= kN · mm)$		0.112 984 8	pound inch	lb_f · in
Bending Moment, Torque	newton millimeter $\quad$ N · mm $(= mN · m)$		112.984 8	pound inch	lb_f · in
Spring Rate (Linear)	newton per mm $\quad$ N/mm $(= kN/m)$		0.175 126 8	pound per inch	lb_f/in
Torsional Spring Rate	newton millimeter per radian $\quad$ N · mm/rad		112.984 8	pound inch per radian	lb_f · in/rad
Plane Angle	degree		57.295 780[b]	radian	rad
Stress, Modulus of Elasticity	pascal $\quad$ Pa (N/m²) $\\$ kilopascal $\quad$ kPa $\\$ megapascal $\quad$ MPa		6894.7573[c] $\\$ 6.894 757 3 $\\$ 0.006 894 757 3	pound per square inch $\quad$ lb_f/in² (= psi)	
Density of Material e.g. for Steel	kilogram per cubic meter $\quad$ kg/m³ $\\$ 7850 kg/m³		27 679.90	pound per cubic inch $\\$ ~ 0.283	lbₘ/in³ $\\$ lbₘ/in³
Acceleration "g" due to Gravity (by International Agreement)	9.806 650 m/s² $\\$ 9.806 650 m/s²		0.3048§ $\\$ 0.0254§	32.174 $\\$ 386.09	ft/s² $\\$ in/s²
Natural Frequency (Hz = cycles/s) $\frac{1}{2\pi}\sqrt{\frac{g}{F_s}}$ where F_s = static deflection = load/spring rate	$f = \sqrt{0.248/F_s(m)}$ $\\$ $f = \sqrt{248/F_s(mm)}$			$f = \sqrt{9.78/F_s(in)}$ $\\$ $f = \sqrt{9.78/F_s(in)}$	

[a] $4.448\ 221\ 6 = 0.453\ 592\ 37 \cdot 9.806\ 650$

[b] $57.295\ 780 = 180/\pi$

[c] $6894.757\ 3 = \dfrac{4.448\ 221\ 6}{0.000\ 645\ 16}$

Chapter 1

Fundamental Considerations

1. Energy

Definitions

A mechanical spring is defined as an elastic body which has the primary function to deflect or distort under load, and to return to its original shape when the load is removed.

The external work imparted to the spring by the product of the two factors "load" and "deflection" is stored by the spring as potential energy and is subsequently recovered (except for frictional losses) as kinetic energy. The complete cycle may require only a fraction of a second or up to several years; two typical examples are whether the spring actuates an engine valve or a circuit breaker.

For most spring applications, the deflection F can be considered as proportional to the load P, provided the elastic limit of the material is not exceeded. Not all springs, however, have linear load deflection diagrams. Notably in this class are disk or Belleville and volute springs. Where accurate load deflection characteristics are a factor, changes in coil diameter and non-axial force components must be considered. If the spring is considered as having a constant load-deflection rate R, with load P expressed in terms of N (newton) and deflection F expressed in terms of m (meter), then the energy stored by the spring at deflection F or load P can be expressed in terms of J (joule = N . m) as follows:

$$\text{Energy} = \frac{P\,F}{2} = \frac{P^2}{2\,R} = \frac{R\,F^2}{2} \qquad \text{(J)}$$

However, in this Manual, F is expressed as mm and R as N/mm; this requires that the denominator in the energy formula be multiplied by 10^3 so as to convert mm into m, thereby allowing the energy still to be expressed as J. The stored energy will then be expressed as follows:

$$\text{Energy} = \frac{P\,F}{2 \times 10^3} = \frac{P^2}{2\,R \times 10^3} = \frac{R\,F^2}{2 \times 10^3} \qquad \text{(J)}$$

The need for this conversion arises wherever lengths are expressed in mm, as in d, D, F, L, and also where compound terms involving a length dimension are used, such as R (N/mm) and E, G, S (MPa = N/mm^2).

There are four mathematical symbols in this Manual where the length factor has not been converted to mm. They are the symbols for density - γ (kg/m^3), standard acceleration of gravity - g (9.81 m/s^2), and velocity under dynamic loading - c and v (m/s). Therefore, the reader must be alert to apply the correct conversion factor where a formula contains both converted and unconverted factors, as in Sections 2 and 3 of this Chapter.

The preceding formulae are based on the fact that the deflection starts from the free length or free position depending upon the type of spring. Generally the spring is given some initial deflection F_1 upon assembly into its application which develops the load P_1. From this point, the increase of the load to its maximum value P_2 will cause the spring to be deflected through a distance or "stroke" $(F_2 - F_1)$ to the maximum deflection F_2. The subsequent decrease in load to P_1 will return the spring to its assembled length. In such instances, the energy to be absorbed by the spring during the cycle is:

$$\text{Energy} = \frac{P_1 + P_2}{2 \times 10^3}(F_2 - F_1) = \frac{(P_2^2 - P_1^2)}{2\,R \times 10^3} \qquad \text{(J)}$$

$$= \frac{R(F_2^2 - F_1^2)}{2 \times 10^3} \qquad \text{(J)}$$

Energy Capacity of Different Spring Types

The general problem for the designer is to know the overall energy capacity required of the spring and to relate this to the maximum stress in the spring. It has been established in fundamental studies that the energy capacity of a spring for a given maximum stress increases in direct proportion to the volume of the active spring material. For each type of spring, the energy capacity per unit volume of active material represents the basic index for the efficient utilization of the spring material.

This Manual deals with helical compression and extension springs, in which the spring wire is principally subjected to torsional stress, and with torsion and flat spiral springs, in which the spring wire is principally subjected to bending stress. The two formulae expressing the specific energy capacity for these spring types are as follows:

For springs subject to torsional stress:

$$\frac{Energy}{Volume} = U \frac{S^2}{G}$$

For springs subject to bending stress:

$$\frac{Energy}{Volume} = U \frac{S^2}{E}$$

where:

$U = 1/2$ for material under uniform stress, therefore most efficient utilization of material possible; approached by a thin-walled tube used as a torsion bar spring (under torsional loading) and by a ring spring (under compression)

$U = 1/4$ for round wire torsion bar springs and coil springs

$U = 1/6.5$ for square wire coil springs

$U = 1/6$ for rectangular cross sections in uniform circular bending

$U = 1/8$ for round cross sections in uniform circular bending

The formulae indicate that the specific energy capacity increases with the square of the maximum stress which the material can withstand without permanent set.

Specific energy capacity is also inversely proportional to the modulus of elasticity of the material. It is for these reasons that there is a constant search for materials and fabricating processes which will permit springs to be highly stressed. In many applications, density of material is an important factor in mass reduction.

As for the type of springs to incorporate the highest possible energy capacity, the formulae are not conclusive in themselves, because the maximum design stresses vary with the type of spring and material, as the diagrams in the later chapters of this Manual show. A comparison of different configurations on the basis of volume of metal can be made by selecting suitable stress values for each configuration and solving the appropriate energy/volume equations. When other than metallic materials are considered for mass reduction, the comparisons should be made by converting the energy/volume equations to energy/mass equations.

The principal spring types treated in this manual can be arranged in descending order of their specific energy capacities:

1. Round wire compression springs.
2. Flattened round wire or rectangular wire compression springs.
3. Round wire extension springs.
4. Square wire compression springs.
5. Torsion springs and flat spiral springs with rectangular wire section.
6. Square wire extension springs.
7. Torsion springs and flat spiral springs with round wire section.

Space Limitations

Frequently it is valuable to determine early in the design of a particular spring configuration the maximum volume of spring material which can be used within a given space.

For a flat spiral spring, the solution depends only upon the required clearance between coils and on the space required for the inner fastening device.

For round wire compression and extension springs, the curves in Fig. 1.1 will determine if the space is approximately adequate. The curves represent "space efficiency factors" V/V_s, where V is the volume of active spring material (the inactive end coils are ignored), and V_s is the volume of the cylindrical space in which the spring is contained when solid. The curves are plotted against "spring index" ($C = D/d$).

Two types of spring application are considered. One type is springs under a static load throughout their life with only infrequent further deflection, such as a steam boiler safety valve spring. In such a spring, the entire (round) wire surface is assumed to be under uniform torsional stress. For this spring the space efficiency factor is:

$$V/V_s = \frac{\pi C}{(C + 1)^2}$$

Other springs are expected to withstand a great number of load cycles which subject the wire to large stress ranges, as in most engine valve springs. In such a spring, where the maximum stress range is the criterion for failure, the higher stresses near the inside of the coil, due to wire curvature and direct shear, become more critical. This is usually taken into account by multiplication of the torsional stress with a "stress correction factor" greater than 1.0 and dependent upon the spring index C. the "Wahl factor" K_w, which is widely used for this purpose, is expressed by the formula:

$$K_w = \frac{4 C - 1}{4 C - 4} + \frac{0.615}{C}$$

For the spring, which must not fail under these "fatigue loading" conditions, the space efficiency factor is:

$$V/V_s = \frac{1}{K_w^2} \frac{\pi C}{(C + 1)^2}$$

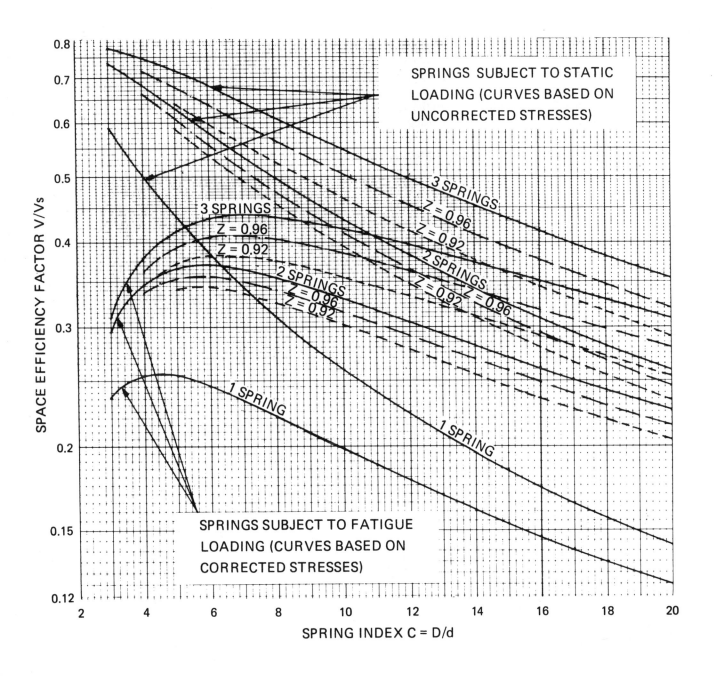

Fig. 1.1—Space efficiency of round wire compression and extension springs

After the spring index is estimated, the space efficiency factor may be found from Fig. 1.1. When this factor is multiplied by the volume available for the solid spring, the result will be the maximum volume of active spring material which may be used in the available space.

With decreasing index, the space efficiency factor will be increased, but the maximum deflection of the spring will be reduced.

Nested Springs

If the volume of active material, which can be built into one spring within the available space, is insufficient to satisfy the energy requirements for the spring, as expressed in the formula for specific energy capacity, then the designer should consider the use of a second (and possibly a third) spring inside the first one and acting in parallel with it. The springs in such a nest should be wound right and left, alternately, to avoid jamming.

The most economical design will be approximated if all springs have equal values of solid length (L_s), free length (L_θ), and spring index (C), since their maximum stresses will then be equal. This means that the wire diameters of the springs will be in the same proportion as their mean coil diameters, and their energy capacities will have the same ratio as the squares of the wire

2.3

diameters. Rather than have equal maximum stresses, it may be more efficient to have the smaller wires assume a somewhat higher stress level because of their greater tensile strength.

Assuming no radial clearance either between the nested springs or between the outer spring and the housing (with diameter D_o) in which it is to operate, the wire diameters and mean coil diameters can be expressed in terms of D_o, as follows:

Subscripts 1, 2, and 3 in the following equation denote dimensions of the outer, center, and inner springs, respectively.

$$d_1 = \frac{D_o}{C + 1} \qquad D_1 = \frac{C\,D_o}{C + 1}$$

$$d_2 = \frac{D_o}{C + 1} \cdot \frac{C - 1}{C + 1} \qquad D_2 = \frac{C\,D_o}{C + 1} \cdot \frac{C - 1}{C + 1}$$

$$d_3 = \frac{D_o}{C + 1}\left[\frac{C - 1}{C + 1}\right]^2 \qquad D_3 = \frac{C\,D_o}{C + 1}\left[\frac{C - 1}{C + 1}\right]^2$$

If V represents the total volume of active material in all springs of the nest, then the space efficiency factor for static loading is:

For two springs:

$$V/V_s = \frac{\pi\,C}{(C + 1)^2}\left[1 + \left[\frac{C - 1}{C + 1}\right]^2\right]$$

For three springs:

$$V/V_s = \frac{\pi\,C}{(C + 1)^2}\left[1 + \left[\frac{C - 1}{C + 1}\right]^2 + \left[\frac{C - 1}{C + 1}\right]^4\right]$$

For fatigue loading conditions, these formulae will again have to be multiplied by $1/K_w^2$. In Fig. 1.1, the solid lines show the space efficiency factors for spring nests without clearance between the springs.

For practical purpose, it is necessary to provide radial clearances within the nest. This is expressed by a factor Z, the ratio of actual diameters to diameters which would allow no clearance. Values of Z may be established from consideration of tolerances and diameter changes.

With the assumptions that C, L_s, and L_θ are equal for all springs, the space efficiency factor for static loading then becomes:

For two springs:

$$V/V_s = \frac{\pi\,C}{(C + 1)^2}\left[1 + Z^3\left[\frac{C - 1}{C + 1}\right]^2\right]$$

For three springs:

$$V/V_s = \frac{\pi\,C}{(C + 1)^2}\left[1 + Z^3\left[\frac{C - 1}{C + 1}\right]^2 \right.$$
$$\left. + Z^6\left[\frac{C - 1}{C + 1}\right]^4\right]$$

Values of Z may be as low as 0.92 for springs of less than 25 mm outside diameter. They may increase by approximately 0.01 with every 25 mm of increase in outside diameter. Thus, a second spring with 125 mm outside diameter should have ample clearance within the outer spring if it is designed with:

$$d_2 = 0.96\,\frac{D_o}{C + 1} \cdot \frac{C - 1}{C + 1}$$

$$D_2 = 0.96\,\frac{C\,D_o}{C + 1} \cdot \frac{C - 1}{C + 1}$$

The dashed lines in Fig. 1.1 indicate the space efficiency values for spring nests with clearance between the springs obtained by using $Z = 0.92$ and $Z = 0.96$.

If clearance must be provided between the outer spring and the housing, the space efficiency will be further reduced since it will then be necessary to use:

$$d_1 = Z\,\frac{D_o}{C + 1}$$

$$d_2 = Z^2\,\frac{D_o}{C + 1} \cdot \frac{C - 1}{C + 1}$$

$$d_3 = Z^3\,\frac{D_o}{C + 1}\left[\frac{C - 1}{C + 1}\right]^2$$

$$D_1 = Z\,\frac{C\,D_o}{C + 1}$$

$$D_2 = Z^2\,\frac{C\,D_o}{C + 1} \cdot \frac{C - 1}{C + 1}$$

$$D_3 = Z^3\,\frac{C\,D_o}{C + 1}\left[\frac{C - 1}{C + 1}\right]^2$$

Fig. 1.1 confirms that the use of more than one round wire spring is more beneficial in a space of large diameter (permitting a high spring index C) than in a space of small diameter. Introduce a second spring only for a spring index greater than 4, and a third spring only for a spring index greater than 6.

Changing from a lower to a higher spring index is not efficient for a single spring, but it is when a second spring is added. In particular, nested springs of high index are necessary if a greater flexibility is required. In a given space, the higher spring index is equivalent to a smaller

2.4

wire size, and generally indicates fewer manufacturing difficulties and higher permissible stresses.

The Space Efficiency Factor is a rather useful tool when the designer must fit a spring (or several springs) having specified spring rate, load, and stress at that load into a limited space of specified outside diameter D_o and solid spring length L_s. The designer could start by choosing a value for the spring index $C = D/d$, and then go through steps 1-6 for the single spring, assuming it is to undergo static loading:

1. $d = \dfrac{D_o}{(C + 1)}$

2. $N_a = \dfrac{G\,D_o}{8\,R\,C^3\,(C + 1)}$

3. $L_s = N_a d$ if this proves excessive, choose a higher value of C

4. $S_s = \dfrac{8\,C\,P\,(C + 1)^2}{\pi\,D_o^2}$ if this proves excessive, choose a lower value of C

5. $V_s = \dfrac{\pi\,L_s\,D_o^2}{4}$

6. $V = V_s$ times Space Efficiency Factor from Fig. 1.1

If there is no value of C which will satisfy the specified maximum values of both L_s and S_s, the designer will then use two springs, with a higher value of C. For the two springs (here assumed to have no radial clearance and equal values of L_s, L_θ, and C) he will go through steps 7-17:

7. $d_1 = $ same formula as for d

8. $d_2 = \dfrac{d_1\,(C - 1)}{(C + 1)}$

9. $N_{a1} = \dfrac{2\,G\,D_o\,(C^2 + 1)}{8\,R\,C^3\,(C + 1)^3}$

10. $L_s = N_{a1}\,d_1\,(= N_{a2}\,d_2)$ if this proves excessive, choose a higher value of C

11. $R_1 = \dfrac{G\,D_o}{8\,N_{a1}\,C^3\,(C + 1)}$

12. $N_{a2} = N_{a1}\,\dfrac{C + 1}{C - 1}$

13. $R_2 = R_1\left[\dfrac{C - 1}{C + 1}\right]^2$

14. $P_1 = \dfrac{P\,R_1}{R}$

15. $S_1 = \dfrac{8\,C\,P_1\,(C + 1)^2}{\pi\,D_o^2}\,(= S_2)$ if this proves excessive, choose a lower value of C

16. $V_2 = $ same formula as with single spring

17. $V = V_s$ times Space Efficiency Factor from Fig. 1.1

A numerical example follows using the following specified parameters:

Max. available outside diameter	D_o	= 125 mm
Max. available solid length	L_s	= 127 mm
Load at solid length	P_s	= 13 000 N
Spring rate	R	= 52 N/mm
Max. stress (uncorrected) at solid length	S_s	= 860 MPa
Modulus of elasticity	G	= 79 300 MPa
Type of loading	static	
Spring index	C	$= \dfrac{D}{d}$

A—Single Spring:

3. $L_s = N_a d = \dfrac{G\,d^4}{8\,D^3\,R}\,d = \dfrac{G\,D_o^2}{8\,R\,C^3\,(C + 1)^2}$

$L_s = \dfrac{79\,300 \times 125^2}{8 \times 52\,C^3\,(C + 1)^2}$

$= \dfrac{2.98 \times 10^6}{C^3\,(C + 1)^2} \leq 127$

$C^3\,(C + 1)^2 \geq 23\,460;\ C \geq 7.1$ (by trial solution)

4. $S_s = \dfrac{8\,C\,P\,(C + 1)^2}{\pi\,D_o^2}$

$S_s = \dfrac{8 \times 13\,000\,C\,(C + 1)^2}{\pi \times 125^2} \leq 860$

$C\,(C + 1)^2 \leq 406;\ C \leq 6.7$ (by trial solution)

Since no value of C will satisfy the requirements for both L_s and S_s, two springs must be considered.

B—Two Springs: by trial select C = 8

7. $d_1 = \dfrac{125}{8 + 1} = 13.9$

$D_1 = 8 \times 13.9 = 111$

8. $d_2 = 13.9\,\dfrac{8 - 1}{8 + 1} = 10.8$

$D_2 = 8 \times 10.8 = 86$

9. $N_{a1} = \dfrac{2 \times 79\,300 \times 125 \times 65}{8 \times 52 \times 8^3 \times 9^3} = 8.30$

2.5

10. $L_s = N_{a1} d_1 = 8.30 \times 14.7 = 122$ (OK)

11. $R_1 = \dfrac{79\,300 \times 125}{8 \times 8.30 \times 8^3 \times 9} = 32.4$

12. $N_{a2} = 8.30 \dfrac{8+1}{8-1} = 10.7$

13. $R_2 = 32.4 \left[\dfrac{8-1}{8+1}\right]^2 = 19.6$

14. $P_1 = \dfrac{13\,000 \times 32.4}{52} = 8100$

15. $S_1 = \dfrac{8 \times 8 \times 8100 \times 9^2}{\pi \times 125^2} = 855 \ (= S_2)$ (OK)

16. $V_s = \dfrac{122\,\pi \times 125^2}{4} = 1.50 \times 10^6$

17. $V = 1.50 \times 10^6 \times 0.498 = 747 \times 10^3 \ mm^3$

Helical springs of rectangular wire are used in special applications where a greater volume of material is required in a limited space, and they may provide a solution to a problem of high loads in a small space.

A helical spring of cylindrical or conical form, made from a round bar of tapered cross section, is feasible and finds an application as a vehicle suspension spring. The variable bar cross section can impart a variable rate to the spring as the smaller diameter portion of the bar closes solidly during the early stages of spring compression. This feature can provide improved ride "characteristics" to a loaded vehicle. The use of a papered bar, combined with variable coil pitch, reduces the linear space and spring mass requirements. Manufacturing methods have been developed to form the tapered round bar prior to coiling.

The tapered bar helical springs can be designed with the tapered bar section at one or both ends of the spring, depending upon the application and the characteristics desired. Due to the added variables involved, the design of these springs becomes complicated but can be handled by developing modifying factors on the standard design formulae and then experimentally verifying the spring characteristics. Another approach is to develop new formulae to encompass the added variables and constraints imposed on the variable rate and the stress distribution along the bar length.

2. Static Versus Dynamic Loading

The term "static loading" implies a force application of sufficiently low velocity that the spring is not deflected beyond an equilibrium position, and deflections and stresses are stabilized throughout the spring. In practice, a force is usually applied until the deflection is limited by some external means. If this position is reached at a low velocity, no surge within the spring will occur and it may be considered as statically loaded. When a mass W (kg) is gradually lowered on a vertical helical spring of rate R (N/mm), assuming that the standard acceleration of gravity g = 9.81 m/s^2 prevails, and since force equals mass times acceleration, the mass W (kg) will exert a vertical force P = W g = W x 9.81 (N) on the spring, when at rest, and a deflection F = P/R (mm) will result.

The term "dynamic loading" implies an application in which the velocity of the moving spring end is appreciable. This is true for all impact loadings as well as for sinusoidal loadings under high frequencies. Because of the inertia of the spring coils, the whole spring deflection is not equally distributed among the coils so that some of them must undergo a stress range greater than the corresponding stress range with static loading. For example, let the mass W be suddenly released from a point where it just touches the top of the free vertical helical spring of rate R. In this case, the spring will undergo a maximum deflection F_1 before it comes to rest at the equilibrium position under deflection F.

When the mass descends a distance F_1 (mm), its potential energy decreases by $W\,g\,F_1/10^3$ (J); assuming no damping, the strain energy in the spring increases by $R F_1^2/(2 \times 10^3)$ (J).

Therefore:

$$\frac{W\,g\,F_1}{10^3} = \frac{R\,F_1^2}{2 \times 10^3}; \quad F = \frac{W\,g}{R}; \quad F_1 = \frac{2\,W\,g}{R} = 2\,F$$

It then follows that, for this instance of dynamic loading, the maximum deflection and, consequently, the maximum stress, is twice as large as would result from the gradual (static) application of the same mass. When the mass is released from some point higher than the top of the free spring, then the maximum stress is more than twice the static stress.

Two basic instances of dynamic loading can be distinguished. In one instance the stroke is practically constant. This applies to the driving spring of a machine gun, since the striking part is finally stopped by a stiff buffer acting in parallel to the driving spring. The other instance is that of a buffer spring, which absorbs almost the full kinetic energy of the striking part and therefore has a variable stroke. In the first instance, the spring itself may take a greater deflection than that prescribed by the striking part owing to the kinetic energy of its coils. In addition, at the minimum compressed length, the whole deflection of the spring is not equally distributed among the coils, so that some of them must necessarily take a higher than average stress. Therefore, the conventional formulae give too low a maximum stress, and special formulae are required for the maximum dynamic stress.

An example of dynamic loading will serve to illustrate the relationships. A precompressed spring is suddenly compressed by a heavy mass moving with constant velocity v over a given stroke. The motion of the front end of the spring causes a surge wave which is propagated along the spring wire with surge velocity c, while the coils themselves move axially with velocity v. This wave may be called "single wave," as long as no reflection occurs at the rear end of the spring. The amplitude of this single wave can be expressed by one of the following:

1. Change of load, ΔP (N)
2. Change of stress, ΔS (MPa)
3. Change of deflection per coil, ΔF (mm)

It has been found that all these changes of load, stress, and deflection per coil are proportional to the velocity of loading v (m/s). They may be expressed as follows:

$$\frac{\Delta P}{v} = R\,T_s \times 10^3 = \frac{\pi\,d^3}{8\,D \times 10^3}\sqrt{2\gamma G} \quad \left(\frac{N}{m/s}\right)$$

$$\frac{\Delta S}{v} = \frac{\Delta P}{v} \cdot \frac{\Delta S}{\Delta P} = \frac{1}{10^3}\sqrt{2\gamma G} \quad \left(\frac{MPa}{m/s}\right)$$

$$\frac{\Delta F}{v} = \frac{T_s}{N_a \times 10^3} = \frac{\pi\,D^2}{d \times 10^3}\sqrt{\frac{2\gamma}{G}} \quad \left(\frac{mm}{m/s}\right)$$

where:

T_s = travel time of the wave over the whole wire length, or "surge time" s

γ = 7850 kg/m³ for steel

G = 79 300 MPa

then:

$$\frac{\Delta P}{v} = 13.86\,\frac{d^3}{D} \quad \left(\frac{N}{m/s}\right)$$

$$\frac{\Delta S}{v} = 35.3 \quad \left(\frac{MPa}{m/s}\right)$$

$$\frac{\Delta F}{v} = \frac{1.40\,D^2}{d \times 10^3} \quad \left(\frac{mm}{m/s}\right)$$

The additional stress ΔS is independent of the spring dimensions; for any compression spring made of steel, each m/s of impact velocity increases the stress by 35.3 MPa.

After reflection of the single wave from the fixed end of the spring, the incident and the reflected waves are superimposed by ordinary addition of the stress amplitudes. Therefore, the maximum stress built up in the spring depends on the stress amplitude of the single wave as well as on the number of superimpositions of the single wave over the duration of the compression stroke. While the stress amplitude depends upon the impact velocity only, the number of superimpositions is a function of the surge time T_s, that is, the spring dimensions.

With slow dynamic loading during compression time, there are many superimpositions of the surge wave of small amplitude. Therefore, the dynamic stress will not noticeably exceed the static stress.

However, with rapid dynamic loading, the compression time can be of the order of the surge time. There are few superimpositions, but the amplitude is high and, consequently, the dynamic stress can exceed the static stress substantially. In such instances, it may be advisable to tune the surge time T_s with regard to the compression time by proper selection of the spring dimensions.

In most instances of dynamic loading, the analysis of stress is not as simple as in the example given. However, in general, two groups of factors appear to be decisive for the maximum stress built up and the spring life expected:

1. The "loading function" v(t), that is, the time-velocity function of the moving end of the spring, with stroke, compression time, maximum velocity, and maximum acceleration as important parameters.

2. The spring, characterized by its material, maximum static stress, surge time T_s, and inherent damping.

The influence of the loading function v(t) on the life of springs has been strikingly proved by fatigue tests: The average life of springs under sinusoidal loading may be reduced as much as 50:1 under impact loading, with the same stroke and frequency. However, the importance of each individual life factor remains to be explored by more fundamental testing.

Gun springs are among the springs which are subjected to severe dynamic loadings. The following methods of design have been successfully tried:

1. Single wire springs—Proper adjustment of the surge time T_s to the given compression time reduces the maximum dynamic stress.

2. Stranded wire springs, coiled from 3-wire or 7-wire strand. Their greater inherent damping improves the spring life.

3. Two-section springs, that is, a front spring in series with a somewhat stiffer rear spring. In comparison with a single spring, a two-section spring gives more possibility of adjusting the spring assembly to the given loading function since a more equalized distribution of stresses over the wire length can be obtained at the moment of maximum compression. This results in lower peak stresses and better life than with a single spring.

3. Natural Frequency

With dynamic loading of compression springs, the inertia of the spring coils, in addition to their elasticity, becomes effective. When moving one end of a precompressed free spring in an axial direction, the disturbance of the equilibrium at this end is propagated along the spring wire in a transient wave with a definite surge velocity.

$$C = L \sqrt{\frac{R}{W}} \qquad (1)$$

where:

mass of active part of spring $= W = \dfrac{\pi^2 d^2 N_a D \gamma}{4g}$ (2)

spring rate $= R = \dfrac{Gd^4}{8D^3 N_a}$ (3)

Length of active part of spring $= L = \pi D N_a$ (4)

Substituting W(2), R(3), L(4) in (1)

$$C = \frac{d}{D} \sqrt{\frac{Gg}{2\gamma}} \qquad (5)$$

Here all length units must be used uniformly in terms of meter, as follows:

$g = 9.806\ 650$ m/s^2 (6)
$G = 79300 \times 10^6$ N/m^2 (for Steel) (7)
$\gamma g = 7850 \times 9.806\ 650$ N/m^3 (for Steel) (8)

Substitute g(6), G(7) and γ(8)

in (5)

$$C = \frac{d}{D} \sqrt{\frac{79300 \times 10^6 \times 9.806\ 650}{2 \times 7850 \times 9.806\ 650}} = 2247\ \frac{d}{D}\ (m/s)\ (9)$$

The travel time of the wave over the whole wire length or "surge time" is (L, D, d specified in mm here expressed as m)

$$T_s = \frac{L}{10^3 C} = \frac{\pi D N_a}{10^3 C}\ (s) \qquad (10)$$

with mean value for spring steel

$$T_s = \frac{\pi D^2 N_a}{(2247)\ 10^3 d}\ (s) \qquad (11)$$

Active solid length of the spring $= L_s = N_a d$ (12)

$$\therefore N_a = \frac{L_s}{d}$$

$$\therefore T_s = \frac{\pi}{2.247 \times 10^6} \left(\frac{D}{d}\right)^2 L_s = \frac{1.4}{10^6} \left(\frac{D}{d}\right)^2 L_s\ (s)\ (13)$$

T_s increases with the active solid length and with the index of the spring. A spring of index $D/d = 7$ has a surge time of 69×10^{-6} s/mm of active solid length.

In some instances, another formula can be used to good advantage.

$$T_s = \sqrt{\frac{W}{R \times 10^3}}\ (s) \qquad (14)$$

where W(kg) is the mass of the spring and R (N/mm) is the rate.

When both ends of the spring are supported, as is usual, one cycle of vibration comprises just twice the surge time T_s, therefore, the natural frequency of the spring, measured in Hertz, is:

$$f = \frac{1}{2T_s} = \frac{2.247 \times 10^6 d}{2 D^2 N_a}$$

$$= \frac{358 \times 10^3 d}{N_a D^2}\ (Hz) \qquad (15)$$

$$\text{or } f = \frac{358 \times 10^3 d^2}{L_s D^2}\ (Hz) \qquad (16)$$

$$\text{or } f = \frac{1}{2 \sqrt{\dfrac{W}{10^3 R}}} = 15.8 \sqrt{\frac{R}{W}}\ (Hz) \qquad (17)$$

The natural frequency decreases when the spring index and the active solid length are increased. A spring of index $D/d = 7$ and an active solid length of 250 mm has a frequency of 29 Hz.

The natural frequency of valve springs, as used in automotive engines, tends to be somewhat lower than that calculated from the above formula. Actual measurements reveal that the frequency may be as much as 10% less than that calculated, due, evidently, to the end effects introduced by more active coils during a spring surge.

NOTE: It is necessary to distinguish clearly between the natural frequency of the spring itself, as described above, and the natural frequency of a mass W_1 (kg) attached to spring of much smaller mass and rate R (N/mm).

The natural frequency of such a mass W_1 is:

$$f = \frac{1}{2\pi} \sqrt{\frac{R \times 10^3}{W_1}}$$

The mass W_1, when attached to the spring, will cause a static deflection

$$F = \frac{W_1 g}{R}$$

The natural frequency of the mass W_1, relative to the spring deflection F (measured in cycles/minute)

$$f = .159 \times 60 \sqrt{\frac{8 \times 10^3}{F}} = \frac{945}{\sqrt{F}} \text{ (cycles/minute)}$$

4. Residual Stresses

Residual stresses (sometimes called "trapped stress") may be set up in springs in a variety of ways including heat treatment, shot peening, and hot and cold presetting. As a rule, the presence of residual stresses of proper sign makes possible the utilization of higher working stresses; in springs, this is of considerable advantage since the energy storage capacity increases as the square of the working stress. For coiled springs, the shot peening and presetting methods of inducing residual stresses are most practical.

Shot Peening

The shot peening process consists essentially of subjecting the spring to a stream of shot moving at high velocity. The peening action of the shot sets up beneficial biaxial compressive stresses in a thin surface layer and also results in cold working this layer. The depth of this cold-worked layer depends, among other things, upon the size and impingement velocity of the shot used, the direction of impingement relative to the surface, and the spring material. Shot peening, when properly conducted, results in relatively large increases of the stress range through which the spring can safely operate under fatigue or repeated loading conditions.

In commercial use, two types of methods are most adaptable. In the earlier method, air is used for the propelling agent as in the common sand blast equipment. The second method is mechanical, the shot being fed on a rapidly rotating wheel having radial blades. Thus, the energy of the wheel is transferred to the shot by the blades and by centrifugal force, the shot in turn impinging on the parts. Tests have shown that either method can pro-

duce the same results in increasing fatigue life, but the mechanical method is more economical for high production.

The increase in fatigue life is accomplished in two ways. One consists in stressing the surface layers in compression. Fatigue failures are due to tensile stresses and, if a compressive stress can be set up in advance where subsequently the load application will produce a tensile stress, the part will carry considerable load before it is actually stressed in tension. The second effect is the cold working and strengthening of the surface layers which raises the physical properties where the highest stresses occur. This second effect is of minor importance compared with the first. Because of the twofold beneficial results imparted by shot peening, this surface treatment has done more to increase the life of springs than any alloy steel or other process ever employed.

Shot peening can be successfully carried out on wire sizes in excess of 0.75 mm. But it will be effective only if it is thoroughly and evenly applied over the entire surface, with particular emphasis on those parts of the surface which will undergo the highest load stresses. Thus, on compression springs the inside of the coil is most important, whereas on torsion springs, it is the outside that is important. Time of exposure to the shot is one of the prime factors to obtain effective peening, as indicated by the curve in Fig. 1.2. For each type of spring, a knowledge of this time factor, with a definite type and size of shot, is needed.

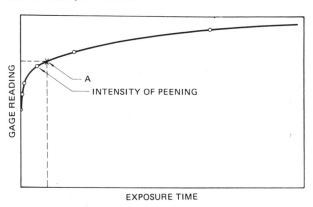

Fig. 1.2—Intensity determination curve

The effects of shot size and peening time can be checked by means of Almen strips and other observations. The height of the arc in the Almen strip is not in itself a measure of fatigue life, but it furnishes a check on uniform operation of the shot peening machine. The SAE Manual on Shot Peening (HS 84) details the methods for controlling the machine with the help of Almen strips and for measuring coverage. The importance of proper peening coverage cannot be overemphasized.

Since shot peening is a mechanical working process, it can be affected by heat. If springs are heated sufficiently after shot peening (Fig. 1.3), the benefits will be reduced and the endurance limit will drop, eventually to that of the wire as originally received. The use of shot peened springs in applications involving high temperatures necessitates careful planning. If the spring is heated above 430°C during manufacture, the maximum resistance to settling at temperatures usually encountered in automotive practice is obtained. But in this event, the beneficial shot peening stresses are completely removed, and no increase will occur in fatigue life due to the shot peening operation. Therefore, compromises must be made if an endurance is to be realized which exceeds that of the untreated wire.

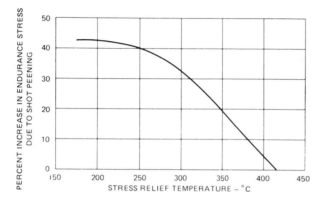

Fig. 1.3—Reduction in effectiveness of shot peening by subsequent heating or stress relieving

Presetting

Residual stresses are most commonly set up in helical compression springs by means of a presetting operation, known also as scragging or cold setting when performed at ambient temperatures. When the presetting operation is performed at an elevated temperature it may be called hot pressing, heat setting, or warm setting.

The need for presetting depends upon the design stresses, the application and its conditions and requirements. The use of presetting is most beneficial when design stresses are at or near the yield point, and settling or sag prevents the spring from performing as required.

Presetting is an operation that is performed during the manufacturing of helical compression springs in which the spring is compressed beyond the yield point of the material. The yielding of the surface layers of the wire, which occurs during the presetting, produces beneficial residual stresses, thus increasing the elastic limit of the spring and thereby reducing the amount of settling or sag in subsequent service. The spring is initially coiled to a free length greater than the final desired free length and is then compressed to a point where the spring may be at or near the solid length (Fig. 1.4, point B). On releasing the load, a permanent set OC is produced while the

spring behaves nearly elastically along line BC. In special instances several additional compressions to point B may be carried out.

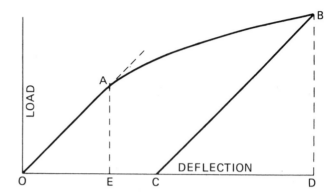

Fig. 1.4—Typical load-deflection diagram of helical spring during presetting

When presetting on helical compression springs at some elevated temperature, the operation known as hot pressing, heat setting, or warm setting may be employed to produce greater resistance to relaxation than cold setting.

Heat setting consists of compressing the spring on a fixture, subjecting the compressed spring to a temperature higher than the desired operating temperature for a time suitable to insure complete penetration of the heat, and then cooling to room temperature before releasing.

For a description of warm setting, see Chapter 4, Section 5.

Hot pressing consists of heating the spring in its free or relaxed position to some temperature for sufficient time to insure complete penetration; then while the spring is at the temperature, it is compressed to some height below the installed or operating position and released.

Fig. 1.5A represents the stress distribution over the cross section at a load which gives a stress S somewhat below the yield point.

For purposes of illustration, it is assumed that no strain hardening occurs, even under the higher presetting load (Fig. 1.4, point B) with which Fig. 1.5B deals. Here, yielding occurs at stress S, and the distribution of stress will be as indicated by the shaded area f-h-i-k. The nominal stress calculated from the presetting load by the usual formula is equal to S_n, and without yielding the stress distribution resulting from this presetting load would be as indicated by the triangular area f-l-k. Actually, yielding has started at the surface (wire diameter d) and has progressed inward to a diameter:

$$d_1 = d \sqrt[3]{4 - 3S_n/S_y}$$

In Fig. 1.5B, the ratio S_n/S_y has been chosen as 31/24. In that instance, d_1 becomes d/2.

When the presetting load is reduced to zero (Fig. 1.4, point C), residual stresses are set up as indicated by the

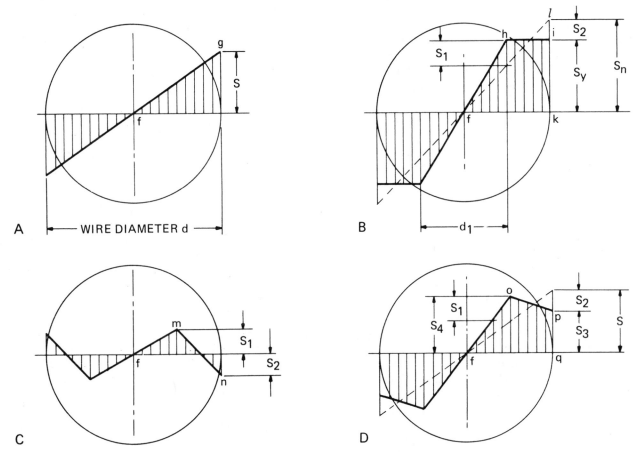

Fig. 1.5—Effect of presetting on stress distribution over cross section of helical spring

shaded area in Fig. 1.5C. These are obtained by subtracting the ordinates of the dashed line f-1 from those of the broken line f-h-i in Fig. 1.5B. The reason for this is that on removal of the load, the spring will behave approximately elastically. Therefore, removal of the load is equivalent to subtracting the elastic stress distribution f-1 from the plastic distribution f-h-i.

If the spring is loaded subsequent to presetting, the resultant stress distribution is found by adding the load stress distribution to that of Fig. 1.5C. For example, a load may be applied to the spring which gives a stress S, the same as that shown in Fig. 1.5A. The resultant stress distribution under that load is indicated by the shaded area f-o-p-q in Fig. 1.5D. It will be seen that, due to the presetting operation in this instance, the peak stress has been reduced from a value S (Fig. 1.5A) to a lower value S_4 (Fig. 1.5D).

Still dealing with a round wire helical compression spring of large index and still assuming that no strain hardening occurs, it may be considered that the material yields at an essentially constant stress equal to the shearing yield point and that complete yielding, over the entire cross section, will occur theoretically at a load equal to 1.33 times the load at which yielding first starts. This would indicate that, for these theoretical conditions, the maximum presetting load would be about 33% above the load corresponding to the torsional yield stress and that, after presetting, the spring would behave elastically up to a load corresponding to a stress 1.33 times the torsional yield point. Because of curvature and strain hardening effects, the ratio of the maximum presetting load to the load at which yielding first starts, tends to be higher than 1.33 in actual springs, and, in some instances, for the smaller spring indexes, it may approach 2. Since the spring will behave elastically nearly up to the presetting load, this means that the elastic limit of a spring may be increased considerably by presetting; much of this increase is attributable to the residual stresses set up in the presetting operation. In actual practice, presetting deflections equal to 2, and in some instances even 3, times the elastic limit deflection may be used (OD = 2 to 3 times OE in Fig. 1.4). In other instances, however, lower values of presetting deflection may be required to avoid excessive distortion of the spring.

In cold wound helical torsion springs, residual stresses are also set up by the coiling operation; tension residual stresses are set up on the inside of the coil and compression stresses on the outside. If the spring is loaded by a torque so as to wind up the coil, it is clear that the residual stresses, in this instance, will also subtract from the maximum load stresses near the outer fibers. To take advantage of the residual stress pattern, torsion springs should therefore be loaded in the windup direction.

Residual stresses set up in cold wound helical compression springs as a consequence of the winding process are usually removed or reduced by a low temperature, stress-relieving (bluing) treatment. These stresses tend to augment the effects of the torsional working stresses in such springs, and hence, their removal is beneficial.

5. Temperature Effects on Springs

Spring design becomes complicated for extreme temperature applications. Generally, the usual spring materials will operate satisfactorily at very low temperatures under static loading conditions. The combination of extreme cold temperature and impact loading represents a more severe situation; however, it may not be catastrophic because of the resiliency of the spring material.

Under high temperature conditions, springs undergo load losses which in most instances are due to a decrease in the torsional elastic limit of the material. Springs may be designed to operate satisfactorily at normal temperatures, but under elevated temperatures these same springs will undergo a slow plastic deformation (commonly known as "relaxation" when the spring is at a fixed length, or a "creep" when spring is under a fixed load) and take a permanent set. The amount of set depends upon the working stress level, heat intensity, and the duration at the elevated temperature. See Fig. 5.7 in Chapter 5.

Thus, in any spring problem involving elevated temperature, the designer should be fully aware of the impending difficulties and make suitable allowances for all the factors involved. In general, the spring stresses should be reduced to the lower practical values. However, there are applications where such modifications are not possible, and in these instances, some form of presetting may be used. See Chapter 1 - Section 4.

The temperature, time, and stress level of any of the presetting methods may be varied to minimize the loss in load when subjected to the stresses and elevated temperatures subsequently imposed in operation. This relaxation or sag may be reduced as much as 50% when compared to a spring that has not been preset.

For recommended guidelines on selecting materials to resist the effects of elevated temperatures, see Table 2.1 in Chapter 2.

If springs are subject to elevated temperatures during operation, residual stresses may be greatly reduced or even lost. For example, shot peening begins to lose its effectiveness at temperatures around 260°C, with complete loss at temperatures around 430°C on steel springs. This indicates that, at these temperatures, residual stresses are considerably reduced, and the benefits of presetting will be reduced correspondingly.

6. Fatigue Durability

Some springs are required to function cyclically with accompanying repetitive stresses. Failure of these springs is commonly known as fatigue. Generally, these springs perform an essential function and should be designed to have high fatigue durability. This applies particularly to valve springs for internal combustion engines and to injector springs for diesel engines. Designing a spring of this type requires a consideration of the stress and stress range, the waveform and harmonic content of the cyclic stress, the environmental conditions, and the specialized surface treatments such as presetting and shot peening.

From fatigue test data on cyclic loaded springs an S-n curve (endurance factor S_E/S_U versus life in number of cycles) can be plotted as shown in Fig. 1.6. This is a typical log-log plot of the endurance factor K_E which is normalized to represent the ratio of the peak alternating stress as a fraction of the minimum ultimate tensile strength of the material. The log-log coordinates permit the test data to be reasonably well represented by straight lines between 10^3 and 10^6 cycles of life. Spring steels develop a "knee" in the curves at about 2×10^6 cycles beyond which infinite life can be expected.

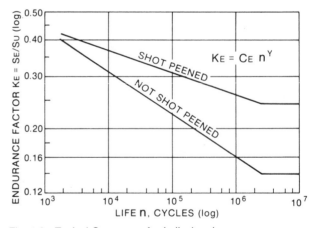

Fig. 1.6—Typical S-n curves for helical springs

Springs under a fluctuating load normally are alternately stressed about a mean level with the actual stress value not going through zero. The S-n curve of Fig. 1.6 refers to a completely reversed alternating stress with a mean value of zero. Therefore, another form of diagram, known as the "Goodman-type Diagram" as illustrated in Fig. 1.7, is employed to correlate the actual stress range with the conventional S-n diagrams by determing an equivalent value of K_E. As an example, illustrated in Fig. 1.7, a typical stress range between the initial stress of K_{S1} reference line and the maximum stress K_{S2} on the "Goodman" line establishes K_E, which is one-half of the overall peak-to-peak cyclic stress range and represents a spring having equivalent life. These two diagrams are the basis for the three fatigue strength diagrams presented in Chapter 5 as Figs. 5.1, 5.2, and 5.3 for use in actual spring design work. They are intended to establish safe stress limits consistent with expected life endurance.

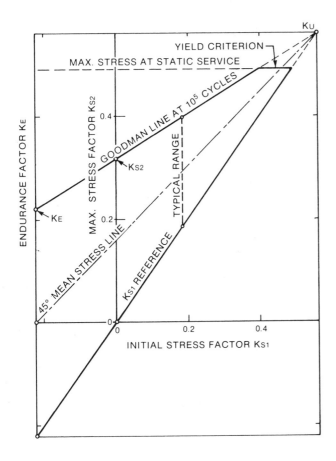

Fig. 1.7—Goodman-type fatigue strength diagram for helical springs

$$K_E = C_E n^Y$$

$$K_E = \frac{K_U(K_{S2} - K_{S1})}{2 K_U - (K_{S2} + K_{S1})} = \frac{S_E}{S_U}$$

$$K_{S2} = \frac{2 K_U K_E + K_{S1}(K_U - K_E)}{K_U + K_E} = A + B K_{S1}$$

$$\text{Min. TS} = \frac{S_1}{A}\left[\frac{P_2}{P_1} - B\right]$$

$$\text{Max. } K_{S2} = C_S n^M$$

$$A = K_{S2} \text{ at } K_{S1} = 0$$

$$B = (K_U - A)/K_U$$

Where:

C_E, C_S, M and Y are constants
K_E = Endurance factor
K_{S1} = Initial stress factor
K_{S2} = Maximum stress factor
K_U = Ultimate strength factor
n = Life cycles
P = Spring load
S_E = Endurance stress
S_1 = Initial stress
S_U = TS = Ultimate stress (min. tensile strength)

Refer to Tables 5.3, 5.4, and 5.5.

In selecting a spring material for prolonged fatigue durability, first consideration should be given to carbon steel valve spring quality wire. In the normal sizes, it usually performs as well as alloy steel and is generally more dependable. In most instances, the use of pretempered steel is advisable to avoid hardening cracks that can occur in material hardened and tempered after coiling. Alloy steels are usually more subject to seams and have a greater tendency toward quench cracks than are carbon steels. Therefore, it is advisable to subject such springs to magnetic particle inspection. While this process is not ordinarily used for springs in the automotive industry, it is seldom omitted on aircraft engine springs. Another technique, more recently developed and commonly used for quality spring manufacturing, is magnetic inspection of spring wire at the coiling machine.

7. Common Causes of Spring Failure

A spring designed and produced with due regard to all the foregoing details should not fail. In spite of this, a rare failure may occur on a properly designed spring. It can usually be traced to the following causes:

1. Surface Imperfections—They occur as hardening cracks, seams, pipe, laps, pits, die marks, and tool marks.

It should be understood that the S-n curves and "Goodman-type Diagrams" are limit lines established from accumulated test data by many investigators. They have no theoretical basis, but are conveniently presented in mathematical and graphical form for design purposes. The lines given tend to be conservative and are intended only to serve as a guide for the designer. In the final application, the designer should consult with the spring manufacturer to establish realistic values for stress limits in cyclic stressed springs.

The following formulae are the basis for developing the fatigue strength diagrams. They may be used in conjunction with computers for spring design calculations instead of using the fatigue strength diagrams directly.

Of these, magnetic particle inspection will locate only the first three. The other imperfections must be found by careful inspection of the raw materials, usually by etching both ends of every bundle of wire and by visual inspection of the finished part. The difficult part of this procedure is that the small sharp pits or marks may be overlooked while the broad, round bottom marks, which are harmless, are easily observed. For critical springs such as automotive valve springs it is common practice to inspect the wire continuously by an electromagnetic device which rejects springs with defects.

2. Corrosion—This is a more common cause of spring breakage than is usually understood by the spring user. In valve springs, it may be caused by condensation of the products of combustion on the springs. A small pit is formed or hydrogen embrittlement has occurred from the oxidation of the steel. It is not generally recognized how fast this process can lead to breakage. Springs which were perfect have failed in 3-6 h in an engine which was started and stopped before it could run long enough to scavenge the crankcase. In applications of this type, it will be necessary to provide corrosion protection for the springs. (Refer to Section 8 on surface protection.)

3. Improper Heat Treatment—It is a threat which must not be overlooked, even though it is not frequently encountered. If springs are overheated, the grain structure will be coarse and the fatigue life poor. Failure to heat to a proper temperature, or to heat long enough to get all the carbides into solution, will result in ferrite patches and low endurance limit. Automatic time-temperature control in modern heat treating departments has almost completely eliminated this once common cause of breakage.

Decarburization—This is the least frequent offender in the entire list of imperfections. Partial decarburization is usually present in spring wire, at least to a slight extent. The degree of permissible decarburization will depend upon the type of material and the application for which it is used. When the examination by a competent metallographer reveals a full ferrite ring around the circumference, the wire is subject to rejection because the surface condition will ultimately lead to premature failure.

The foregoing data cover the general considerations which deal with fatigue durability springs, and some of the special methods which have been developed to meet the conditions where the consequences of breakage may be disproportionally serious.

8. Surface Protection

The most widely used metals in the spring industry are the carbon and low alloy steels. These are also the least corrosion resistant. The wide acceptance of these steels as spring material directs consideration toward choice of the proper type of finish for a particular spring application. The most important factors to be considered are:

1. Corrosive severity of the spring environment.
2. The degree of corrosion protection required to give spring life expectancy.
3. Effect of coating and its method of application on the mechanical properties of the spring. (See guidelines in following paragraphs to minimize hydrogen embrittlement.)
4. Comparative corrosion resistance and the cost of applying the coating.
5. Effect of the protective coating on associated members.
6. Availability of the protective materials and the equipment necessary for the application of this coating.

Protection From Mildly Corrosive Environment

For many applications, springs are coated with one of several slushing oils or rustproofing compounds. A more durable coating, if required, is a phosphate coating with rustproofing compound or organic finish. The organic finish can be compounded with low, medium, or high oil content in order to give the degrees of protection required, though baking may be required to dry the springs coated with compounds containing the higher oil content. An example of an organic coating with a low oil concentration is the asphaltic-based material known as Gilsonite paint or Japan Black, while the medium and high oil concentration paints are alkyd enamels. The addition of zinc chromate pigment to these synthetic enamels may be used to further enhance their corrosive protection.

When springs are protected by applying a zinc or iron phosphate coating, the thickness of the coating depends, among other factors, on the precleaning process. When an emulsion cleaner is used, a very fine coating is formed; an alkaline cleaner produces a thicker, hard, tight coating; if an acid is used, the coating is coarse and very thick. A uniform coating depends on good cleaning and chemical control. This phosphate coating is a basis for the subsequent coating with a rust preventive oil, wax, grease, or organic finish providing good protection from corrosion. The mechanism of these coatings is the ability of the minute pits, surface irregularities, and the crystalline structure of the zinc and iron phosphate crystals to retain the subsequent coating, thereby providing maximum corrosion protection. The phosphating process must be tailored to protect the spring from acid pickling, both in the precleaning sequence and during the phosphate process by controlling the free acid ratio.

When the end use of the spring is in a mildly corrosive atmosphere, a black oxide finish provides some degree of corrosion protection. A black oxide coating provides some of the same mechanics to hold rust preventive materials as the zinc and iron phosphate coatings, but with an overall lesser degree of protection. When black oxide coatings are applied they have the advantage of causing very slight, if any, dimensional changes and no hydrogen embrittlement.

Frequently, it is advantageous to use precoated spring wire where mildly corrosive conditions are encountered. These wires are coated by dipping into molten metal or by electroplating. They may be coated and then drawn to size, or they may be coated at the finished size. Cadmium, tin, and zinc are the metals employed for the wire coating process. The commercial importance of the process is considerable because the parts formed from such wires may be used immediately without having to undergo additional operations normally associated with the plating of finished parts. In the case of extension springs, the use of precoated wire is preferred because the tightly wound coils prevent applying a satisfactory coating after coiling.

Electroplating

Many springs are being plated with a variety of metals of which cadmium and zinc have generally proved to be the most satisfactory.

Cadmium and zinc plate are bluish white metals having good ductility and providing a protective coating on springs for most applications. When properly applied, cadmium or zinc plating does not flake or peel. They protect the base metal by sacrificial galvanic action. The protective life of both zinc and cadmium plate may be extended by adding a chromate coating. The degree of protection will still be dependent upon the thickness of the plate which may vary from 4 to 13 μm. Neither of these metals is resistant to attack from strong acids or alkalis. Both zinc and cadmium are poisonous when vaporized. Neither should be used on springs which come in contact with food or drink.

The end use of the spring and the environment determines which coating is to be applied. When either zinc or cadmium is indicated, zinc may be chosen because it is less expensive; the choice may be cadmium because of the longer time before producing white corrosion products. Cadmium can be baked at a higher temperature for relief of hydrogen embrittlement without the hazard of blistering. Copper plate on springs should be used only for identification purposes as it provides only slight corrosion protection for steels. The use of a copper, nickel, or chromium plate may be indicated for decorative reasons but is not generally applicable. Springs requiring nickel plate may be plated using the Sulfamate, Watts, or electroless process. When the nickel corrodes,

it forms green nickel corrosion products, and the protection depends on the thickness and continuity of the plate. Plated nickel is dense and prevents the escape of hydrogen which enters the steel during the precleaning, pickling, and plating processes. Hard chromium plating of springs is generally unsatisfactory because of the heavy thickness (50 μm or more) required to give corrosion protection. Chromium plate is hard with low ductility; therefore, it is subject to peeling and cracking while causing reduction in spring fatigue life.

When springs are exposed to severe environmental conditions such as salt water, acid, and caustic atmosphere or solutions, the most resistant coating is lead alloy. Lead coatings should be 25 μm in thickness to afford maximum protection. Heavy tin coatings, exceeding 25 μm in thickness, should be used for springs coming in contact with foodstuff because of the resistance of tin to weak organic acids and its low toxicity. Lead and tin coatings are electropositive to steel (more noble in the electromotive series); therefore, these coatings will only protect the steel spring when the coating is continuous and of a sufficient thickness. Lead and tin plates are soft materials; therefore, abrading of the coating must be avoided in order to provide the necessary continuous thick protective coating.

Techniques for Minimizing Hydrogen Embrittlement

Hydrogen may be introduced either during the cleaning steps in preparation for plating or it may be introduced during the plating process. If springs can be fabricated without producing heavy scale or film, the first hazard will be greatly reduced.

Residual stresses should be reduced as much as possible before plating.

If scale removal is required, a mechanical process is preferred. Unless the parts would distort, they may be blasted with shot, grit, glass, aluminum oxide, etc., or they may be tumbled or vibrated with various abrasive media.

When either a sulfuric or hydrochloric acid solution is used to remove scale from springs, the use of a suitable organic pickle inhibitor will eliminate excessive pickling and tend to reduce hydrogen embrittlement. Springs with a hardness over Rc 40 should not be descaled in acid.

The resistance of the passage of hydrogen from the steel increases as the thickness of the plate increases; therefore, excessive plate thickness should be avoided as well as the conditions producing hydrogen embrittlement. The following procedure should be followed to achieve this. The anodic or cathodic cleaning and pickling cycle should be as short as possible to reduce possible cracking of the spring in highly stressed areas. The zinc or cadmium plate should be initially deposited at a low amperage until the springs are completely coated. The amperage should then

TABLE 1.1—CORROSION PROTECTION FOR SPRINGS

Criteria for Determining Protection to be Used	Recommended Protection in Order of Preference	Dimensional Buildup, Minimum/Surface, μm	Methods of Applying Protection	Special Advantages, Precautions, and Limitations
Directly exposed to weather, not making close fits	Phosphate coat and paint, or paint alone	25	Dip, spray, or brush	Close fits will peel paint
Directly exposed to weather, making close fits	Cadmium or zinc plate	13	Electroplate	Corrosion protection good. Relief of hydrogen embrittlement essential after electroplating[a]
Not directly exposed to weather, making close fits	Cadmium or zinc plate	4	Electroplate	Same as above
Not directly exposed to weather, making close fits	Phosphate coat plus oil	8	Dip	Corrosion protection poorer than phosphate coat plus paint
Relatively mild corrosive conditions	Precoated spring wire	Thickness is variable and is dependent on size	Hot dip or electroplate	Fair all purpose coating applied to spring wire suitable for mildly corrosive conditions
Where corrosive conditions are very mild	Oxide black	Less than 5	Alkaline oxidizing solution, 30–60 min at 140°C	
Mildest corrosive conditions	Slushing oil or rust-proofing compounds	Dependent on method of application and/or viscosity of material	Dip or spray	Good for protection prior to installation and for short period after installation

[a]See text for alternate methods of plating and recommendation for relief of hydrogen embrittlement.

be raised, as the plating operation is continued until the specified thickness is achieved.

In spite of these precautions, springs which have been cadmium or zinc plated have been exposed to a severe hydrogen embrittlement environment during the preplate and plating steps. To produce a satisfactory spring, conditions must be set up to force the hydrogen to pass from the steel spring through the cadmium or zinc plate. The time and temperature required to relieve hydrogen embrittlement may vary from 1 h at 180°C, to as much as 24 h or longer at 220°C, depending on the tensile strength, hardness, residual stress, size of the spring, and thickness of the zinc or cadmium plate. This operation should be performed as soon as possible after plating.

The following processes for applying metallic coatings may be used with no exposure to hydrogen or substantially less exposure than occurs with conventional electroplating: mechanical plating, peen plating, vacuum plating, and low-embrittling electroplating.

Effect of Finish on Environment

The choice of finish depends also on the conditions encountered in use. Each individual application presents a different problem since the type of protective finish suitable for one may be wholly inadequate for another. For example, consider the question, "Should zinc- or cadmium-coated springs be used in hydraulic systems using petroleum hydraulic fluids?" While the zinc- or cadmium-coated spring will function perfectly for years in this medium, conditions can develop causing the petroleum hydraulic fluid to break down producing acid, sludge, and varnish by-products. These breakdown products are caused by temperature and time. When a breakdown occurs, it will be accelerated by the presence of zinc or cadmium. Zinc or cadmium metal will act as a catalyst speeding up the formation of acids, sludges, and varnishes. These conditions may cause little or no etching of the steel spring but will prevent normal operation of the hydraulic system due to the sludge and varnish deposits.

It is essential to consider all of the factors involved in order to select the finish best suited for a particular purpose. If possible, the spring maker should be consulted before final decision, so as to coordinate the finish with the manufacturing procedure.

Table 1.1 can be used as a general guideline for selecting finishes.

Chapter 2

Spring Materials

1. Material Selection

Among the factors governing the selection of a material for a spring are the desired load on the spring, the stress range through which the spring operates, mass and space limitations, expected fatigue life of the part, the environment in which the spring will operate with respect to temperature or corrosive conditions, and the severity of deformation encountered in its fabrication. As loads are increased and space is limited, springs must be made from higher tensile strength materials. If the load cycles also increase as the stress levels rise, the spring material must have correspondingly better fatigue properties. Music spring wire has the highest tensile strength of all the common engineering materials, and valve spring wire has the best fatigue properties.

Short descriptions of the types of wire and strip most generally considered for use in the manufacture of springs will be found in the following sections. Complete details and tolerances of the materials are covered in the reference specifications. The SAE specifications also give the requirements of the finished springs made from these materials.

2. Cold Wound Spring Material

Spring wire is produced from hot rolled rods by cold drawing through carbide dies to obtain the required size, surface finish, dimensional accuracy, and mechanical properties. By varying the chemistry, the amount of cold reduction and other mill practices including heat treatment, a wide diversity of mechanical properties and finishes is made available.

The several types of wire used in spring manufacture are made from a variety of chemical compositions and are either annealed, hard drawn, or pretempered to definite mechanical properties. The performance of springs is dependent upon the mechanical properties of the material used or the properties which may be developed as a result of subsequent treatments. Some of these properties are given in Table 2.1. Tensile strength values are shown in Tables 2.2-2.20.

3. Ferrous Wire

Hard Drawn Carbon Spring Wire—SAE J113, UNS K06501 and Special Quality High Tensile Carbon Spring Wire—SAE J271

The mechanical properties are developed by cold drawing patented hot rolled material. Patenting may be defined as heating to above the critical range followed by rapid cooling to transform at an elevated temperature in the 455-465 °C range. This operation produces a tough uniform structure which is suitable for severe cold reduction without actual or incipient breakage. To replace patenting, a new process has been developed wherein the hot rolled rods are given a controlled cooling treatment on the hot bed prior to coiling into bundles. The resulting structure and physical properties are equivalent to those obtained by patenting. This treatment is limited to a maximum of 13 mm rod diameter. The subsequent cold drawing imparts a fibrous structure and increases the tensile strength of the wire without appreciable loss in ductility. Hard drawn wire is intended to withstand more severe deformations than oil tempered grades. For the most part, hard drawn wire is intended for use in springs subject to static loads, low stress, or infrequent stress repetitions. Tensile strength values of SAE J113, Class 1 and Class 2 hard drawn wires are shown in Tables 2.2 and 2.3. SAE J271 is similar to SAE J113 except the former has restricted wire tolerances and is used for applications requiring higher stresses. See Table 2.4 for typical tensile strength values.

Music Spring Wire—SAE J178, UNS K08500

Music spring wire represents the highest quality of hard drawn steel spring wire. Its manufacture involves careful selection of heats of steel or portions thereof, special hot rolling practice, and very closely controlled heat treatment and cold drawing operations. The attainment of the high mechanical properties of this grade is dependent upon the chemistry of the steel, the high degree of craftsmanship employed in processing, and the amount of cold reduction (greater than that used for SAE J113) after the last of several patenting operations. This greater degree of cold work involves an increased number of drafts, a lower percentage of reduction per draft and slower wire

TABLE 2.1—PROPERTIES OF COMMON SPRING MATERIALS

| Name | UNS Number[a] | Specification | | Density kg/m³ | Modulus of Elasticity | | Available Wire Sizes mm | Maximum Temperature[b] °C |
		SAE	Comparable ASTM		Tension (E) MPa	Shear (G) MPa		
High Carbon Steels								
Hard drawn	K06501	J113	A227	7850	205 000	79 300	0.50—14	110
Hard drawn valve	K06701	J172	—	7850	205 000	79 300	2.2—6.5	110
Hard drawn special quality	K08200	J271	—	7850	205 000	79 300	0.5—14	110
Music	K08500	J178	A228	7850	205 000	79 300	0.10—6.5	140
Oil tempered	K07001	J316	A229	7850	205 000	79 300	0.50—16	160
Oil tempered valve	K06701	J351	A230	7850	205 000	79 300	1.6—6.5	160
Annealed	G10650	—	—	7850	205 000	79 300	Thru 5	110
	G15660	—	—	7850	205 000	79 300	Over 5	110
Alloy Steels								
Chromium Vanadium	K15048	—	A231	7850	205 000	79 300	0.5—12	230
Chromium Vanadium valve	K15047	J132	A232	7850	205 000	79 300	0.5—12	230
Chromium Silicon	K15590	J157	A401	7850	205 000	79 300	0.8—11	250
Chromium Silicon valve	—	—	A877	7850	205 000	79 300	0.8—10	250
Stainless Steel								
Austenitic (18-8)	S30200	J230	A313	7900	190 000	68 900	0.22—10	290
17-7 PH	S17700	J217	—	7810	200 000	75 800	0.5—10	340
Copper Alloys								
Phosphor Bronze	C52100	J463	B103	8800	105 000	43 400	d	c
	C51000	J463	B159	8860	105 000	43 400	0.10–12	c
Beryllium Copper	C17200	J463	B194	8260	130 000	48 300	d	c
	C17200	J463	B197	8260	130 000	48 300	0.08—12	c
Silicon Bronze	C65500	J463	B99	8530	105 000	43 100	0.10—13	c
Brass	C26000	J463	B36	8530	105 000	37 900	d	c
	C27000	J463	B134	8470	105 000	37 900	0.10—6.5	c
Nickel Alloys								
Inconel 600	N06600	J470	—	8500	215 000	75 800	0.10—12	430
Inconel X-750	N07750	J470	—	8300	215 000	75 800	0.10—14	430
Monel 400	N04400	J470	—	8830	180 000	65 500	0.10—12	230
Monel K500	N05500	J470	—	8470	180 000	65 500	0.10—12	260

[a]Unified Numbering System for Metals and Alloys, SAE HS 1086 (ASTM DS-56).
[b]Temperatures represent the approximate point at which a compression spring will experience 10% load loss when loaded to 700 MPa corrected stress for 100 h.
[c]Data for this material are not available.
[d]Sheet and strip.

TABLE 2.2—HARD DRAWN CARBON STEEL SPRING WIRE, SAE J113, CL1

| Diameter[a] mm | Tensile Strength[b] MPa | | Diameter[a] mm | Tensile Strength[b] MPa | |
	Min	Max		Min	Max
0.50	1950	2230	2.80	1480	1680
0.55	1920	2200	3.00	1450	1650
0.60	1900	2180	3.50	1420	1620
0.65	1880	2160	4.00	1400	1600
0.70	1850	2130	4.50	1380	1580
0.80	1800	2080	5.00	1350	1550
0.90	1780	2060	5.50	1320	1520
1.00	1750	2020	6.00	1300	1480
1.10	1720	1980	7.00	1250	1420
1.20	1700	1950	8.00	1200	1380
1.40	1680	1920	9.00	1180	1350
1.60	1650	1900	10.00	1150	1320
1.80	1600	1850	11.00	1120	1300
2.00	1580	1800	12.00	1100	1280
2.20	1550	1780	14.00	1050	1220
2.50	1500	1700			

[a]Preferred sizes.
[b]Tensile strength values for intermediate sizes may be interpolated.

drawing speeds in order to control heat buildup. Music wire is used in the more severe types of spring applications which require high tensile strength (see Table 2.5) and exacting surface control to obtain the necessary fatigue properties.

Oil Tempered Carbon Spring Wire—SAE J316, UNS K07001

The mechanical properties for this grade are developed by heat treating the wire at its finished size. The general processing to this point is the same as for hard drawn wire. The heat treatment consists of heating the wire in the strand form to a temperature above the critical range, quenching in oil and then tempering by passing it through a molten lead bath maintained at the proper temperature to produce the desired mechanical properties. The resulting tempered martensitic structure provides greater freedom from relaxation under repeated and continued stress

TABLE 2.3—HARD DRAWN CARBON STEEL SPRING WIRE, SAE J113, CL2

Diameter[a] mm	Tensile Strength[b] MPa	
	Min	Max
0.50	2240	2500
0.55	2220	2480
0.60	2200	2460
0.65	2180	2440
0.70	2150	2420
0.80	2120	2380
0.90	2100	2350
1.00	2050	2300
1.10	2000	2250
1.20	1980	2220
1.40	1920	2160
1.60	1880	2120
1.80	1850	2080
2.00	1820	2050
2.20	1780	2000
2.50	1750	1950
3.00	1700	1900
3.50	1650	1850
4.00	1600	1800
4.50	1550	1750
5.00	1520	1720
5.50	1500	1700
6.00	1450	1650
7.00	1420	1620
8.00	1380	1580
9.00	1350	1550
10.00	1320	1520
11.00	1300	1500
12.00	1250	1450
14.00	1200	1350

[a]Preferred sizes.
[b]Tensile strength values for intermediate sizes may be interpolated.

TABLE 2.4—SPECIAL QUALITY HARD DRAWN STEEL WIRE, SAE J271

Diameter[a] mm	Tensile Strength[b] MPa Typical
0.50	2600
0.60	2500
0.80	2400
1.00	2300
1.20	2200
1.60	2100
2.00	2000
2.50	1950
3.00	1900
4.00	1800
5.00	1750
6.00	1700
8.00	1600
10.00	1500
12.00	1450
14.00	1400

[a]Preferred sizes.
[b]Tensile strength values for intermediate sizes may be interpolated.

TABLE 2.5—MUSIC WIRE, SAE J178, UNS K08500

Diameter[a] mm	Tensile Strength[b] MPa		Diameter[a] mm	Tensile Strength[b] MPa	
	Min	Max		Min	Max
0.10	3000	3300	0.90	2200	2450
0.12	2900	3200	1.00	2150	2400
0.16	2800	3100	1.10	2120	2380
0.18	2750	3050	1.20	2100	2350
0.20	2700	3000	1.40	2050	2300
0.22	2680	2980	1.60	2000	2250
0.25	2650	2950	1.80	1980	2220
0.28	2620	2920	2.00	1950	2200
0.30	2600	2900	2.20	1900	2150
0.35	2550	2820	2.50	1850	2100
0.40	2500	2750	2.80	1820	2050
0.45	2450	2700	3.00	1800	2000
0.50	2400	2650	3.50	1750	1950
0.55	2380	2620	4.00	1700	1900
0.60	2350	2600	4.50	1680	1880
0.65	2320	2580	5.00	1650	1850
0.70	2300	2550	5.50	1620	1820
0.80	2250	2500	6.00	1600	1800
			6.50	1530	1750

[a]Preferred sizes.
[b]Tensile strength values for intermediate sizes may be interpolated.

application as compared to hard drawn wire. Oil tempered wire is also more suitable for precision forming and coiling operations than hard drawn wire because of the close control of tensile strength and its superior straightness. Tensile strength and hardness values of Class 1 and Class 2 oil tempered wires are shown in Tables 2.6 and 2.7.

Oil Tempered Carbon Valve Spring Wire—SAE J351, UNS K06701

Valve spring wire represents the highest quality of oil tempered wire. It is used for springs subject to dynamic stress applications with maximum life expectancy. To meet this requirement, valve spring wire has the highest degree of uniformity with respect to surface, structural soundness, and mechanical properties. The production of this quality requires rigid inspection, selection, and continuous control of the product from the steel melting furnace to the finished wire. The surface of this wire will have no areas of detrimental carbon depletion. It is free of seams, scratches, die marks, pits, or other surface defects that would impair the fatigue life of the spring. Wire tensile strength and hardness values are shown in Table 2.8.

TABLE 2.6—OIL TEMPERED CARBON STEEL SPRING WIRE, SAE J315, CL1

Diameter[a] mm	Tensile Strength[b] MPa		Hardness	
	Min	Max	Min	Max
			Rockwell 15N	
0.50	2050	2250	88.0	90.0
0.55	2020	2220	88.0	90.0
0.60	2000	2200	88.0	90.0
0.65	1980	2180	88.0	90.0
0.70	1950	2150	87.5	89.5
0.80	1900	2100	87.5	89.5
0.90	1850	2050	87.0	89.0
1.00	1800	2000	86.5	88.5
1.10	1780	1980	86.0	88.0
1.20	1750	1950	86.0	88.0
			Rockwell 45N	
1.40	1700	1900	55.0	60.0
1.60	1650	1850	54.0	59.0
1.80	1620	1820	53.0	58.0
2.00	1600	1800	52.5	57.5
2.20	1580	1780	52.0	57.0
2.50	1550	1750	51.5	56.5
2.80	1520	1720	50.5	55.5
			Rockwell C	
3.00	1500	1700	45	50
3.50	1450	1620	44	48
4.00	1400	1580	43	48
4.50	1380	1550	42	48
5.00	1350	1520	41	47
5.50	1320	1500	41	47
6.00	1300	1480	40	45
7.00	1280	1450	40	45
8.00	1250	1420	39	44
9.00	1220	1400	38	44
10.00	1200	1380	37	43
11.00	1180	1350	36	42
12.00	1150	1320	35	41
14.00	1120	1300	35	41
16.00	1120	1300	35	41

[a]Preferred sizes.
[b]Tensile strength values for intermediate sizes may be interpolated.

TABLE 2.7—OIL TEMPERED CARBON STEEL SPRING WIRE, SAE J316, CL2

Diameter[a] mm	Tensile Strength[b] MPa		Hardness	
	Min	Max	Min	Max
			Rockwell 15N	
0.50	2240	2440	90.0	92.0
0.55	2220	2420	90.0	92.0
0.60	2200	2400	89.0	91.0
0.65	2190	2390	89.0	91.0
0.70	2180	2380	89.0	91.0
0.80	2150	2350	88.5	90.5
0.90	2100	2300	88.0	90.0
1.00	2050	2250	87.5	89.5
1.10	2020	2220	87.5	89.5
1.20	2000	2200	87.0	89.0
1.40	1950	2150	87.0	89.0
			Rockwell 45N	
1.60	1900	2100	58.0	63.0
1.80	1880	2080	57.0	62.0
2.00	1850	2050	56.0	61.0
2.20	1800	2000	55.5	60.5
2.50	1750	1950	55.0	60.0
2.80	1720	1920	55.0	60.0
3.00	1700	1900	54.5	59.5
			Rockwell C	
3.50	1650	1850	49	54
4.00	1600	1800	48	53
4.50	1550	1750	48	53
5.00	1520	1720	47	52
5.50	1500	1700	46	51
6.00	1480	1680	45	50
7.00	1460	1660	44	49
8.00	1440	1640	44	49
9.00	1420	1620	44	49
10.00	1400	1600	43	48
11.00	1380	1580	43	48
12.00	1350	1550	42	47
14.00	1320	1520	41	46
16.00	1300	1500	41	46

[a]Preferred sizes.
[b]Tensile strength values for intermediate sizes may be interpolated.

Hard Drawn Valve Spring Wire—SAE J172, UNS K06701

A hard drawn grade of valve spring wire is produced with all the restrictions as to processing and controls of quality equal to the oil tempered grade. The final mechanical properties are the result of the amount of cold work which follows the final patenting treatment. The fabrication of this material is similar to the hard drawn carbon spring wire described in the first paragraph of Section 3. See Table 2.9 for tensile strength values of the wire.

Annealed Wires (SAE 1065, UNS G10650 Thru 5.0 mm; SAE 1566, UNS G15660 Over 5.0 mm)

Annealed wire is produced for certain applications where there is a high degree of severity of deformation in the fabrication of springs and formed parts. This wire is not commonly produced to definite tensile properties, as the parts formed are subsequently heat treated. This grade may be furnished lightly drawn after annealing in order to aid in dimensional control during forming.

Alloy Steel Spring Wires—SAE J132, UNS K15047; SAE J157, UNS K15590

Alloy steel wire is used in the manufacture of springs which operate at moderately elevated temperatures up to 230°C. Because of their resistance to relaxation under these conditions, their high tensile strengths, and their torsional elastic limits, the most commonly used alloys are SAE J132, Oil Tempered Chromium Vanadium Valve

TABLE 2.8—OIL TEMPERED STEEL VALVE SPRING WIRE, SAE J351, UNS K06701

Diameter mm		Tensile Strength MPa		Hardness	
Over	Thru	Min	Max	Min	Max
				Rockwell 45N	
—	2.5	1650	1800	52	57
2.5	3.5	1600	1750	51	56
				Rockwell C	
3.5	4.0	1580	1720	46	51
4.0	4.8	1550	1700	45	50
4.8	5.5	1520	1680	44	49
5.5	—	1500	1650	44	49

TABLE 2.9—HARD DRAWN CARBON STEEL VALVE SPRING WIRE, SAE J172, UNS K06701

Diameter mm		Tensile Strength MPa	
Over	Thru	Min	Max
—	3.5	1600	1750
3.5	4.0	1580	1720
4.0	4.8	1550	1700
4.8	5.5	1520	1680
5.5	—	1500	1650

TABLE 2.10—CHROMIUM VANADIUM STEEL SPRING WIRE, SAE J132, UNS K15047

Diameter mm		Tensile Strength MPa		Hardness		Reduction of Area
Over	Thru	Min	Max	Min	Max	% Min
				Rockwell 15N		
—	0.5	2050	2200	88.5	89.7	—
0.5	0.6	2020	2180	88.3	89.5	—
0.6	0.8	2000	2150	88.0	89.3	—
0.8	1.0	1950	2100	87.5	88.8	—
1.0	1.2	1900	2050	87.0	88.5	—
				Rockwell 45N		
1.2	1.6	1800	1950	57.9	61.4	—
1.6	2.0	1750	1900	56.4	59.4	—
2.0	2.5	1700	1850	55.0	57.9	—
				Rockwell C		
2.5	3.0	1600	1750	48	51	45
3.0	4.0	1550	1700	47	50	40
4.0	5.0	1500	1650	46	49	40
5.0	6.0	1450	1600	45	48	40
6.0	8.0	1400	1550	43	47	40
8.0	10.0	1350	1500	42	46	40
10.0	12.0	1320	1480	41	45	40
12.0	—	1300	1450	41	45	40

TABLE 2.11—CHROMIUM SILICON STEEL SPRING WIRE, SAE J157, UNS K15590

Diameter[a] mm	Tensile Strength[b] MPa		Hardness		Reduction of Area
	Min	Max	Min	Max	% Min
			Rockwell 15N		
0.8	2060	2240	88.5	90.0	—
0.9	2050	2230	88.5	90.0	—
1.0	2040	2220	88.5	90.0	—
1.1	2030	2210	88.5	90.0	—
1.2	2020	2200	88.0	88.5	—
1.4	2010	2190	88.0	89.5	—
			Rockwell 45N		
1.6	2000	2180	59.5	63.0	—
1.8	1980	2160	59.0	62.5	—
2.0	1960	2140	59.0	62.0	—
2.2	1940	2120	58.5	61.5	—
2.5	1920	2100	58.0	61.5	45
2.8	1910	2090	58.0	61.0	45
3.0	1900	2080	57.5	61.0	45
			Rockwell C		
3.5	1860	2040	51.5	54.0	45
4.0	1830	2000	51.0	53.5	40
4.5	1800	1960	50.5	53.0	40
5.0	1780	1940	50.5	53.0	40
5.5	1750	1920	50.0	52.5	40
6.0	1720	1900	49.5	52.5	40
7.0	1700	1880	49.5	52.0	40
8.0	1680	1850	49.0	52.0	40
9.0	1660	1820	48.5	52.0	40
10.0	1640	1800	48.5	51.5	40
11.0	1620	1780	48.0	51.0	40

[a]Preferred sizes.
[b]Tensile strength values for intermediate sizes may be interpolated.

Spring Quality Wire and Springs, and SAE J157, Oil Tempered Chromium Silicon Alloy Steel Wire and Springs. See Tables 2.10 and 2.11 for tensile strength and hardness values of oil tempered wire. These alloys may also be obtained hard drawn or annealed, and in commercial quality or valve spring quality.

4. Stainless Steel Wire—SAE J230 and SAE J217

Stainless steels, SAE J230 in either UNS S30200 or UNS S30400 grades, are produced to mechanical properties by cold drawing. In this condition, these grades have good corrosion resistance and also good temperature resistance up to 260°C. SAE J217, UNS S17700, Spring Wire and Springs, covers a grade of stainless normally produced cold drawn which, after forming, is age hardened by heating to 480 ± 5°C for 1 h. In this condition, it has mechanical properties equal to music wire. It has the corrosion-resistant properties of UNS S30200 and, in addition, greater resistance to relaxation up to about 340°C. All metallic coatings, such as lead or copper, must be removed prior to heat treatment. Springs made from stainless wire must be passivated after heat treating to ensure maximum corrosion resistance. All of these stain-

TABLE 2.12—HARD DRAWN STAINLESS STEEL WIRE, SAE J230, UNS S30200 (18-8)

Diameter[a] mm	Tensile Strength[b] MPa		Diameter[a] mm	Tensile Strength[b] MPa	
	Min	Max		Min	Max
0.22	2240	2440	1.60	1750	1950
0.25	2220	2420	1.80	1720	1920
0.28	2200	2400	2.00	1680	1880
0.30	2180	2380	2.20	1650	1850
0.35	2150	2350	2.50	1620	1820
0.40	2120	2320	2.80	1580	1780
0.45	2100	2300	3.00	1550	1750
0.50	2080	2280	3.50	1480	1680
0.55	2050	2250	4.00	1420	1620
0.60	2020	2220	4.50	1360	1550
0.70	1980	2200	5.00	1300	1500
0.80	1940	2140	5.50	1260	1460
0.90	1900	2100	6.00	1220	1420
1.00	1880	2080	7.00	1160	1360
1.10	1850	2050	8.00	1100	1300
1.20	1820	2020	9.00	1000	1200
1.40	1800	2000	10.00	900	1100

[a]Preferred sizes.
[b]Tensile strength values for intermediate sizes may be interpolated.

TABLE 2.14—SPRING BRASS WIRE, SAE J461, UNS C26000 AND UNS C27000

Temper	Tensil-Strength, MPa	
	Min	Max
Half-Hard	550	650
Three-Quarter-Hard	625	725
Hard	700	800
Extra Hard	800	900
Spring	825	—

TABLE 2.13—HEAT TREATED STAINLESS STEEL WIRE, SAE J217, UNS S17700 (17-7 PH)

Diameter[a] mm	Tensile Strength MPa Typical
0.50	2400
0.60	2380
0.80	2320
1.00	2300
1.20	2250
1.60	2200
2.00	2100
2.50	2050
3.00	2000
4.00	1950
5.00	1900
6.00	1850
8.00	1800
10.00	1750

[a]Preferred sizes.

less steel spring materials are magnetic at spring hardness. Tensile strength and hardness values for hard drawn wire are shown in Tables 2.12 and 2.13.

5. Copper Alloys

Spring Brass—SAE J461, UNS C26000 (70% Copper—30% Zinc)

This material is most often supplied as strip or wire and is recommended for light-duty springs. Mechanical properties are not usually related to size but rather to the terminology quarter-hard, half-hard, three-quarter-hard, hard, extra hard, spring, and extra spring, which indicate the mechanical properties available irrespective of material section, as shown in Tables 2.14 and 2.15.

Phosphor Bronze—SAE J461, UNS C51000 and UNS C52100 (80% Copper—8% Tin—0.3% Max Phosphorus)

Sheet and Strip—This material is superior to spring brass in mechanical properties and equal in electrical conductivity. Mechanical properties are shown in Table 2.16.

Wire—Spring temper wire has the size-tensile relationship shown in Table 2.17.

TABLE 2.15 — SPRING BRASS SHEET AND STRIP, SAE J461, UNS C26000

Temper	Size Section mm	Tensile Strength MPa		Hardness			
				Min	Max	Min	Max
	Over/Thru	Min	Max	RB		R30T	
Half-Hard	0.50/0.90 0.90/– 0.30/0.70 0.70/–	395	460	60 63 — —	74 77 — —	— — 56 58	— — 66 68
Three-Quarter-Hard	0.50/0.90 0.90/– 0.30/0.70 0.70/–	440	510	72 75 — —	70 82 — —	— — 65 67	— — 70 72
Hard	0.50/0.90 0.90/– 0.30/0.70 0.70/–	490	560	70 81 — —	84 86 — —	— — 70 71	— — 73 74
Extra Hard	0.50/0.90 0.90/– 0.30/0.70 0.70/–	570	635	85 87 — —	89 91 — —	— — 74 75	— — 76 77
Spring	0.50/0.90 0.90/– 0.30/0.70 0.70/–	625	690	89 90 — —	92 93 — —	— — 76 76	— — 78 78
Extra Spring	0.50/0.90 0.90/– 0.30/0.70 0.70/–	655	715	91 92 — —	94 95 — —	— — 77 77	— — 79 79

TABLE 2.16—PHOSPHOR BRONZE SHEET AND STRIP, SAE J461, UNS C52100

Temper	Size Section mm	Tensile Strength MPa		Hardness			
				Min	Max	Min	Max
	Over/Thru	Min	Max	RB		R30T	
Half-Hard	1.0/– 0.75/– 0.50/1.0 0.25/0.75	475	580	76 — 69 —	91 — 88 —	— 67 — 63	— 78 — 75
Hard	1.0/– 0.75/– 0.50/1.0 0.25/0.75	585	690	91 — 89 —	97 — 95 —	— 76 — 73	— 81 — 80
Extra Hard	1.0/– 0.75/– 0.50/1.0 0.25/0.75	670	770	95 — 93 —	100 — 98 —	— 78 — 77	— 83 — 82
Spring	1.0/– 0.75/– 0.50/1.0 0.25/0.75	725	820	97 — 95 —	102 — 100 —	— 79 — 78	— 84 — 83
Extra Spring	1.0/– 0.75/– 0.50/1.0 0.25/0.75	760	840	98 — 96 —	103 — 101 —	— 80 — 79	— 84 — 83

2.25

TABLE 2.17—PHOSPHOR BRONZE WIRE AND ROD, SAE J461, UNS C51000

Diameter mm		Minimum Tensile Strength, MPa		Minimum Elongation % in 50 mm
Over	Thru	Wire	Rod	
—	0.60	1000	850	—
0.60	1.50	925	800	—
1.50	3.00	900	750	—
3.00	6.00	850	725	3.5 (rod only)
6.00	9.00	825	700	5.0
9.00	12.00	725	625	9.0

Beryllium Copper—SAE J461, UNS C17200
(98% Copper—2% Beryllium)

This material is commonly used because of its excellent fatigue resistance and its ability to be formed in a soft or cold worked condition and subsequently aged or precipitation hardened to spring temper at the relatively low temperature of 315 °C. Mechanical properties of the various mill tempers after heat treat, independent of size, are shown in Table 2.18.

Silicon Bronze—ASTM B99, UNS C65500
(95% Copper—3% Silicon)

There are several different chemical grades of silicon copper alloy wires. UNS C65500 strip is also used for flat parts requiring spring properties. They are less costly than the phosphor-bronze group and have about the same tensile-cold worked temper relationships. The tensile strength of spring grade wire is 900 MPa minimum and 725 MPa minimum for strip and is a function of the chemistry of the alloy rather than size.

6. Nickel Alloy Wires, SAE J470

UNS N06600 (Inconel 600)
(75% Nickel—16% Chromium—8% Iron)

This is used chiefly because of its corrosion resistance and its strength and resistance to oxidation at elevated temperatures. Its nominal operating range is from a low of -40 °C to a high of 345 °C when springs are designed with average stresses, such as 380-585 MPa in torsion or 760-930 MPa in bending. As cold drawn, the tensile strength range is 1070-1280 MPa. The nominal stress relieving temperature is 425-480 °C.

UNS N07750 (Inconel X-750)
(73% Nickel—15% Chromium—7% Iron—2.5% Titanium—0.8% Columbium—0.8% Aluminum)

This is an age hardenable alloy that has outstanding resistance to relaxation at elevated temperatures up to 600 °C. This grade is furnished in two tempers. The lower temperature aged product has a tensile strength range of 930-1140 MPa which after aging at 730 °C for 16 h will have a tensile strength of 1170 MPa and maximum resistance to relaxation. If greater strength is desired, an aging treatment of 845 °C for 4 h will develop 1520 MPa tensile strength but with some deterioration of relaxation properties.

UNS N04400 (Monel 400)
(65% Nickel—35% Copper)

This is used because of its high strength, ductility, and resistance to alkalis and acids. It is also the lowest cost nickel grade spring alloy. In the cold drawn spring grade, its tensile strength varies with the amount of cold work in the range of 900-1100 MPa, and it should be stress relieved after coiling at 275-315 °C. It is best used for

TABLE 2.18—BERYLLIUM COPPER, MECHANICAL PROPERTIES OF MILL TEMPERS AFTER HEAT TREAT, SAE J461, UNS C17200

Temper Before Hardening	Hardened Temper	Tensile Strength				Hardness, Min		
		Strip		Wire				
		MPa		MPa				
		Min	Max	Min	Max	RC	R30N	R15N
Solution Treated (A)	AT	1140	1350	1140	1350	36	56	78
Quarter-Hard (¼ H)	¼ HT	1200	1410	1200	1410	38	58	79
Half-Hard (½ H)	½ HT	1270	1480	1270	1480	39	59	79.5
Three-Quarter-Hard (¾ H)	¾ HT	—	—	1310	1520	40	60	80
Hard (H)	HT	1310	1520	—	—	40	60	80

moderately loaded springs which are subject to corrosive attack such as marine or salt water conditions. It is reasonably stable in the temperature range of -70 to +200°C.

UNS N05500 (Monel K-500)
(66% Nickel—31% Copper—3% Aluminum)

This is used in place of regular monel because it can be age hardened to provide higher mechanical properties. When aged at 595°C for 16 h followed by slow cooling to 480°C, this grade of wire has a tensile range of 1100-1240 MPa. It is nonmagnetic and can be used at temperatures to 235°C.

Superalloys

In addition to the above, several specialized grades of nickel or cobalt-base alloys containing varying amounts of chromium-tungsten-molybdenum and/or silicon have been developed. These materials, termed "superalloys," are all age hardening grades and have tensile properties in wire form that can be used to make springs that are dimensionally stable up to 815°C and at stress levels to 550 MPa. Details of these materials are given in SAE J467, Special Purpose Alloys ("Superalloys"). Further information should be obtained from the manufacturers.

7. Spring Steel for Flat Spiral Springs

Flat wire and strip are terms, sometimes used interchangeably, which identify two different products. There is an area of overlap where a cross section may be produced either as a flat wire or as a strip. The processing is different, and since this may have an effect upon the application, the two shall be described briefly below.

Flat Wire

Flat wire is produced by passing round wire through the rolls of a flattening mill. The wire may be left with a natural round edge formed by the flow of material as it passes between the rolls, or it may be shaped by a Turks Head or edger to produce a flat or radius edge.

In the hardened and tempered condition, the desired mechanical properties for a given end use are obtained by quenching and tempering in the strand. The required combination of strength, hardness, toughness, and ductility can be produced by varying the carbon content and the conditions of quenching and tempering.

In ordering flat wire, it is necessary to specify the edge required, unless the approximately round edge produced by cold work is acceptable.

Strip

The ability of high carbon steel strip to attain high hardness and elastic limits makes it suitable for many special applications. Spring steel strip is produced in a variety of chemistry combinations and mechanical properties.

Selection of the proper grade can only be made after determination is made of the hardness, strength, and fatigue requirements, plus the ductility needed for forming. Spring steel strip is produced by subjecting hot rolled strip to a cleaning operation and a combination of cold rolling and thermal treatments. This process transforms a coarse grained material of uneven dimensions and a rough surface to one that is capable of being formed, responding to heat treatment, and having uniform dimensions and a smooth surface free of defects. Selection of the raw material and the mill practice employed are key factors in determining the quality of the finished product, such as uniformity in response to forming and heat treatment, and ductility.

Hardened and tempered cold rolled steel strip is produced to meet a range of hardnesses or ultimate tensile strengths. This operation is performed on the strip to secure the highest mechanical properties. Material in this condition is capable of only moderate forming.

Hardened and tempered strip may be supplied in any of the following finishes: black tempered, scale-less tempered, bright tempered, tempered and polished, or tempered, polished, and colored.

Untempered steel strip may be supplied with different edges as listed below:

1. Round or square, a prepared edge.
2. Natural mill edge.
3. Approximately square edge produced by slitting.
4. Approximately round edge produced by edge rolling.
5. Approximately square edge produced by edge rolling or filing to eliminate slitting burr.
6. Square edge produced by edge rolling when width tolerances and finish are not as exacting as in Item 1.

Tempered strip is supplied with an edge as described in Item 1.

8. Hot Coiled Spring Materials

Spring steel bars are available in carbon and alloy grades. The bars are generally used in the as-rolled condition, but may be centerless ground before coiling. Table 2.19 lists some materials commonly used in hot coiled springs. Given a steel heat of minimum hardenability and a mild oil quench, the "H" steel sizes listed should

TABLE 2.19—MATERIALS COMMONLY USED IN HOT COILED SPRINGS

| Material | Tensile Properties | | | Torsional Properties | | | Application—Based on Hardenability (See Chapter 2, Section 8) |
| | Ultimate Strength MPa | Elastic Limit MPa | Modulus E[a] MPa | Ultimate Strength MPa | Elastic Limit MPa | Modulus G[a] MPa | |
	Min[b]	Min[b]		Min[b]	Min[b]		
Carbon Steels							
SAE 1085 and SAE 1095	1200	900	200 000	780	520	76 000	up to 10 mm bar size
Carbon Boron Steel							
SAE 15B62H	1450	1280	200 000	1000	720	76 000	up to 25 mm bar size
Alloy Steels							
SAE 5150H	1450	1280	200 000	1000	720	76 000	up to 10 mm bar size
SAE 5160H	1450	1280	200 000	1000	720	76 000	up to 20 mm bar size
SAE 9260H	1450	1280	200 000	1000	720	76 000	up to 10 mm bar size
SAE 51B60H	1450	1280	200 000	1000	720	76 000	up to 30 mm bar size
SAE 4161H	1450	1280	200 000	1000	720	76 000	up to 60 mm bar size
SAE 6150H	1450	1280	200 000	1000	720	76 000	up to 10 mm bar size

[a]Generally accepted values: reduced from nominal values of 205 000 MPa and 79 300 MPa to offset various factors related to hot coiled springs.
[b]Carbon steel minimum values are based on size shown. Core hardness may be lower than surface hardness for size shown.
"H" steel minimum values are based on a through hardened bar of 444 HB (1450–1650 MPa Ultimate Tensile), Ref. SAE J413a. Typical hardness ranges are 444–495 HB and 461–514 HB.

provide a minimum as-quenched core hardness of 58 Rc. Sizes for the "H" steels are based on SAE J1268 JUN80. A minimum core hardness of 58 Rc as-quenched is considered desirable to produce a spring that can be fully preset to operate at maximum stress levels. An automotive suspension spring is typical of the high stress, minimum weight approach.

When springs are operated at lower stresses, a softer core can be permitted and larger bar sizes used. The carbon steel sizes shown are based on this type of application. Selection of a material involves many variables and requires a thorough knowledge of spring requirements, materials, and the heat treat system involved.

9. Corrosion Resistance

To ensure that a new spring design will provide adequate performance, it is necessary to evaluate not only the mechanical requirements of the application, but also the environmental conditions in which the spring must function. Corrosion is the major environmental factor which will promote spring breakage. It destroys and removes metal from the surface of the spring by chemical or electrochemical methods in an irregular fashion which causes an overall reduction in the spring stock size, compounded by localized areas of intensive metal removal. These effects seriously reduce both the static strength and the fatigue strength of the spring. Since the corrosive losses reduce the spring stock size, the resulting strength reduction will vary inversely to the section size and, therefore, gain increasing importance as the size decreases.

To protect the spring against the corrosive environment, one must either isolate the surface from the environment, nullify the environmental effects, or use a spring material that is resistant to the chemical attack of the corrosive agents. Isolation of the surface may be accomplished by removing the spring from the environment or covering the surface with a protective coating. Applied finishes are discussed in Chapter 1, Section 8.

The environmental effects can be nullified on electrochemical forms of corrosion by attaching to the spring a sacrificial metal which is chemically more active than the spring metal. This can be done by plating the spring surface with a sacrificial metal, that is, cadmium or zinc for steel springs, or by providing an electrical conductive link between the spring and the sacrificial metal.

The most common corrosion resistant spring materials are nonferrous metals or stainless steels. To be truly corrosion resistant, these metals must be selected on the basis of the corrosive environment in which they are to operate. Among the nonferrous materials used are phosphorbronzes, brasses, and copper nickel alloys. However, due to lower elastic properties, it is necessary to use a larger volume of these metals to give performance comparable to ferrous metals. For certain severely corrosive conditions, stainless steels offer very satisfactory performance.

The materials have very nearly the same mechanical properties as the standard ferrous spring materials and will permit satisfactory design with the least amount of metal and a minimum of space.

Because of economic conditions, it is not always possible to use anti-corrosive materials in spring parts subject to such attack. It thus becomes necessary to analyze each application to determine the type and degree of protection that is necessary to ensure adequate performance.

10. Material Handbooks

Additional information on materials and further details of specifications can be obtained from the following sources:

1. Society of Automotive Engineers, Inc., 400 Commonwealth Drive, Warrendale, PA 15096. SAE Handbook includes material compositions and properties.

2. American Society for Testing and Materials, 1916 Race Street, Philadelphia, PA 19103. Annual Book of Standards: Part 3 covers steel strip and wire, Part 6 covers copper alloys, Part 10 covers tests of metals.

3. Huntington Alloys Products Division, The International Nickel Co., Huntington, WV 25720. Handbooks of high-nickel alloys.

4. Copper Development Association, Inc., 405 Lexington Avenue, New York, NY 10017. Standards Handbook for copper alloys.

Chapter 3

Cold Wound Helical and Spiral Springs

Helical compression and extension springs are normally cold wound when the material is less than 10 mm in diameter, and hot coiled when it is greater than 16 mm (see Chapter 4). When the cross section is between 10 mm and 16 mm both processes are common.

A. HELICAL COMPRESSION SPRINGS

1. General

A helical compression spring is an open helical spring which offers resistance to a compressive force. It is made in various forms and from different shapes of wire depending upon a wide variety of uses. While it is necessary in some cases to employ square, rectangular, or special shaped wire, the use of round wire predominates in the manufacture of most compression springs. The most common form has the same diameter for its entire length and is known as a straight spring. Variable diameter springs are used to some extent.

The main objective of a spring design is to obtain a spring which will be reasonably economical for a given application, will fit into the available space, and will give satisfactory performance. The space governs the dimensional limits of operating length and outside and inside diameters. In addition, some other requirements, such as load, deflection, and maximum solid length, need to be known at the start. These dimensions, together with the load and deflection requirements, determine the stresses in the spring.

When a helical compression spring is loaded, the coiled wire is stressed in torsion. The stress is greatest at the surface of the wire; as the spring is deflected, the load varies, causing a range of operating stress. Stress and stress range govern the life of the spring. The higher the stress range, the lower the maximum stress must be to obtain comparable life. Relatively high stresses may be used when the stress range is low or if the spring is subjected to static loads only. The stress at solid length must be sufficiently high to permit presetting if this manufacturing operation is included. However, if the spring is not preset, then the solid stress should be kept low enough to avoid permanent set, since springs are frequently compressed solid during installation.

The material selection depends upon the spring application and such factors as cost, allowable stress, fatigue durability, corrosion resistance, and resistance to set or load loss at elevated temperatures. If the space is limited, or if optimum fatigue durability is required, shot peening and spring presetting may be considered as means of increasing the allowable working stress. If a spring is subject to elevated temperatures and/or high level stresses, then presetting by either heat setting or hot pressing may be necessary to reduce the relaxation in service. See Chapter 1, Sections 4 and 5. Spring materials are fully described in Chapter 2, which should be referred to for material selection.

2. Spring Details

Terminology and Types of Ends

A typical helical compression spring is illustrated in Fig. 3.1 to explain the spring terminology. This figure shows the configuration of closed and ground ends of a compression spring. This minimizes the eccentricity of loading at the ends and reduces the installation space required.

Typical types of end configurations which may be employed are illustrated in Fig. 3.2. The unground ends may be used for reasons of economy, but they give eccentric loading with some increase in maximum spring wire stress and space required. The plain ends similarly produce eccentric loading and additionally present a handling problem due to springs tangling together.

Wire Diameter (d)

The desired size of wire will be determined by the elastic limit and the modulus of the material, loads, range of operation, and type of application of the spring. Round wire provides the most economical compression spring design and should be used whenever possible. At times, however, it is necessary to resort to the use of square, rectangular, or flat wire in order to obtain the desired characteristics within a limited space.

If the design necessitates a rectangular section, a rolled wire (with round edges) is more economical and can quite often be rolled from round wire in stock. For these wire sections, rectangular wire formulae are applicable, although the load capacity, rates, and stresses will be somewhat altered due to the large edge radii.

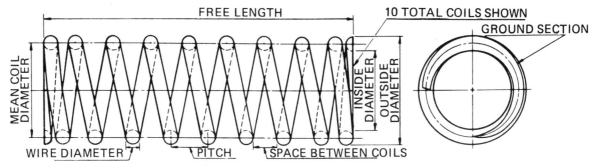

Fig. 3.1—Spring details and terminology

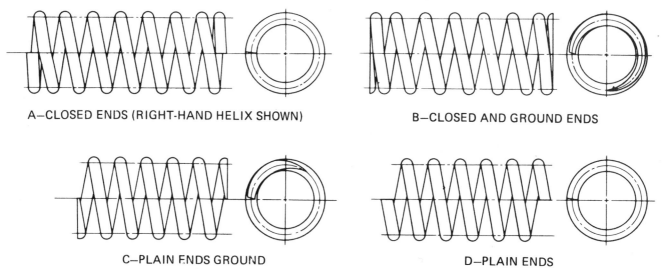

A—CLOSED ENDS (RIGHT-HAND HELIX SHOWN)

B—CLOSED AND GROUND ENDS

C—PLAIN ENDS GROUND

D—PLAIN ENDS

Fig. 3.2—Typical ends of helical compression springs

Coil Diameter (D)

The following procedure is recommended when specifying the coil diameter in the itemized spring specifications.

1. If the spring works over a rod, the minimum inside diameter of the spring should be specified.
2. If the spring works inside a housing, the maximum outside diameter of the spring should be specified.
3. If the spring is confined both internally and externally, the limiting dimensions may be stated in addition to the mean coil diameter.

Some consideration should be given to the spring index (C = ratio of mean coil diameter to wire diameter) in the design. An unusually low index is difficult to manufacture, while a high index will cause large variations in coil diameter. Indexes 3.5-15 are commercially practical to manufacture, but indexes in the range of 5.5-9 are preferred, particularly for close tolerance springs and

those subjected to cyclic loading. For details on coil diameter tolerances, refer to Part C of this chapter.

The spring index is generally not selected at will as it depends upon other spring parameters. Considering only the active portions of a compression spring, there is a direct relationship between the ratio of maximum deflection to solid length and the stress and index. This is expressed by the formula:

$$\frac{F_s}{L_s} = \frac{\chi S_s C^2}{G}$$

Free Length (L_θ)

The free length is an overall dimension measured parallel to the axis of the spring when it is in the free or unloaded state. If definite loads are specified for the manufacture of the spring, the free length should be an approximate dimension which may be varied to meet the load requirements. On the other hand, if a load with a tolerance is not specified, then it is advisable to specify the free length with a tolerance. Detailed information of free length tolerances is given in Part C of this chapter.

Number of Coils (N)

The number of coils should be specified as a reference figure, and it should be stated whether it refers to active (N_a) or total coils (N_t). The maximum number of coils will be restricted when there is a solid length specification. When specifying the number of coils, it is advisable to use total coils rather than active coils. Total number of coils is counted from tip to tip. On springs of great length, it is advisable to specify the pitch or number of coils per 25 mm rather than the total number of coils. Springs with closed ends or with closed and ground ends have one inactive coil at each end. Springs with plain ends are considered to have virtually no inactive end coils unless they are fitted into specially shaped spring seats. Springs with plain ends ground are considered to have about one-half inactive coil at each end; the actual number depends upon the free pitch of the spring and the thickness at the tip after grind. Tolerances to number of coils are outlined in Part C of this chapter.

Helix of Coil

In applications such as one spring operating inside another, it is necessary to coil the springs so that the helices are in opposite directions, right and left. If a spring operates over a screw thread, the direction of the helix should be opposite to that of the thread. In some instances, the direction of the helix may be chosen for part number or source identification. In such special cases, the direction of coil helix should be specified; otherwise, it will be taken as optional.

Solid Length (L_s)

It is desirable that the solid length be specified as a maximum dimension, allowing the manufacturer any tolerance under this dimension determined by variation in wire diameter and the amount of grind on the ends.

Load (P)

It is the force required to compress the spring to a specified loaded length. It may be used as a resisting force to absorb the energy of a blow or shock, or to exert a force as in the case of a valve gear. In some cases, load is static and is used to exert force at a given point. Sometimes a spring load is utilized for an energy source to actuate parts of a device rather than to offer resistance.

The axis of a helical compression spring should be positioned vertically when the spring is being checked for load. The readings should be made during the unloading which will produce a lower load due to the spring hysteresis.

Load should be specified at a definite compressed length of the spring and not at some deflected distance from the free length, as the free length should be a variable dimension. Tolerances should be applied to the load and not to the length whenever possible. Load tolerances are explained in Part C of this chapter.

Rate (R)

It is the change of load per unit length of spring deflection as the spring is being compressed. Rate should be determined between 20 and 60% of total deflection when test lengths are not otherwise established. If even coil spacing or pitch is used, a constant rate develops. In a compression spring with closed ends, the rate is not constant throughout the entire deflection of the spring. This is due to a portion of the active coil adjacent to the closed ends closing down as the deflection nears the solid compressed length of the spring, thus increasing the actual rate of the spring due to the decrease in the number of active coils.

It is possible to design a spring with a definitely increasing rate by using a variable pitch of coils or a variable diameter spring; however, a helical compression spring cannot be designed with a decreasing rate. Rate tolerances are outlined in Part C of this chapter.

B. HELICAL EXTENSION SPRINGS

1. General

An extension spring is a particular type of helical spring, usually close wound, which offers resistance to a tension force applied through suitable end forms. Simple hooks and loops serve the purpose most frequently. A distinguishing characteristic of the conventional extension spring is the presence of initial tension, a force in the close wound coils which must be overcome by a pull applied at the ends of the spring before the coils will separate.

Music wire, hard drawn spring wire of various kinds, and pretempered carbon and alloy wire are the materials used in extension springs. Annealed or untempered wire which requires hardening and tempering after coiling is rarely used for conventional extension springs, because during the hardening process the springs lose their initial tension.

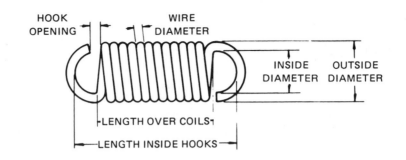

TYPICAL ENDS USED ON EXTENSION SPRINGS

MACHINE CUT
PLAIN ENDS

SINGLE FULL LOOP
OVER CENTER

 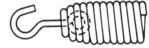

SMALL OFFSET
HOOK AT SIDE

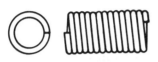

MACHINE HALF HOOK
OVER CENTER

DOUBLE FULL LOOP
OVER CENTER

CONED END WITH
SWIVEL BOLT

HALF LOOP
OVER CENTER

FULL LOOP
AT SIDE

CONED END WITH
SWIVEL HOOK

LONG ROUND END
HOOK OVER CENTER

SMALL EYE
AT SIDE

THREADED PLUG TO FIT
PLAIN END SPRING

V–HOOK
OVER CENTER

EXTENDED EYE
OVER CENTER

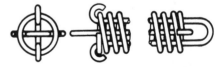

DRAWBAR SPRING

Fig. 3.3—Extension springs

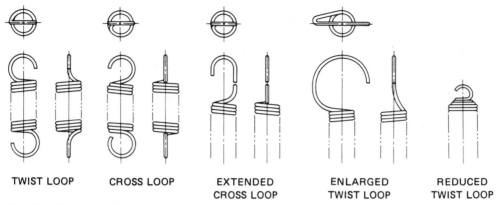

TWIST LOOP   CROSS LOOP   EXTENDED CROSS LOOP   ENLARGED TWIST LOOP   REDUCED TWIST LOOP

Fig. 3.4—Typical machine made loops

2. Types of Ends

The sketches in Fig. 3.3 show some of the many types of ends which have been used on extension springs. Details such as hook opening or restraint of the loop within the body diameter should be specified on the drawing because the word descriptions in Fig. 3.3 are not universally standard. There is no limit to the designs of loops which may be developed for a particular installation. In general, the machine hooks and machine loops are the preferred designs of those shown in Fig. 3.3.

Many extension springs of small wire size can be looped automatically on a coiling machine. Fig. 3.4 illustrates typical loops which can be made complete on modern spring making machinery.

The cross loops made by this technique are free from the mutilated portions which reduce the endurance of this style of loop as made by other techniques.

3. Stress Concentrations in End Coils

When sharp bends are made in forming the end hooks or loops of extension springs, stress concentrations are produced which may cause failures. An adequate estimate of the stress concentration at the bend in the end hook or loop may be found by multiplying the bending or shear stress obtained with the ordinary design formulas by the ratio r_0/r_1, where r_0 is the radius of the center line of the bend and r_1 is the inside radius of curvature of the wire at the bend, as illustrated in Fig. 3.5. For best results, this ratio should not be permitted to exceed 1.25; in other words, the inside radius of curvature should be at least twice the diameter of the wire.

Coned ends with swivel eyes, hooks or bolts, as illustrated in Fig. 3.3, provide effective means for reducing the stress in the end coils of the spring. Another way to avoid stress concentration in the end coils is to obtain

$$S = \frac{8\,PD}{\pi\,d^3}\left(\frac{r_0}{r_1}\right)$$

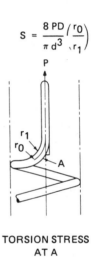

TORSION STRESS AT A

Fig. 3.5—Stress concentration at end hook bend

an extension spring action by combining a compression spring with two yoke-like drawbars. The limit imposed upon deflection by the definite solid length of the compression spring prevents excessive extension and provides a substantial measure of safety for the spring in this arrangement. Extension springs may be made by screwing threaded plugs into the ends of plain machine-cut springs, the threaded plug being equipped with a hook of substantial proportions to minimize the effects of stress concentration.

4. Position of Hooks

The position of hooks relative to each other can be in line, at right angles, or at any other angular position as required. The hooks on springs with many coils may change position somewhat in shipment, and the variation of position is also affected by other variations during manufacture. Hence, in order to maintain the position of hooks relative to each other, it is necessary to exercise special care in manufacturing. If the position of ends relative to each other is important, the drawing of the spring should emphasize the importance of position by a statement as well as by pictorial representation.

5. Specifications

The drawing specifications for an extension spring should include the coil diameter, the wire diameter, the free length, some reference to number of coils, and full particulars concerning the type of ends. Coil diameter is usually specified as outside diameter. Free length should be measured inside hooks. Either the number of coils in the body of the spring or the length over the coils may be specified, but only as an approximate figure. In computing the length over coils, it should be recognized that there is always one more wire diameter in the length than the number of coils in a close wound spring. Full hooks and loops are the same diameter as the body coils, and the length required for two ends is approximately the same as two diameters of the body coils.

When load is specified, the extended length inside hooks under the load should be given. The manufacturing tolerance should be applied to the load rather than to the extended length. Initial tension should be controlled by including it in the test load rather than specifying it independently. If load and tolerance are specified at a particular extended length, the free length inside hooks should be shown as an approximate dimension.

Extension springs normally do not have a definite stop to their deflection such as the solid compressed length of a compression spring. The drawing specifications, therefore, should include a statement of the maximum extended length which must be attained without encountering permanent set.

6. Initial Tension

The initial tension that can be wound into an extension spring depends upon the index and the winding technique. Fig. 3.6 shows a range of practical values of initial tension in terms of stress. The corresponding load should be figured from the load-stress formula without any correction factor. It is difficult to coil springs with a very small amount of initial tension, because some springs will then have some open wound coils. Very high values of initial tension may require special coiling techniques and may increase cost.

Normally, the initial tension varies slightly from coil to coil, so it is necessary to extend the spring an appreciable amount before overcoming all of the initial tension.

A practical method of measuring the initial tension in a spring is as follows:

1. Extend spring some unit of length until all coils are open, then measure load.
2. Extend spring another equal unit of length and measure load.
3. Subtract the first load from the second load to obtain the scale for the selected unit of length.
4. Subtract the scale for the selected unit of length from the first load to determine the initial tension.

After separation of all the coils, the slope of the load-deflection diagram is the same as for an open wound spring.

7. Spring Design

The design of extension springs, apart from the need for making allowance for initial tension, differs little from the design of compression springs. The end construction affects the deflection of the springs, and spring calculations should include an allowance for the effect of end deflection. In the case of full hooks or loops at each end, the total number of active coils for rate computations can be taken as the number of body coils plus one to allow for end effect.

The stress levels for extension springs are substantially lower than the stress levels for compression springs. There are several reasons for this. In the first place, the torsional stress in the body of the spring must be low to avoid high bending stresses in the loops. Second, extension springs are normally made without residual stress, so the limiting factor is the torsional elastic limit rather than the high apparent stress which results from taking out set in a compression spring. However, in cases where higher stresses are needed, it is possible to coil the spring with excess initial tension and then stretch it, removing initial tension and inducing residual stresses. If this practice is used, the remaining initial tension cannot be high, and it may be necessary to use coned ends or inserted hooks to avoid excess stress in the loops. Without residual stress, the maximum stress may be a compromise with the requirement for initial tension. Normally, springs are given a low temperature heat treatment to obtain the maximum elastic limit, but this heat treatment may reduce the initial tension in an extension spring by 25-50% while increasing the elastic limit by a similar amount.

Still another reason for lower stresses is the hazard that an extension spring may be extended beyond its design length either in assembly or by careless handling. Loops on the side will cause a higher stress than loops over center which apply the load axially.

Considering all of these uncertainties, a reasonable uncorrected stress level is 35% of tensile strength for ferrous materials and 30% of tensile strength for nonferrous and austenitic stainless steel for static or mild service. Prediction of fatigue life is very difficult because of fur-

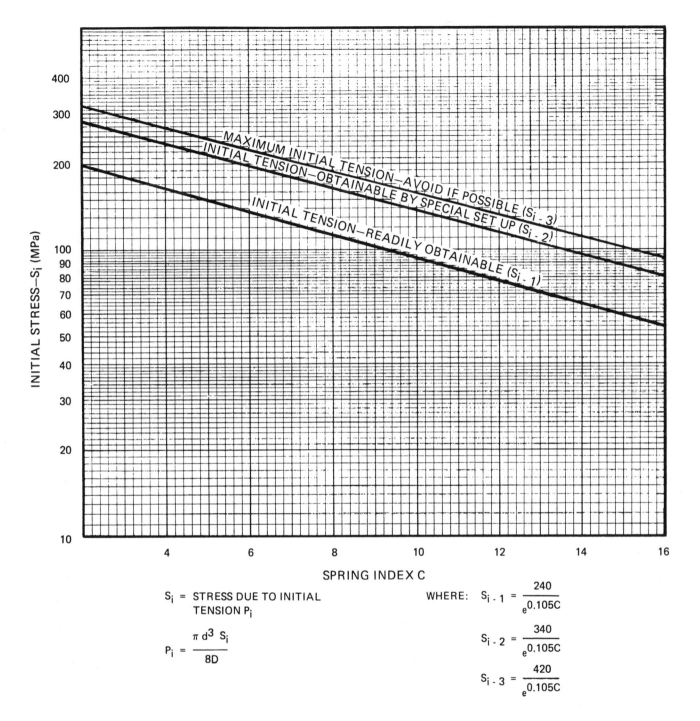

S_i = STRESS DUE TO INITIAL TENSION P_i

$$P_i = \frac{\pi d^3 S_i}{8D}$$

WHERE: $S_{i\text{-}1} = \dfrac{240}{e^{0.105C}}$

$S_{i\text{-}2} = \dfrac{340}{e^{0.105C}}$

$S_{i\text{-}3} = \dfrac{420}{e^{0.105C}}$

Fig. 3.6—Initial tension in extension springs

ther complications, such as surging caused by impact loads and increased stress which results from bowing of the spring, if it operates in the horizontal position. Extension springs have many advantages, but experimental work and testing are necessary to utilize their maximum potential.

C. TOLERANCES FOR HELICAL COMPRESSION AND EXTENSION SPRINGS

1. General

The tolerances in this section apply to cold coiled springs made of round wire. When applied with sound

2.37

judgment, they represent economical limits for good commercial springmaking practice. Closer tolerances can be maintained, but usually with increased cost.

The perfect universal tolerances system for cold coiled helical compression and extension springs has not been devised. That it will ever be is unlikely.

There are many reaons for this situation. One is the peculiar interdependence that each dimension of a spring has upon its other dimensions and upon its load-deflection characteristic. Another reason is that certain aspects of specifying spring tolerances can have a great influence upon the cost of springs. If the designer is unaware of them, or chooses to disregard them, he may design considerable extra cost into a spring without increasing its value.

The wire which spring makers use in the manufacture of cold coiled helical springs is purchased to standards or specifications that include tolerance for the diameter and mechanical properties of the wire. Variation within the limits of these tolerances causes significant variation in the dimensions and load-deflection characteristics of springs.

This variation in wire is one reason why spring makers recommend that designers apply realistic tolerance only to mandatory requirements and specify other requirements only as approximate, suggested, or limiting values.

The designer should specify either free length, solid length, or loads at specific lengths. If more than one of these is specified, the secondary requirements will have a wider tolerance which includes the effect of the spring rate tolerance.

When the above practice is followed, the spring maker has the opportunity to make compensating adjustments in the noncritical dimensions in order to maintain the mandatory requirements. Without this freedom for adjustment, the use of special controls or salvage operations may be required, which would increase manufacturing costs.

The magnitude of the manufacturing tolerances required for a spring of a given design depends upon many factors. Sometimes a customer allows insufficient space for a satisfactory spring. This results in a spring design that is overstressed and allows very little manufacturing variation. Tolerances in such a case are severely restricted by the application, and cost must become a secondary matter.

Generally, cost is a major consideration, and springs must be produced in a most economical manner. Specified tolerances need to be generous enough to allow acceptable springs to be made using ordinary methods.

Those types of wire made to close tolerances, such as music wire and carbon steel valve spring quality wire, tend to reduce variation in spring dimensions and permit the holding to closer tolerances than is the case when the commercial grades of wire are used.

Compression springs designed to high stresses, within safe stress limits, tend to vary less than springs designed to low stresses.

The degree of nonconformance which the customer's receiving inspection will accept can be a vital factor in the cost-tolerance relationship. When the spring maker is aware of the customer's inspection and acceptance practice, he can adjust his control measures accordingly to meet the required quality level at lowest possible cost.

The above is particularly true for large production runs which allow time for process refinement and the effective application of statistical control. When this can be done, it tends to reduce variation in spring dimensions to the minimum, and very close tolerances can be met.

Many drafting rooms use standard drawing forms having a tolerance box for machine dimensions. These box tolerances are almost always impracticable for springs. It is strongly recommended, when drawing forms with box tolerances are used for springs, that the tolerance box be crossed out and that realistic spring tolerances be specified for mandatory requirements.

Figs. 3.7-3.13 give tolerances for dimensions, load, and rate for compression and extension springs.

The diameter and out-of-round tolerance for the common grades of spring wire can be found in the SAE Handbook under the section entitled Spring Wire and Springs.

2. Coil Diameter

Coil diameter tolerance may be specified on either the inside or outside coil diameter. The tolerances in Figs. 3.7 and 3.8 are functions of the wire diameter and the ratio of mean diameter to wire diameter. The tolerances are to be considered as manufacturing tolerances and do not take into account changes in diameter due to applied loads. Fig. 3.7 gives coil diameter tolerance for springs having wire diameters in the range 0.30-9.50 mm. Fig. 3.8 gives coil diameter tolerance for springs having wire diameters in the range 9.5-16.0 mm.

3. Free Length

The free length tolerances specified in Figs. 3.9 and 3.10 can be maintained, in most cases, unless changes in free length are required to compensate for other variables which affect load.

Therefore, when a load with tolerance is not specified, it is advisable to specify the free length with tolerance for control and inspection purposes. Otherwise, it is recommended that the free length be a reference dimension.

In the case of compression springs, the free length is an overall dimension measured parallel to the axis of the spring. For extension springs, the free length refers to the

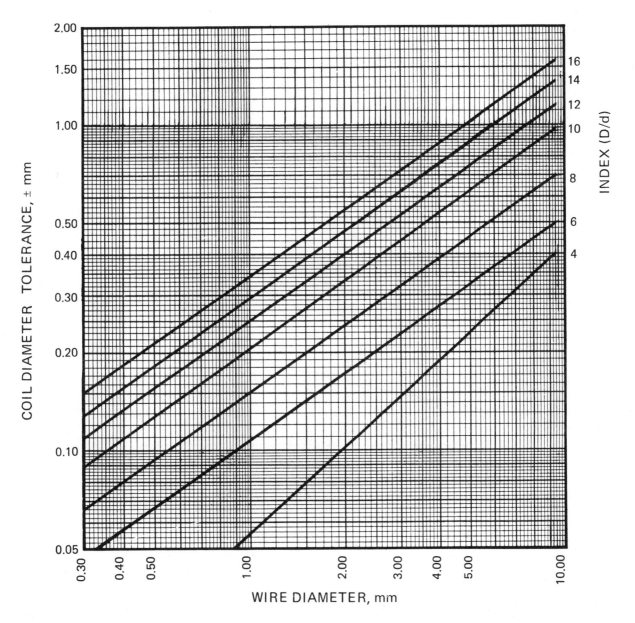

Fig. 3.7—Coil diameter tolerance—compression and extension springs for wire diameters 0.30 to 9.50 mm. Round off index to nearest whole number. Interpolate when the rounded-off value is an odd number. Use tolerance for 0.30 mm wire diameter when wire diameter is less than 0.30 mm.

length inside to inside of the hooks. Free length tolerances for compression springs (expressed as mm per mm of free length) are shown in Fig. 3.9 as functions of the number of active coils per mm of free length and the spring index.

Free length tolerances for extension springs expressed as functions of the free length appear in Fig. 3.10.

4. Load

Load tolerances for compression springs, as functions of the nominal free length tolerance of Fig. 3.9 and the deflection from free length, are shown in Fig. 3.11.

The normal load tolerance for an extension spring is equal to the product of the appropriate tolerance factor from Fig. 3.12A and the appropriate multiplying factor from Fig. 3.12B. Tolerances determined in the above manner are functions of the spring index, the ratio of free length to deflection, and the wire diameter.

5. Rate

When spring rate is a functional requirement, it should be specified with tolerance. Normal rate tolerances are shown in Fig. 3.13 as functions of the number of active coils.

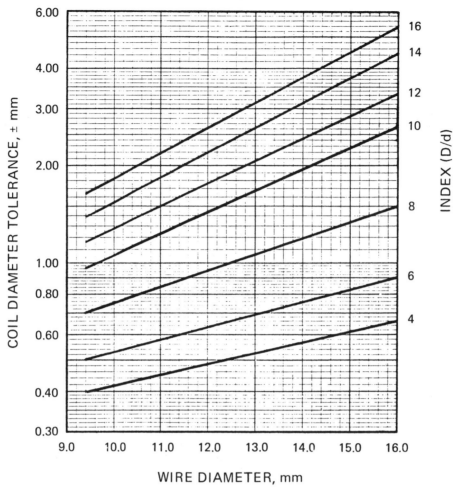

Fig 3.8—Coil diameter tolerance—compression and extension springs for wire diameters 9.5 to 16.0 mm. Round off index to nearest whole number. Interpolate when rounded-off value is odd number.

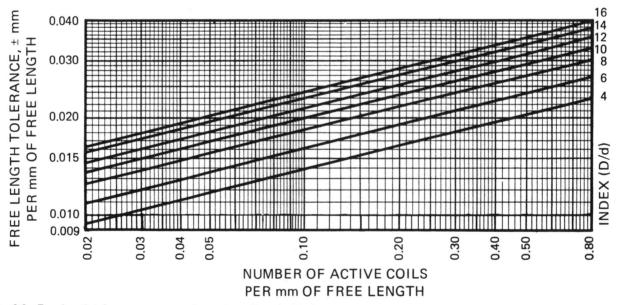

Fig. 3.9—Free length tolerance—compression springs. Round off index to nearest whole number. Interpolate when rounded-off value is odd number. These are tolerances for springs with ends closed and ground. For springs with ends closed but not ground, multiply by 1.7.

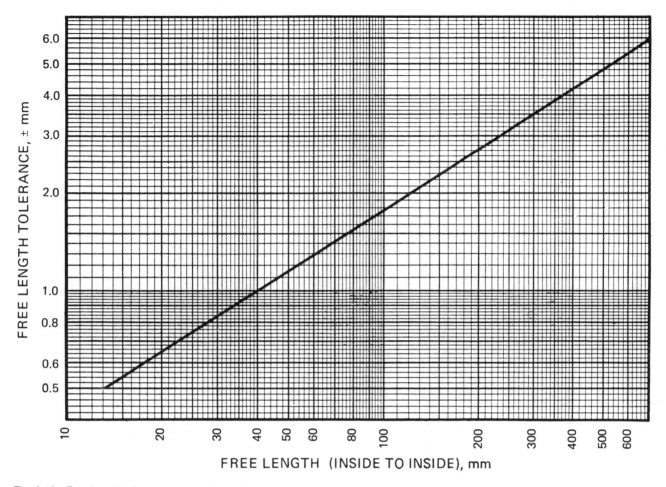

Fig. 3.10—Free length tolerance—extension springs

6. Solid Length

If the working range of a compression spring approaches solid compression, it is advisable to specify the solid length as a maximum. If the spring ends are closed and ground, the lowest specified maximum solid length should be not less than the product of the maximum allowable number of coils and the maximum allowable wire diameter.

If the ends are not ground, the lowest specified maximum solid length should be not less than the product of the maximum allowable number of coils, plus one, and the maximum allowable wire diameter.

When the wire diameter is less than 2.50 mm, it is advisable to add to the solid length an additional amount equivalent to one-half wire diameter to compensate for thick ends which occasionally occur during production end-grinding operations. Failure to make this allowance could require a costly sorting operation after grinding.

Finishes such as paint or plating, when applied to springs, significantly increase the solid length of the springs and must be allowed for when calculating solid lengths.

7. Number of Coils

Tolerance on the number of coils is given in Table 3.1 for compression springs and in Table 3.2 for extension springs. The tolerance is expressed in degrees as a function of the number of active coils.

8. Squareness and Parallelism of Ends

The squareness of the ends of compression springs having ends that are closed and ground should be within 3 deg of the spring axis when measured in the unloaded position.

The ends shall be parallel within a limit equal to twice the squareness of the ends.

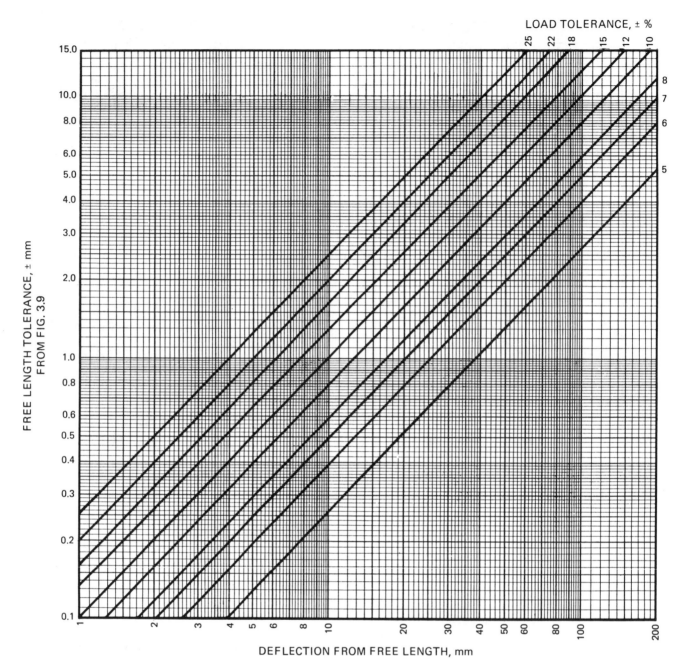

Fig. 3.11—Load tolerance—compression springs, ±%. Enter chart from bottom with deflection from free length to loaded length and from left with free length tolerance of Fig. 3.9. Round off percent load tolerance values to next larger whole number. Interpolate when rounded-off value is odd and between 8 and 25%.

TABLE 3.1—NUMBER OF COILS TOLERANCE OF COMPRESSION SPRINGS

Active Coils	Tolerance ± deg
3-10	45
For each additional 10 coils, add	30

TABLE 3.2—NUMBER OF COILS TOLERANCE OF EXTENSION SPRINGS

Active Coils	Tolerance, ± deg	
	Close Wound	Open Wound
3	30	90
4-10	45	90
For each additional 10 coils, add	15	30

2.42

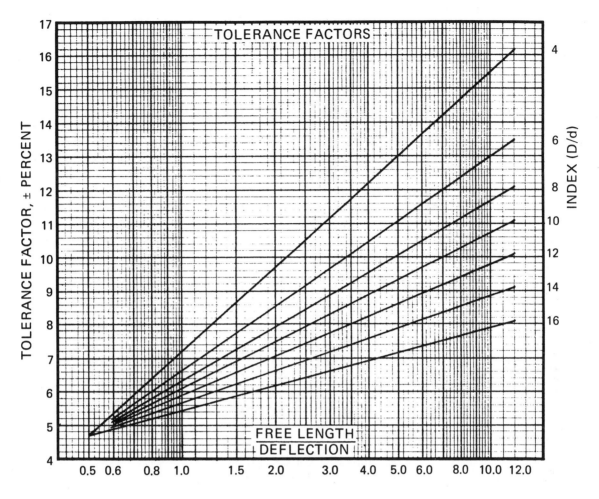

Fig. 3.12 A

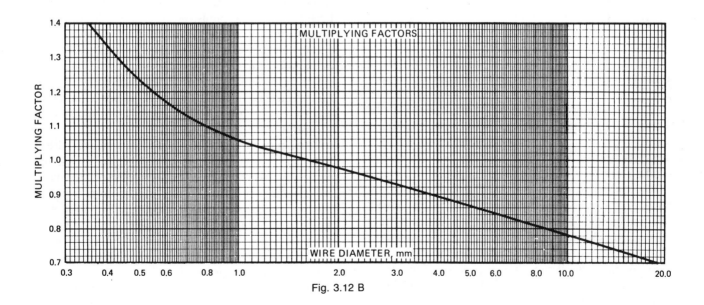

Fig. 3.12 B

Fig. 3.12—Load tolerance — extension springs, ± %. To find load tolerance, multiply tolerance factor from Fig. 3.12A by multiplying factor from Fig. 3.12B.

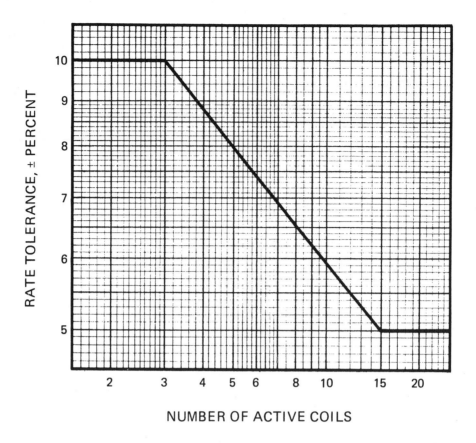

Fig. 3.13—Rate tolerance—compression and extension springs. Rate and rate tolerance should be specified only when rate is functional. When rate specification is necessary, range of deflection over which it is to apply must be clearly identified. Deflection ranges for rate control should fall within 20-60% limit of total deflection because rate is likely to be variable outside this range.

D. HELICAL TORSION SPRINGS

1. General

Torsion springs store energy or offer resistance to an applied torque when subjected to an angular deflection. This deflection results in an increase in the number of coils in the spring and a relative decrease in the spring diameter. The term "torsion spring" may be confusing, since compression and extension springs are subjected to torsional stresses, while torsion springs are subjected to bending stresses.

The contour of a torsion spring is usually cylindrical; however, it can be conical, hourglass, or barrel-shaped. The use of hourglass or barrel-shaped torsion springs makes it possible to obtain an increasing rate. Assembly of these springs over a rod of the same size as the inside diameter of the smallest coil results in some of the spring material becoming inactive during deflection, thus increasing the spring rate.

Torsion springs may be either close wound or open wound. Open wound springs are recommended wherever possible, since close wound springs produce frictional loads which are difficult to evaluate. When an open wound torsion spring is used in a confined space, the length of this space must be adequate to accommodate both the increased number of coils due to winding of the spring and the variations in the original number of coils in manufacture. Close wound torsion springs increase in length as they are wound up. Clearance must be allowed for this increase in length and for variations in the springs; otherwise, the springs will bind, and breakage or setting will result.

Torsion springs may be coiled from annealed wire and fully heat treated after coiling for certain applications, particularly where unusually difficult end coil formations are specified. It is by no means uncommon to produce heavy torsion springs from large round or square bars which are coiled hot, as described in detail in Chapter 4 which relates to hot coiled springs. However, it is by far the most general practice to coil torsion springs from either hard drawn or oil tempered wire.

Square and rectangular wire cross sections are frequently used in torsion springs due to their greater material efficiency compared with that of round wire. However, the unit cost of square and rectangular wires is generally greater than that of round wire, and, in consequence, the use of round wire springs usually results in a lower net material cost. It is to be appreciated that "keystoning" will result when square and rectangular wires are coiled and that allowances for the effect should be made, as discussed in Chapter 5, Section 6.

2. Effect of Direction of Coiling and Beneficial Residual Stress

Torsion springs should be designed so that torque is applied in a direction which tends to reduce the coil diameter. Cold wound helical torsion springs made from hard drawn or pretempered wire must be coiled to a smaller diameter than that desired in the finished spring, to allow for recoiling when released from the coiling arbor. This causes a tensile residual stress to be set up on the inside of the coil and a compressive residual stress on the outside. If the spring is then loaded by a torque so as to reduce the coil, it is clear that these residual stresses will be subtracted from the maximum load stresses near the outer fibers. These beneficial bending stresses are analogous to the effect of presetting on torsional stress distribution over the cross section of helical springs. To take advantage of these residual stress patterns, torsion springs should therefore be loaded in the same direction they were coiled.

3. Support of Torsion Springs

Torsion springs can be wound without appreciable set until the inside diameters of the coils are only slightly larger than the inside diameter to which they were originally coiled before recoiling from the arbor. Winding to any smaller diameter would be equivalent to recoiling from a smaller mandrel than that used for the original coiling and would produce substantial set. For this reason, it is necessary to support adequately torsion springs which are coiled in such a direction that the diameter is reduced by the applied torque. The most common way to effect adequate support is to insert a rod through the spring, and experience shows that the diameter of this rod should be approximately 90% of the smallest inside diameter to which the spring is reduced when under maximum load or travel. Insufficient support on the inside is conducive to early failure.

The size of the supporting rod, as indicated above, is affected by the decrease in the coil diameter due to the angular deflection of the spring. The reduced mean diameter may be determined by the following formula:

$$D' = \frac{D\,N_a}{N'_a}$$

where:

D = free mean coil diameter
D' = reduced mean coil diameter
N_a = free number of active coils
N'_a = increased number of active coils

4. Types of Ends and End Stresses

The variation in torsion spring ends is almost limitless, but a few of the more common types are illustrated in Fig. 3.14. The double torsion type of spring shown is necessary for some applications, but it is expensive to make. In many cases, it may be less costly to use two springs instead, one left-hand and the other right-hand.

When torsion springs are used with eyes at the ends or with bends off the coil, special care should be taken to keep stresses low. When bends are made at smaller radii and then loaded, the stress becomes tensile at the inside of the bend and may be much greater at that point than the stress in the coil. Due to the sharp curvature, the neutral axis moves in toward the center of the bend and the tension stress becomes that of a cantilever loading multiplied by a constant greater than unity, depending upon the degree of curvature. A close approximation of the peak stress at such bends is given by:

$$S_e = S\,\frac{r_0}{r_1}$$

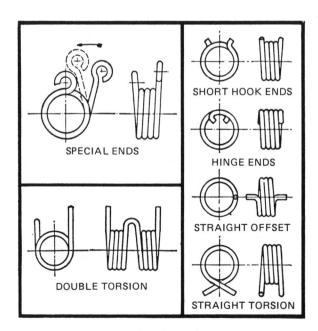

Fig. 3.14—Details of helical torsion springs

SPECIAL ENDS

DOUBLE TORSION

SHORT HOOK ENDS

HINGE ENDS

STRAIGHT OFFSET

STRAIGHT TORSION

TABLE 3.3—HELICAL TORSION SPRING FORMULAE

	Round Wire	Square Wire	Rectangular Wire
$M =$	$\dfrac{E\,\theta\,d^4}{3672\,N_a\,D}$	$\dfrac{E\,\theta\,t^4}{2160\,N_a\,D}$	$\dfrac{E\,\theta\,b\,t^3}{2160\,N_a\,D}$
$S =$	$\dfrac{10.2\,M}{d^3}$	$\dfrac{6.0\,M}{t^3}$	$\dfrac{6.0\,M}{b\,t^2}$
$M/\theta =$	$\dfrac{E\,d^4}{3672\,N_a\,D}$	$\dfrac{E\,t^4}{2160\,N_a\,D}$	$\dfrac{E\,b\,t^3}{2160\,N_a\,D}$

where:

b = axial dimension
t = radial dimension

where:

S_e = peak stress, MPa
S = stress from torsion spring formula, MPa
r_o = radius of bend to center line of wire, mm
r_1 = radius at inside of ends, mm

5. Specifications

Torsion spring applications are seldom alike and have no standard forms. It is recommended that all inquiries and orders for torsion springs be accompanied by drawings clearly illustrating the types of ends, the diameter of the rod over which the spring must work (if one is used), the length of space available, and the position of the ends at maximum working deflection. In addition, the torque in newton-millimeters at loaded position and the angular position of the ends under this torque should be given. Particularly important is whether the spring should be left- or right-hand wound.

6. Design Formulae

The constants in the formulae of Table 3.3 are derived from the theory of the bending of straight bars. Tests may show appreciable differences from calculations as discussed in Part D, Section 8 of this chapter.

If the length of the ends is appreciable in comparison to the body of the spring, this length may be converted into the number of coils and included with N_a in the load deflection formulae.

7. Design Stresses

The maximum design stresses allowed for static loading with various commercially available spring materials can be established as a percentage of the tensile strength of the material, as shown in Table 3.4. The end stress should also be considered in any torsion spring design, as discussed in Part D, Section 4 of this chapter.

8. Commercial Tolerances

In the accompanying outline of tolerances for torsion springs, no load tolerances are listed. There are several reasons for this. Standards have not been set up as to what size arbor a spring of given ID is to be tested on. If and when the arbor size is agreed upon, then the amount of coil tension must be set, as this also affects friction and hence input and output of the springs. So far, no accepted standard test machine has been developed for universal use as has been done in weighing devices used on extension or compression springs. In order to test torsion springs, the vendor and customer must develop their

TABLE 3.4—ALLOWABLE DESIGN STRESSES AS A PERCENTAGE OF TENSILE STRENGTH FOR VARIOUS MATERIALS

SAE No.	Type of Wire	% of Tensile Strength
J132	Oil tempered alloy wire	77
J316	Oil tempered carbon wire	76
J178	Music	78
J113	Hard drawn carbon steel	68
	Hard drawn stainless type 302	65
	Phosphor Bronze spring wire	70

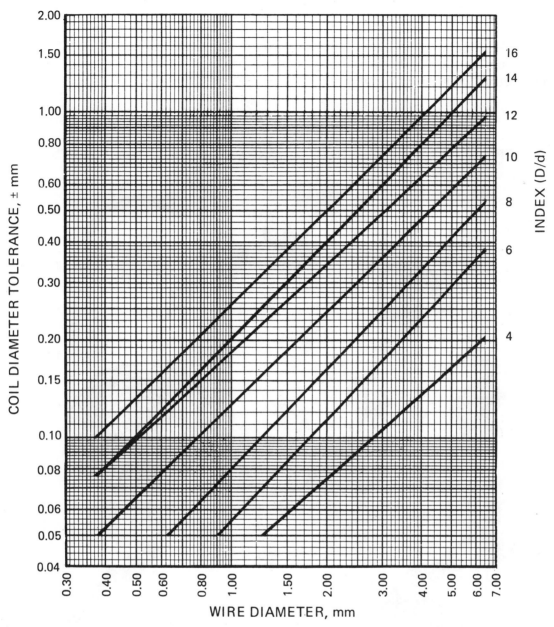

Fig. 3.15—Commercial coil diameter tolerances for helical torsion springs with wire diameters 0.30 to 7.00 mm. Round off indexes to nearest whole number. Interpolate when rounded-off value is odd number.

own testing technique on the spring in question, making such allowances for friction, end position, diameter, and wire size as are inherent in the product. Such a procedure makes load testing of each spring type come under a different method. In the present state-of-the-art, this will have to be endured, as there is no remedy yet in sight.

Coil Diameter

Coil diameter tolerances can be specified on either the inside or the outside diameter of the coils, depending upon the importance of the respective dimensions to the uses. These tolerances apply to torsion springs made of 7.00 mm or smaller wire size on torsion coiling machines. They are given as functions of the ratio of the mean diameter D to the wire diameter d, as shown in Fig. 3.15. These tolerances are to be considered as manufacturing tolerances and do not take into account changes in diameter resulting from angular deflection.

Wire Diameter

For wire diameter tolerances, refer to the SAE Handbook under the section entitled Spring Wire and Springs.

2.47

Position of Ends

The ends or position of the arms with respect to the coil of a torsion spring are subject to the tolerances shown in Table 3.5 up to and including a D/d ratio of 16.

TABLE 3.5—TOLERANCE FOR POSITION OF ENDS

Total Number of Coils	Tolerance, ± deg
Up to 3	8
Over 3 thru 10	10
Over 10 thru 20	15
Over 20 thru 30	20
Over 30	25

Fig. 3.16—Flat spiral springs

E. FLAT SPIRAL SPRINGS

1. General

A flat spiral spring consists essentially of flat spring material wound on itself with open space between the coils in the free position. These springs are fundamentally made to deliver a torque, but this may be translated to linear action by means of linkages. They provide a compact means for storing energy and, therefore, are often used to produce force or motion in preference to other types of springs.

Flat spiral springs are very widely used for locks and window counterbalances in automobiles and for maintaining pressure on carbon brushes of electric motors and generators. Typical examples are shown in Fig. 3.16.

Another closely related type of spring is identified as a clock or motor spring. In this style of spring, the length of material is so great that it is confined in a case or drum, and a majority of the coils are contacting each other in the unwound position. As the spring is wound, the active length changes continuously so the load-deflection curve is nonlinear. The design of this style of spring is a specialty which will not be covered in this manual. It is recommended that the manufacturer be consulted in design problems of this spring style.

2. Spring Materials

The material for flat spiral springs is described as flat wire in Chapter 2. By far the greatest proportion is made from oil tempered wire, although for certain applications, the springs are formed from untempered or annealed material and are fully heat treated after forming. Special applications occasionally require the use of stainless steel, nickel alloys, or nonferrous materials.

3. Design Formulae

The formulae used for this type of spring are similar to those found under the torsion spring section:

$$\theta = \frac{687\ ML}{Ebt^3} \qquad S = \frac{6M}{bt^2}$$

Variations may be expected from the calculated deflection values for several reasons. If the material is rolled from round wire so the edges have small radii, the load may be about 5% weaker than predicted. If the outer end is free to swivel about a pin, the load may be weaker than with a clamped end. If the active length varies as the spring contacts the arbor, or if the coils touch, the rate will increase. If only one load test is specified, it is customary for the manufacturer to vary the free position to compensate for these deviations. If two load tests are specified, it is necessary to vary the developed length to meet a rate requirement.

The arbor size and free outside diameter do not affect the load-deflection characteristics very much, but they do affect the total available deflection. If the arbor size is large and the outside diameter is small, the overlaps may be so close together that very little deflection is available. A rigorous equation for total deflection would be very complex. The following approximation equation is useful as a guide:

$$\text{Number of available turns} < \sqrt{\frac{D_a^2 + 1.27Lt - D_a}{2t}} - \frac{2L}{\pi(D_o + D_a)}$$

The equation assumes uniform spacing of successive coils and does not account for the extra thickness of material at the first overlap. Springs made from tempered material normally have a slight increase in spacing as the coils get larger. This gives less total deflection than uniform spacing. Therefore, the total deflection will be less than the formula indicates.

Because of the large amount of residual stress, the apparent stress for static applications may be as high as the tensile strength of the material. Fig. 3.17 shows the stress versus thickness for steel springs with maximum residual stress, which is normal when the springs are made from tempered material.

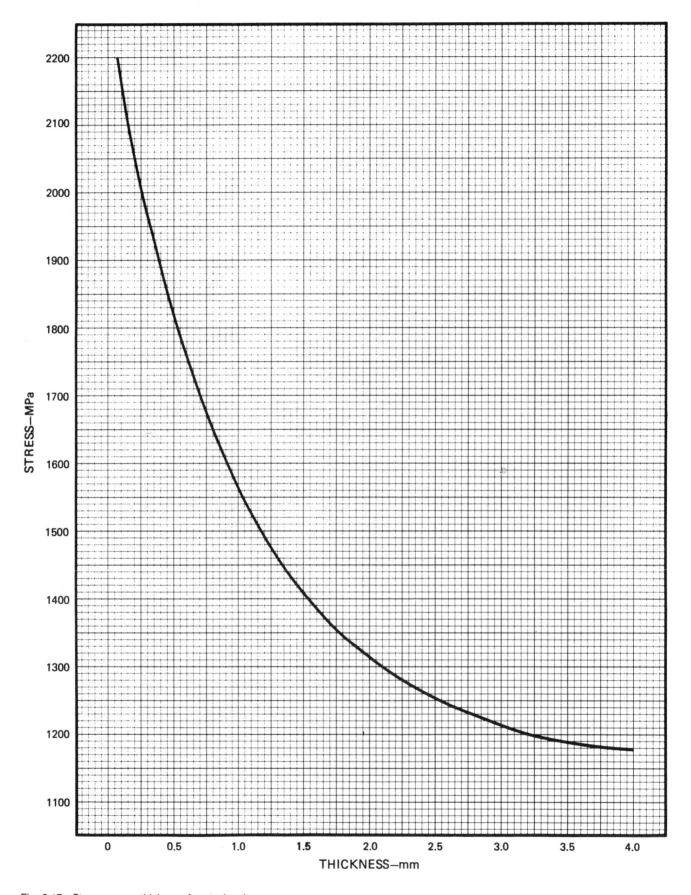

Fig. 3.17—Stress versus thickness for steel springs

2.50

Chapter 4

Hot Coiled Helical Springs

1. General

Definitions

Helical compression and extension springs are normally cold wound when the material is less than 10 mm in diameter (see Chapter 3), and hot coiled when it is greater than 16 mm. When the cross section is between 10 and 16 mm, both processes are common.

Hot coiled springs find extensive use in many fields of heavy industry and particularly in automotive, railway, agricultural implement, construction equipment, and ordnance applications. Basically, they are designed according to the general helical spring formulas of Chapter 5, but some of their features require special attention.

Spring steel bars are available in carbon and alloy analyses. The bars are generally used in the as-rolled condition, but may be centerless ground before coiling. Table 2.2 lists some materials commonly used in hot coiled springs.

2. Heat Treatment

In a typical hot coil spring operation the bar is heated, coiled, and quenched in a continuous operation. Care must be used to prevent excessive temperature before coiling, while maintaining adequate temperature for quenching. After quenching, the spring is tempered to produce the specified hardness. It is important that the finished spring have a fine grained structure and a surface free from excessive pitting, roughness, or other defects.

It is sometimes desirable to heat treat a spring that has cooled below the critical temperature for quenching. This may occur when a spring receives secondary operations after coiling or as part of a salvage operation. This is sometimes referred to as double heat treat and involves heating the spring to the proper temperature, quenching,

and tempering. This process will produce satisfactory springs, but care must be used to prevent excess scale and decarb during two heating operations.

Additional information can be found in SAE J412, "General Characteristics and Heat Treatments of Steels" and SAE J413, "Mechanical Properties of Heat Treated Wrought Steels."

3. Shot Peening

Shot peening is used to increase the fatigue strength of springs, thereby helping to reduce their size and cost. Refer to Chapter 1, Section 5 for details. Hot coiled springs need shot peening to remove scale in addition to its other benefits.

4. Presetting

Hot coiled compression springs normally receive a manufacturing operation commonly called presetting. This process consists of coiling a spring to a free length greater than specified free length. Following heat treatment, the spring at room temperature is compressed solid or to a specified preset length, to produce yielding. For high stress springs, the preset stress (corrected with Wahl factor) is usually 1100-1150 MPa. Presetting is covered in more detail in Chapter 1, Section 5.

5. Warm Setting

In order to reduce the "sag" or "settling" of helical suspension springs which occurs when they are subjected to vehicle loading over time, it has become common practice to warm set—or hot press—the springs at an elevated temperature (usually in the range of 120°C-260°C depending on the particular spring design). One theory holds that the major benefit of this operation results from

TABLE 4.1—PREFERRED
DIAMETERS OF ROUND
STEEL BARS (ANSI B32.4)

mm	mm	mm	mm
9.0	18	30	55
9.5	19	32	60
10.0	20	35	65
11.0	21	36	70
12.0	22	38	75
13.0	23	40	80
14.0	24	42	85
15.0	25	45	90
16.0	26	48	95
17.0	28	50	100

TABLE 4.2—CROSS SECTION TOLER-
ANCES FOR HOT ROLLED CARBON AND
ALLOY STEEL ROUND BARS

Specified Diameter, mm		Tolerance, Plus and Minus, mm	Out of Round, mm
Over	Thru		
—	8	0.13	0.20
8	10	0.15	0.22
10	15	0.18	0.27
15	20	0.20	0.30
20	25	0.23	0.34
25	30	0.25	0.38
30	35	0.30	0.45
35	40	0.35	0.52
40	60	0.40	0.60
60	80	0.60	0.90
80	100	0.80	1.20

TABLE 4.3—CROSS SECTION TOLERANCES FOR TURNED AND POLISHED
ROUND BARS

Specified Size, mm		Tolerance, Minus Only, mm			
Over	Thru	Maximum of Carbon Range 0.28 or Less	Maximum of Carbon Range Over 0.28 to 0.55 Incl	Maximum of Carbon Range to 0.55 Incl Annealed or Stress Relieved after Cold Finishing	Maximum of Carbon Range Over 0.55. Also all Grades Quenched and Tempered or Normalized and Tempered, before Cold Finishing
Carbon Steel					
—	40	0.05	0.08	0.10	0.12
40	60	0.08	0.10	0.12	0.15
60	100	0.10	0.12	0.15	0.18
100	150	0.12	0.15	0.18	0.20
150	200	0.15	0.18	0.20	0.22
200	225	0.18	0.20	0.22	0.25
225	—	0.20	0.22	0.25	0.28
Alloy Steel					
Coils					
—	25	0.05	0.08	0.10	0.12
Cut Lengths					
—	40	0.08	0.10	0.12	0.15
40	60	0.10	0.12	0.15	0.18
60	100	0.12	0.15	0.18	0.20
100	150	0.15	0.18	0.20	0.22
150	200	0.18	0.20	0.22	0.25
200	225	0.20	0.22	0.25	0.28
225	—	0.22	0.25	0.28	0.30

TABLE 4.4—CROSS SECTION
TOLERANCES—STEEL BARS (ROUND),
CARBON AND ALLOY: COLD DRAWN,
GROUND AND POLISHED AND TURNED,
GROUND AND POLISHED

Specified Diameter, mm		Tolerance, Minus, mm	
Over	Thru	Cold Drawn, Ground and Polished	Turned, Ground and Polished
—	25	0.03	0.03
25	50	0.04	0.04
50	75	0.05	0.05
75	100	0.08	0.08
100	150	—	0.10[a]
150	200	—	0.12[a]

[a] For nonresulphurized carbon steels (steels specified to max sulphur limits under 0.08%) or for any steel quenched and tempered (heat treated), normalized and tempered, or any similar double treatment prior to turning: the tolerance is increased by 0.03 mm.

an increase in the amount of strain hardening that occurs when the spring is stressed past the proportional limit (point "A" in Figure 1.4). Increasing the temperature lowers the proportional limit to some stress lower than point "A", and therefore if the spring is still stressed to point "B", the amount of strain hardening that occurs is greater. This increase in strain hardening will reduce the dynamic or static settling (load loss) that occurs over the useful life of the spring.

A second theory is that a more effective beneficial residual stress pattern is set up over the bar cross section when a spring is warm set at elevated temperature.

Also refer to the general treatment of presetting at an elevated temperature which is given in Chapter 1, Section 4.

It should be noted that a final (cold) presetting operation is still necessary.

In general, warm setting will decrease the load loss by more than 50%, depending on the working stress level.

6. Specifications and Tolerances

Bar Diameter

Round bars are available as preferred sizes as described in American National Standard, ANSI B32.4, "Preferred Metric Sizes for Round, Square and Hexagon Metal Products." Excerpts from this standard are shown in Table 4.1. Since spring performance is predicated on several factors including the bar diameter, adherence to preferred sizes is not generally feasible and the desired

actual size is to be specified. Generally hot rolled steel bars are specified for hot coiled springs. Table 4.2 shows tolerances for hot rolled steel bars in the range of diameters generally required. Other processed forms of round steel bars are available such as cold drawn, turned and polished, ground and polished, and turned, ground, and polished. These are generally specified when close tolerances are required; to control decarburization; for reduction or elimination of surface imperfections; and availability in limited quantities. See Tables 4.3 and 4.4.

Bar Length

Bars are commonly purchased to exact length. Suggested sheared length tolerances for hot rolled carbon and alloy steel are found in Table 4.5.

Length tolerances of special straightened machine cut bars that are approximately one half of the tolerances in Table 4.5 may be found in ASTM A29. For other tolerances consult the manufacturer.

TABLE 4.5—LENGTH TOLERANCES FOR
HOT ROLLED CARBON AND ALLOY
ROUND STEEL BARS

Specified Diameter, mm		Length Tolerance, Plus Only, mm	
Over	Thru	Over Thru 1500 3000	1500 3000
—	25	12	20
25	50	16	25
50	100	25	40

Coil Diameter

The coil diameter can be expressed in terms of the mean coil diameter (D) which is used in the rate and stress formulae. However, coil diameter tolerances should be specified on either the inside diameter (ID) or the outside diameter (OD) of the coils, depending upon the importance of the respective dimensions to the user. Tolerances are shown in Table 4.6, based on coil diameter and spring length.

For motor vehicle suspension springs, it is customary to specify the ID in order to facilitate the coiling of a family of springs on a single arbor. Where tangent tail ends are specified, additional clearance must be provided for the tangent tail end where the straight portion of the bar extends beyond the outside diameter, since the last 12-25 mm of coiled bar length does not continue to wind in a uniform diameter. See Table 4.6.

TABLE 4.6—COIL DIAMETER TOLERANCES

| For Specified or Computed Outside Diameter, mm | Inside or Outside Diameter Tolerance, Plus and Minus, mm | | | | |
| | For Free Spring Length, mm | | | | |
	Up to 250	Over 250 thru 450	Over 450 thru 650	Over 650 thru 850	Over 850 thru 1050
75.0 thru 110.0	0.8	1.3	2.5	3.6	4.6
Over 110.0 thru 150.0	1.3	2.5	3.6	4.6	5.6
Over 150.0 thru 200.0	2.5	3.6	4.6	5.6	6.6
Over 200.0 thru 300.0	3.6	4.6	5.6	6.6	6.6

Spring Lengths

Spring lengths are to be measured after preloading (see "Preload Length") as the distance parallel to the axis between the end surfaces, or else between two reference points specified on the spring drawing.

Free Length—Free length is the length when no external load is applied. When load is specified, free length is used as a reference dimension only. When load is not specified, free length tolerance equals ± (1.5 mm + 4% of free-to-solid deflection).

Solid Length—(see also "Number of Coils")—Solid length is the length when the spring is compressed with an applied load sufficient to bring all coils in contact; for practical purposes, this applied load is taken to equal approximately 150% of the load beyond which no appreciable deflection takes place. See Table 4.7 for the formulae to calculate the solid length. When nominal solid length is specified, the tolerances shown in Table 4.8 apply.

Preset Length—In the presetting operation (see "Presetting") the spring is usually compressed solid. However, if the stress at solid length is so high that the spring would be excessively distorted, the presetting operation may only be carried to a specified preset length. If more than one preset compression is desired, this must be specified on the drawing.

Preload Length—Preloading is the operation of deflecting the spring to the preload length in order to remove temporary recovery of free length before the spring is checked for load and rate.

If the spring was preset during the manufacturing process to the solid length, the preloading may also be carried to the solid length; but it may be restricted to a preload length slightly greater than the solid length, provided the maximum deflection during subsequent service will not go below the preload length.

If the spring was preset to a specified preset length greater than the solid length, the preloading should be restricted to a preload length greater than the preset length.

However, the preload length must not exceed the minimum spring length possible in the mechanism for which the spring is designed. In suspensions, this is called the "length at metal-to-metal position." The metal-to-metal contact will occur in the suspension mechanism when rubber bumpers are disregarded. The spring deflection from the specified loaded length to the metal-to-metal position is called "Clearance."

Loaded Length—Loaded length is the length while the load is being measured; it is a fixed dimension, with the tolerance applied to the load.

Load

Load is the force in newtons (N) measured on the load testing machine required to deflect the spring to the specified loaded length. It is to be measured during compression of the spring (compression load) and not during release of the spring (release load), unless otherwise specified.

With loaded length fixed, the usual tolerance for motor vehicle suspension springs is expressed in terms of load equivalent to a deflection of ± 5 mm at the nominal rate. Where the demand for greater accuracy warrants the cost of additional presetting or other operations, the load tolerance may be specified as low as ± 1.50 mm at the nominal rate.

In the springs for general automotive use, the load tolerance (with loaded length fixed) typically equals ±(1.50 mm ± 3% of free-to-solid deflection) at the nominal rate. This tolerance is limited to springs where the free length does not exceed 900 mm or six times the free-to-solid deflection, and is not less than 0.8 times the OD.

Rate

Rate is the change of load per unit length of spring deflection (N/mm).

In the springs for motor vehicle suspension, the rate is expressed in terms of the load increase per 25 mm

TABLE 4.7—FORMULAE FOR TOTAL COILS AND FOR NOMINAL SOLID LENGTH

End Configuration	Total Coils (N_t)	Nominal Solid Length (L_s)
Both ends taper rolled	$N_a + 2$	$1.01\,d\,(N_t - 1) + 2t$
Both ends with tangent tail	$N_a + 1.33$	$1.01\,d\,(N_t + 1)$
Both ends with pigtail	$N_a + 1.50$	$1.01\,d\,(N_t - 1.25)$
Taper rolled plus tangent tail	$N_a + 1.67$	$1.01\,d\,N_t + t$
Taper rolled plus pigtail	$N_a + 1.75$	$1.01\,d\,(N_t - 1) + t$
Tangent tail plus pigtail	$N_a + 1.42$	$1.01\,d\,N_t$

where:
d = bar diameter
t = tip thickness of taper rolled bar
1.01 = factor used to compensate for the cosine effect of the coil helix angle.

The bracketed term in the solid length formula for springs with two pigtail ends may vary between $(N_t - 0.90)$ and $(N_t - 1.60)$, depending on the pigtail details.

TABLE 4.8—SPRING SOLID LENGTH TOLERANCES

Nominal Solid Length, mm		Maximum Deviation of Solid Length Above Nominal Solid Length, mm
Over	Thru	
—	175	1.5
175	250	2.5
250	325	3.0
325	400	4.0
400	475	4.8
475	550	5.5
550	625	6.5

deflection (N/25 mm). It is therefore determined as one half of the difference between the loads measured 25 mm above and 25 mm below the specified loaded length. Tolerance is ±3% with centerless ground or with precision rolled bars, and ±4% when commercial hot rolled bars are used.

In the springs for general automotive use, the rate is determined between 20 and 60% of the total deflection unless otherwise defined. Typical tolerance is ±5%. In non-critical applications, this may be increased to ±10%.

Number of Coils

Total number of coils (N_t) are counted tip to tip, active number of coils (N_a) are specified as the number of working coils at free length. With increasing load, N_a may progressively decrease due to the "bottoming out" effect. Rate is inversely proportional to active coils and will therefore be higher than calculated when it is checked at heights or loads which cause significant bottoming out. It is sometimes necessary for the designer to compensate for the bottoming effect by increasing the calculated number of active coils. If no appreciable bottoming out occurs, the relationships between N_a and N_t are as shown in Table 4.7 which also gives the formulae for nominal solid length.

Since nominal solid length may be exceeded somewhat by actual solid length due to manufacturing variation, a frequent practice is to specify nominal solid length together with a maximum solid length, as shown in Table 4.8.

End Configurations (See Fig. 4.1.)

Four types of configurations are used:

Taper Rolled—This end configuration is defined as having the bar end taper rolled prior to coiling, gener-

ally for 240 deg, to a point thickness (t) of approximately 33% of the bar diameter. After coiling, the tip of the tapered bar is to be in approximate contact with the adjacent coil and must not protrude beyond the outside diameter of the spring by more than 20% of the bar diameter. A squareness limit of 3 deg with the axis of the spring is normal, but where finished end bearing surfaces are required for critical applications, the ends may be ground. This grinding operation is performed perpendicular to the axis of the spring helix. The resulting ground bearing surface must not be less than two-thirds of the mean coil circumference, nor narrower than half the width of the hot tapered surface of the bar. This type of spring is used with full size flat seats.

The advantages of this end type are:

a) Material savings due to elongation during taper rolling.

b) Any partial number of total coils can be designated, because rotational orientation is not required. When maximum stress must be reduced, a partial coil may be added rather than one complete coil (as in the case of a double tangent tail). This can result in a significant material savings.

c) It requires less vertical space than tangent tail ends.

Disadvantages:

a) Taper rolling is an added operation requiring considerable equipment and floor space.

b) Two taper rolled ends require additional labor costs to insure squareness of the ends with the coil center line. The bar must be guided and oriented on both ends during the coiling operation, and this often results in significant scrap or rework problems.

Pigtail—The pigtail end is defined as an untapered end coil formed substantially smaller than the central coils

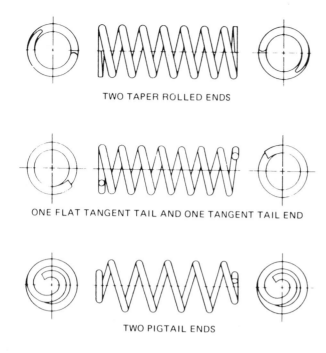

TWO TAPER ROLLED ENDS

ONE FLAT TANGENT TAIL AND ONE TANGENT TAIL END

TWO PIGTAIL ENDS

Fig. 4.1—Typical ends for hot coiled compression springs

of the spring and in such a fashion as to have the outboard bearing surface perpendicular to the axis of the spring helix. This end type is commonly used where the spring must be clamped to the seat to prevent any unseating. It can also be used to fit over a piloted seat without clamping. The pilot seat for a pigtailed end is considerably smaller in diameter than for a non-pigtailed end and may result in a more efficient design. A pigtail end requires the least amount of vertical space and provides for optimum preset conditions. Rotational orientation is not required, therefore, any number of partial coils can be designated. Further, it requires the least amount of inactive material. When forming the pigtail, usually a straight end portion is required to facilitate coiling. The length of this straight portion is dependent on the manufacturer's equipment, but it can have an effect on the size of the pigtail. The major disadvantage of a double pigtail design is the need for additional equipment to form the second pigtail.

Tangent Tail—This end type is defined as an untapered end coil formed as a helix having a pitch substantially equal to the bar diameter and its diameter the same as the central coils. To facilitate coiling, however, a straight end portion approximately 25 mm long is permitted to project tangent to the coil diameter. The resulting helical end requires a spring seat with a corresponding helical ramp. The main advantage of this end type is that it is the simplest and least troublesome method of coiling a spring.

The disadvantages are:

    a) It requires angular orientation during installation.
    b) It requires maximum vertical space.
    c) It requires a greater amount of material than the pigtail or taper rolled ends.
    d) With two tangent tail ends, any design change in the number of coils must be in increments of full coils. The angular relationship between the opposite ends must be a constant for any one particular design application.

Flat Tangent Tail—This end type is similar to the tangent tail end except the last 220 deg (minimum) is coiled perpendicular to the axis of the spring helix. To facilitate coiling, a straight end portion about 25 mm long is permitted to project tangent to the spring diameter. This end type has advantages over the tangent tail end in that it does not require a helical seat and orientation at installation. It also reduces the space requirements. A disadvantage is that it operates with a large amount of inactive material.

Springs can be specified to have any combination of the four types of ends. The combination of two tangent tail ends may involve a complex arrangement for indexing the spring seats, unless the design of every spring is adjusted to an identical number of total coils.

Spring ends and seats are usually so formed as to render approximately two-thirds to one coil inactive at each end.

Direction of Coiling

For most applications, the direction of coiling is unimportant; however, right-hand coiling is preferred because most spring manufacturers are so equipped. When direction of coiling is important, as in the case of concentrically nested springs, it must be specified for each component spring, maintaining opposite directions for adjacent springs. For tangent tail springs, the direction of coiling must conform with the installation conditions.

Uniformity of Pitch

The pitch of coils in a compression spring must be sufficiently uniform so that when the spring is compressed, unsupported laterally, to a length representing a deflection of 80% of the nominal free-to-solid deflection, none of the coils must be in contact with one another, excluding the inactive end coils. This requirement does not apply when the design of the spring calls for variable pitch, or when it is such that the spring cannot be compressed to solid length with lateral support.

When the design of the spring calls for variable pitch, or when it is such that the spring cannot be compressed to solid height without lateral support, the above requirement does not apply.

Concentricity of Coils

At free length, the center of all coils must be concentric with the spring axis within 1.5 mm. This axis is the straight line connecting the centers of the end coils.

Squareness of Ends

Unless otherwise specified, the tapered ends of any spring having an outside diameter to bar diameter ratio of 4 or more, and a free length to outside diameter of 4 or less, shall not deviate more than 3 deg from the perpendicular to the spring axis, as determined by standing the spring on its end and measuring the angular deviation of the outer helix from a perpendicular to the plate on which the spring is standing. In the case of a tangent tail end, the spring must stand on a seat with matching helical ramp. Tolerances for springs outside these limits are subject to special agreement.

Closer Tolerances

The above specifications and tabulations apply to general hot wound springs. However, where large volumes are required, such as automotive suspension coil springs, closer tolerances on some spring parameters are possible. The high volume requirements permit the use of highly automated production equipment and tooling which can be precisely controlled. These same reasons apply to the steel mills which permit them to produce a high quality product to be used for the large volume requirements. These reduced tolerances apply to the following spring parameters:

Coil Diameter—Inside diameters up to 140: ±0.75 mm. Inside diameters over 140: ±1.25 mm.

Bar Diameter—The tolerances detailed in the tabulation of hot coiled springs (Table 4.2) apply for commercial hot rolled bars; tolerance for precision rolled and centerless ground bars is ±0.10 mm.

Bar Length—The bars are commonly purchased to an exact length (for length tolerance see Table 4.5).

Normal Load—The usual load tolerance is ±12% of the specified rate or ±110 N, whichever is greater. Where the demand for greater accuracy warrants the cost of additional presetting or other operations, the load tolerance may be specified as low as ±6% of the specified rate or ±55 N, whichever is greater.

Rate—Rate tolerance is ±3% for springs made from precision rolled or centerless ground bars and ±4% when commercial hot rolled bars are used. The rate is normally checked as the average between 25 mm above and 25 mm below the normal loaded height.

7. Other Hot Coiled Forms

Large extension springs may be made by the hot coiling method, but they cannot be made with initial tension. If (as is usually the case) it is desired to have the coils in contact at no load, then it is impossible to preset the spring and thereby induce a favorable residual stress pattern. Therefore, the design stresses for extension springs must be considerably lower than for compression springs of comparable dimensions (see Part B, Chapter 3).

Hot coiled helical torsion springs should be open coiled for the reasons given in Part D, Chapter 3. A round bar, while theoretically not the most economical section, should be used if possible because of availability and facility of spring manufacture. A square bar may be used, if necessary, to secure additional capacity. Design information and formulae for round bar helical torsion springs are included in Chapter 3, Part D of this manual.

Chapter 5

Design of Helical Springs

1. Design Formulae for Round Wire Springs

All spring design is founded on Hooke's Law, which states that within the proportional limit of any material, deflection is proportional to the load. This means that if stress under any load within the proportional limit is divided by the corresponding strain under that load, the result will always be a constant value. This is independent of hardness or any other property of the material except the proportional limit.

In helical compression and extension springs, the wire is stressed in torsion when the spring is loaded. The basic design formulae for round wire compression and extension springs are shown below:

$$P = \frac{G\,d^4\,F}{8\,D^3\,N_a} \qquad \text{(Force N)}$$

$$R = \frac{P}{F} = \frac{G\,d^4}{8\,D^3\,N_a} \qquad \text{(Linear spring rate N/mm)}$$

$$S = \frac{8\,D\,P}{\pi\,d^3} \qquad \text{(Basic Stress MPa)}$$

$$S_c = \frac{8\,D\,P\,K_w}{\pi\,d^3} \qquad \text{(Corrected Stress MPa)}$$

By making suitable substitutions, the formulae may be combined or transposed to show other useful relationships. For example, the stress may be expressed in terms of deflection, rate, and volume of active spring material.

$$S = F\sqrt{\frac{2\,G\,R}{V}}$$

For springs made of steel wire (with $G = 79.3 \times 10^3$ MPa)

$$S = 398F\sqrt{\frac{R}{V}} \qquad \text{(MPa)}$$

2. Design Stresses

When a helical compression spring is deflected by axial loading, the spring wire is twisted, essentially as a straight bar is twisted under torsion. The torque moment which produces the twist is the product of two factors: one is the load acting along the helical spring; the other is the mean coil radius, or the distance from the spring axis to the center of the coiled wire.

In the straight torsion bar spring of circular cross section, the twisting produces a shear stress which is uniform at every point of the bar surface; but in the helical spring coiled from round wire, the stress pattern on the wire surface is more complex. There are two reasons for this occurrence. The torque moment results in a steeper twist angle for the short wire fibers at the inside of the coil than for the long wire fibers at the outside of the coil, and, consequently, it produces a higher shear stress at the inside of the coil. The axial load causes a direct shear stress which adds to the shear stress from the torque moment at the inside of the coil, but subtracts from it at the outside of the coil.

Under static loading conditions, the variations in stress over the cross section of the spring wire can be neglected. The standard stress formula can be used, as it will furnish an average stress over the entire surface of the wire. However, when the spring is subjected to fatigue loading, the standard formula is inadequate because it conveys neither the higher local stress nor the wider stress range prevailing during each cycle at the inside of the coil. Fatigue failures are induced by a combination of high stress level and wide stress range, and it is a common experience that fatigue failures in helical springs occur at the inside of the coil. A failure at some other point can usually be traced to a local flaw in the wire surface or to severe coiling or presetting stresses.

The maximum design stresses allowed for loading with various spring materials which are commercially available for helical springs can be established as a percentage of the minimum tensile strength of the material. Commercial springs made with a moderate amount of presetting will give a reasonable factor of safety when stressed to the uncorrected stress levels given in Table 5.1.

It has been found convenient to take into account the stress conditions at the inside of the coil by using a "corrected stress." This is obtained by computing the stress from the standard formula and then multiplying the result with a stress correction factor KW. The factor increases with greater curvature of the coiled wire. Greater curvature is equivalent to a smaller spring index, which is the ratio of mean coil diameter to wire diameter. The Wahl factor K_w is most commonly used, and its values may be taken from Table 5.2.

TABLE 5.1—ALLOWABLE UNCORRECTED TORSIONAL
STRESS AS A PERCENTAGE OF TENSILE STRENGTH FOR
VARIOUS SPRING MATERIALS UNDER STATIC LOADING

SAE No.	Type of Wire	% of Tensile Strength
J132/J157	Oil tempered alloy steel	50–55
J316	Oil tempered carbon steel	45–50
J178	Music	40–48
J113	Hard drawn carbon steel	42–45
30302	Stainless steel	42
	Nonferrous metals	42

When the design stress is calculated by including the Wahl stress correction factor, the stress values can be increased proportionally providing the spring index is between 4 and 9. For springs with smaller or larger indexes, the maximum design stress should be kept lower than that previously outlined.

The recommended apparent stresses for compression springs may exceed the torsional elastic limits normally found in the wire. Nevertheless, they are applicable since, by suitably presetting, the spring manufacturer can induce offsetting residual stresses. Presetting is a beneficial operation which will minimize the load loss in operation. A detailed discussion on presetting is given in Chapter 1.

Values 20% lower than those previously given should be employed when designing a spring which is not preset. While the 20% lower values for nonpreset springs may seem to be unattractive to the designer, it should be realized that many springs are so designed because they can be produced at a much lower cost and they may perform fully as well as the preset springs.

In general, helical compression springs should be designed so that the stress at solid compression is within the limits previously outlined. Any compression which falls short of deflecting the spring to its solid length will not produce permanent set and therefore will not result in loss of load. Springs designed for long endurance life, such as valve springs, or for operation under elevated temperatures, require special considerations with the use of design tables and diagrams given in this chapter, and as explained earlier in Chapter 1.

3. Fatigue Life

For those springs which are cyclically loaded, it is common practice to obtain basic mechanical properties from S-n fatigue data and from fatigue strength diagrams, as explained by Figs. 1.6 and 1.7 in Chapter 1.

Three fatigue strength diagrams for typical round wire spring design applications are presented in Figs. 5.1, 5.2, and 5.3 to aid in the actual design work in establishing the life cycle expectancy. These diagrams cover various spring materials having superior surface qualities, such as valve spring quality wires, and include the variables of presetting and shot peening. The Wahl correction factor is used for all stress values referred to.

The allowable initial and maximum stresses are determined by multiplying the appropriate stress factor K_S with the minimum tensile strength TS of the wire or bar used to make the spring. The initial stress factor K_{S1} is read from the 45 deg reference line, while the maximum stress factor K_{S2} is read from the appropriate higher diagonal line depending on the life cycles desired. The stress range is established by the difference in the ordinates of the two diagonal lines. The fatigue strength diagram is then used either to establish the permissible maximum stress and stress range which will provide the desired life cycle expectancy, or else, to establish the life cycle expectancy for the maximum stress and stress range which must be considered for the spring application in question.

TABLE 5.2—TABULATED VALUES FOR WAHL STRESS CORRECTION FACTOR K_w

$$K_w = \left[\frac{4C-1}{4C-4} + \frac{0.615}{C} \right]$$

C→	0.0	0.1	0.2	0.3	0.4	0.5	0.6	0.7	0.8	0.9
2	2.058	1.975	1.905	1.844	1.792	1.746	1.705	1.669	1.636	1.607
3	1.580	1.556	1.533	1.512	1.493	1.476	1.459	1.444	1.430	1.416
4	1.404	1.392	1.381	1.370	1.360	1.351	1.342	1.334	1.325	1.318
5	1.311	1.304	1.297	1.290	1.284	1.278	1.273	1.267	1.262	1.257
6	1.253	1.248	1.243	1.239	1.235	1.231	1.227	1.223	1.220	1.216
7	1.213	1.210	1.206	1.203	1.200	1.197	1.195	1.192	1.189	1.187
8	1.184	1.182	1.179	1.177	1.175	1.172	1.170	1.168	1.166	1.164
9	1.162	1.160	1.158	1.156	1.155	1.153	1.151	1.150	1.148	1.146
10	1.145	1.143	1.142	1.140	1.139	1.138	1.136	1.135	1.133	1.132
11	1.131	1.130	1.128	1.127	1.126	1.125	1.124	1.123	1.122	1.120
12	1.119	1.118	1.117	1.116	1.115	1.114	1.113	1.113	1.112	1.111
13	1.110	1.109	1.108	1.107	1.106	1.106	1.105	1.104	1.103	1.102
14	1.102	1.101	1.100	1.099	1.099	1.098	1.097	1.097	1.096	1.095
15	1.095	1.094	1.093	1.093	1.092	1.091	1.091	1.090	1.090	1.089
16	1.088	1.088	1.087	1.087	1.086	1.086	1.085	1.085	1.084	1.084
17	1.083	1.083	1.082	1.082	1.081	1.081	1.080	1.080	1.079	1.079
18	1.078	1.078	1.077	1.077	1.077	1.076	1.076	1.075	1.075	1.074

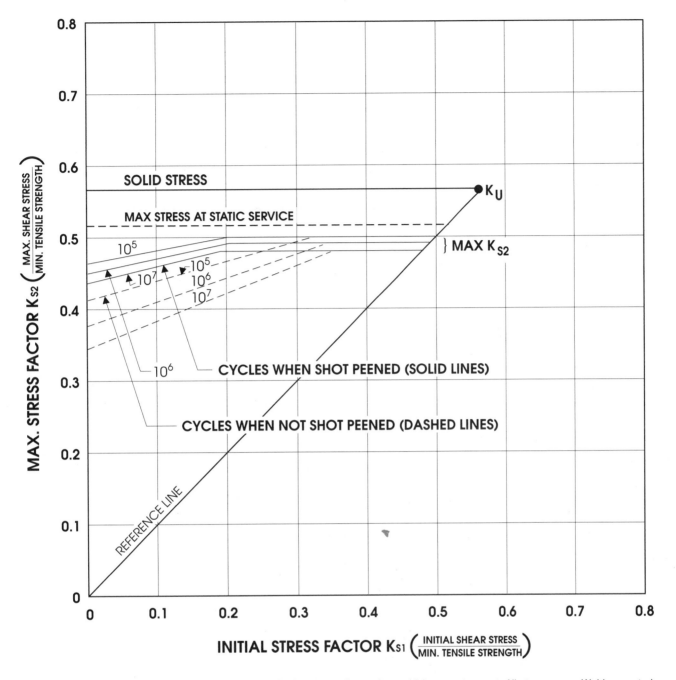

Fig. 5.1 - Fatigue strength diagram for round wire helical compression springs which are not preset. All stresses are Wahl corrected. Diagram applicable to springs which are not preset and to the following materials: Music Steel Spring Wire and Springs—SAE J178; Hard Drawn Carbon Steel Valve Spring Quality Wire and Springs—SAE J172; Oil Tempered Carbon Steel Valve Spring Quality Wire and Springs—SAE J351; Oil Tempered Chromium-Vanadium Valve Spring Quality Wire and Springs—SAE J132

The formulae used as the basis of the fatigue strength diagrams are presented in Chapter 1. When performing spring design calculations with the use of computers, these formulae may be used directly instead of the diagrams. Tables 5.3, 5.4, and 5.5 give the essential parameters required for use with the formulae.

4. Examples of Spring Design Problems

Example 4.1: For a spring with a 28 mm inside diameter, using a 4.5 mm diameter hard drawn wire, determine the stress at a load of 530 N.

$$D = 28 + 4.50 = 32.5 \text{ mm}$$

$$S = \frac{8\,D\,P}{\pi\,d^3} = \frac{8 \times 32.5 \times 530}{\pi \times 4.50^3} = 481 \text{ MPa}$$

TABLE 5.3—PARAMETERS FOR COLD WOUND SPRINGS FOR FATIGUE STRENGTH OF FIGURE 5.1—NOT PRESET

		Not Shot Peened $C_E = 0.6620$, $Y = 0.0622$			Shot Peened $C_E = 0.5021$, $Y = -0.0206$		
Cycles n	Max K_{S2}	K_E	A	B	K_E	A	B
10^5	.500	.324	.410	.268	.396	.464	.171
10^6	.490	.280	.374	.333	.378	.451	.194
10^7	.480	.243	.339	.395	.360	.438	.217

$K_U = .5600$ $C_S = .5546$ $M = -0.0090$

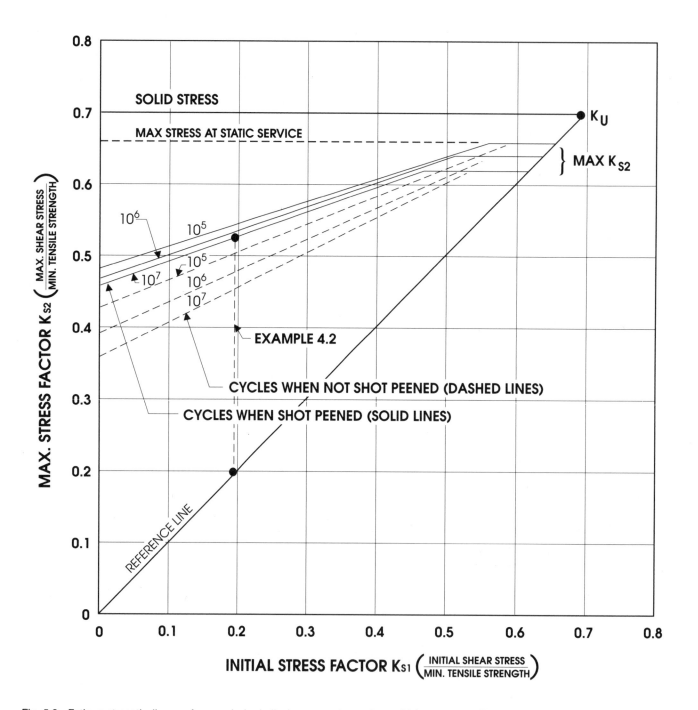

Fig. 5.2—Fatigue strength diagram for round wire helical compression springs which are preset. Diagram applicable to moderately preset springs and to materials mentioned in Fig. 5.1

2.62

TABLE 5.4—PARAMETERS FOR COLD WOUND SPRINGS FOR FATIGUE STRENGTH DIAGRAM OF FIGURE 5.2—PRESET

K_U = 0.7000 C_S = .7757   M = -0.0139

Cycles n	Max K_{S2}	Not Shot Peened C_E = 0.5758, Y = 0.0537			Shot Peened C_E = 0.610, Y = -0.0630		
		K_E	A	B	K_E	A	B
10^5	.661	.310	.430	.386	.372	.486	.306
10^6	.640	.274	.394	.437	.357	.473	.324
10^7	.620	.242	.360	.486	.343	.460	.343

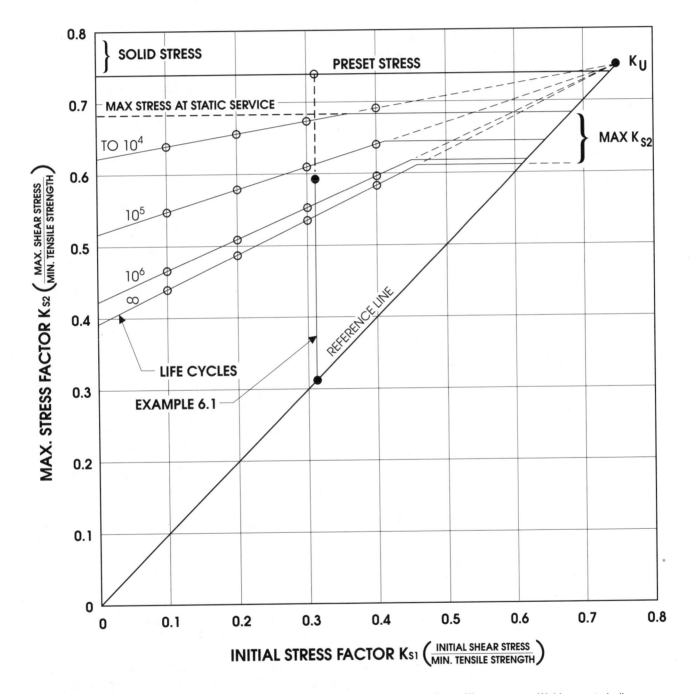

Fig. 5.3 - Fatigue strength diagram for hot coiled round bar helical compression springs. All stresses are Wahl corrected, diagram represents B-10 fatigue life, and is applicable to hot coiled helical compression springs which are shot peened, preset, and using carbon steel or alloy steel bars.

TABLE 5.5—PARAMETERS FOR HOT COILED SPRINGS FOR FATIGUE STRENGTH DIAGRAM OF FIG. 5.3—SHOT PEENED AND PRESET

K_U = 0.740	C_E = 1.808	C_S = 0.830	M = -0.0215	Y = -0.130	
Cycles n	Max K_{S2}	K_E	A	B	
1000	0.715	0.737	0.738	0.002	
10000	0.681	0.546	0.628	0.151	
100000	0.648	0.405	0.523	0.293	
1000000	0.617	0.300	0.427	0.423	
2500000	0.605	0.266	0.392	0.471	

From Table 5.1, the allowable uncorrected stress for SAE J113 hard drawn carbon steel wire is 42% of tensile loading. Since the tensile strength of 4.50 mm wire is 1380 MPa min (see Chapter 2, Table 2.2), the allowable stress is 1380 × 0.42 = 580 MPa. Therefore, the stress of 481 MPa is satisfactory for static loading.

Example 4.2: Using spring dimensions similar to those in Example 4.1, determine the maximum load obtainable using an oil tempered carbon valve spring wire to attain unlimited life when the initial spring load is 270 N and the springs are shot peened and preset.

Tensile strength of 4.50 mm diameter oil tempered carbon valve spring wire is 1550 MPa min for SAE J351 (see Chapter 2, Table 2.8).

$$C = \frac{D}{d} = \frac{32.5}{4.5} = 7.22$$

K_W = 1.206 (Wahl correction factor from Table 5.2)

$$S_{C1} = \frac{8\,D\,K_W\,P}{\pi\,d^3} = \frac{8 \times 32.5 \times 1.206 \times 270}{\pi \times 4.50^3}$$

$$= 296\ \text{MPa}$$

$$K_{S1} = \frac{296}{1550} = 0.191$$

Referring to Figure 5.2, for unlimited life when springs are shot peened and present, or using the formulae in Chapter 1 together with the parameters in Table 5.4:

$$K_{S2} = .460 + .3429\,K_{S1} = .460 + .343 \times .191 = .526$$

Since $\dfrac{P_1}{P_2} = \dfrac{K_{S1}}{K_{S2}}$, therefore $P_2 = 270 \times \dfrac{.526}{.191} = 885\ N$

Example 4.3: Using the data of Example 4.2, determine the number of active coils required to give a load change of 360 N when the deflection is 20 mm.

$$R = \frac{\Delta P}{F} = \frac{360}{20} = 18\ \text{N/mm}$$

$$R = \frac{G\,d^4}{8\,D^3\,N_a} \quad\text{and}\quad N_a = \frac{G\,d^4}{8\,D^3\,R}$$

$$N_a = \frac{79.3 \times 10^3 \times 4.5^4}{8 \times 32.5^3 \times 18} = 6.6\ \text{(active coils)}$$

Thus, if this spring were to be made with closed and ground ends, the total number of coils would be 8.6, with one inactive coil at each end for closing.

5. Design Stresses for Hot Coiled Springs

High stresses can be used in design of hot coiled helical compression springs because presetting is employed and the heat treatment and shot peening processes provide the required control over the physical properties. Based on the material used and its heat treated hardness and tensile strength, the spring can withstand a certain maximum stress for a particular stress range and for an estimated number of fatigue cycles. Load loss or spring set, within permissible limits after a number of cycles of loading is another factor to consider. High cyclic stresses can result in appreciable spring set, so that unsatisfactory spring performance may result even though the spring may not break in service.

The stress limits for hot coiled springs can be determined from the fatigue strength diagram given in Fig. 5.3. This diagram is applicable to springs which are axially loaded without angularly aligned or offset spring seats. When the spring is subjected to excessive bow due to non-aligned spring seats, then the stresses should be decreased proportionally. This diagram is applicable to springs which are preset and shot peened. All stress values are Wahl corrected. The diagram represents B-10 fatigue life or the number of cycles where 10% of the population is estimated to fail (see Section 7).

The allowable initial and maximum stresses are determined by multiplying the appropriate stress factors (K_{s1} and K_{s2}) by the minimum tensile strength for the bar used to make the spring. Basic mechanical properties for the particular material are obtained from S-n fatigue data, as explained in Chapter 1, and also in Section 3 of this chapter where application of the fatigue strength diagrams is explained. Fig. 5.3 is applicable to hot coiled springs and, in addition to stress factors K_{s1} and K_{s2}, it gives the maximum stress factor K_{s2} for solid stress, preset stress, and for maximum stress at static service. Frequently, the solid stress may be the same as the preset stress, in which case the spring is compressed solid to perform the presetting operation. In cases where the design solid stress is in excess of 1150 MPa, the presetting operation can be performed by presetting to a fixed spring length.

6. Fatigue Life of Hot Coiled Springs

Fatigue testing is an accelerated method of examining springs for design adequacy and for quality control purposes. Fatigue life is expressed by the number of deflection cycles a spring will withstand without failure. It can be estimated by the use of the fatigue strength diagram shown in Fig. 5.3. Re-examination of the design will be in order if the fatigue tests result in failures which are confined to one section of the spring.

The procedure to establish suitable parameters for a fatigue test frequently requires the exercise of judgment when the spring application is known to involve random stress amplitudes. As an example, a suspension spring will undergo a large number of cycles of small amplitude near the design load position without failure. Under greater amplitudes, the number of cycles without failure will be reduced, since the maximum stress as well as the stress range are increased, and both are determining factors in the fatigue life of a spring.

The metal-to-metal position (vertical load limit) is frequently used as a maximum deflection position of the spring in a fatigue test; but in heavy truck spring designs this deflection is often considered excessive for the test setup, as it is rarely reached in actual service. Then the maximum deflection position in the fatigue test may be established at compressing the rubber bumper to between one-third and two-thirds of its free height during full jounce.

The length of the test stroke is selected from experience. A frequently used method of establishing the length of test stroke is to add to the compression stroke (from design load to load at maximum deflection position) one-half of this deflection for the release stroke (from design to initial load). This practice may require modification in those cases where it would produce less then 100,000 or much more than 250,000 cycles, according to Fig. 5.3.

There are several considerations as to why the test setup should be preferred to be such that it will result in a preponderance of cycle figures which are neither too short nor too long. At higher stresses (shorter lives), the scatter of the cycle lives is theoretically reduced so that fewer test samples will produce a given degree of precision in the estimated life of the entire population. However, lower stresses (longer lives) give more realistic results, since they duplicate more nearly the actual service conditions and prevent spring settling during the test. Also, comparisons between different groups of springs will be more distinct at lower stresses, since different S-n curves tend to diverge the more they approach the fatigue limit

(or limited value of stress at which 50% of the population would survive a very large number of cycles, say $n = 2.5 \times 10^6$).

In order to establish the fatigue life cycles, which are acceptable in any spring design, it is desirable to have road durability tests run over a prescribed course so that fatigue life test data and actual road durability may be correlated.

Example 6.1: To estimate the expected life of a hot coiled, shot peened, and preset spring which is to meet the following specifications:

End configurations:	Taper rolled plus tangent tail
Inside diameter:	$D_i = 90.0$ mm
Installed load and length:	$P_1 = 6600$ N at 260 mm
Rate:	$R = 51.0$ N/mm
Metal-to-metal clearance:	$cl = 45$ mm
Metal-to-metal limit stress:	$S_2 = 1000$ MPa
Min tensile strength of bar:	$TS = 1570$ MPa
Test stroke (compressed and released from installed):	compressed = 60.0 mm released = 30.0 mm

The computations are as follows:

Bar diameter selected: $d = 15.00$ mm

$$N_a = \frac{G\,d^4}{8\,D^3\,R} = \frac{76.0 \times 10^3 \times 15.00^4}{8 \times 105.0^3 \times 51.0} = 8.15$$

K_w (for $C = 105.0/15.00 = 7.00$) = 1.213

$$S_c/P = \frac{8\,D\,K_w}{\pi\,d^3} = \frac{8 \times 105.0 \times 1.213}{\pi \times 15.00^3} = 0.0961$$

$$L_s = 1.01 \times 15.00 \times (8.15 + 1.67) + 15.0/3 = 153.8 \text{ mm}$$

$$L_\theta = L_I + (P_I/R) = 260 + (6600/51.0) = 389.4$$

$$P_S = P_I + R(L_I - L_S) = 6600 + 51.0(260 - 153.8) = 12020 \text{ N}$$

$P_1 = 6600 - 51.0 \times 30.0 = 5070$ N and
$P_2 = 6600 + 51.0 \times 60.0 = 9660$ N

Tabulation	length L	load P	stress S	factor K_S
Positions:	mm	N	MPa	
Installed—I	260.0	6600	634	0.404
Released—1	290.0	5070	487	0.310
Compressed—2	200.0	9660	928	0.591
Solid	153.8	12020	1155	0.736

The factors (K_{S1} and K_{S2}) may be used with Fig. 5.3 to estimate the life cycles n to be approximately 2.5×10^5; or the formulae in Chapter 1 can be used with the parameters in Table 5.5 to calculate directly the life cycles as follows:

$$K_E = \frac{K_U (K_{S2} - K_{S1})}{2 K_U - (K_{S2} + K_{S1})}$$

$$= \frac{0.740 (0.591 - 0.310)}{2 \times 0.740 - (0.591 + 0.310)} = 0.3591$$

Since $K_E = C_E n^Y$;

$$n = (C_E/K_E)^{1/-Y}$$
$$= (1.808/0.3591)^{1/0.310} = 250,900 \text{ cycles}$$

7. Evaluation of Fatigue Test Results

It must be understood that the fatigue life cycles for any group of springs will vary considerably even under closely controlled test conditions. Moreover, the average

TABLE 5.6—MEDIAN RANKS (PERCENT) FOR SAMPLE SIZES 1—30

Rank Order	Sample Size										Rank Order
	1	2	3	4	5	6	7	8	9	10	
1	50.000	29.289	20.630	15.910	12.945	10.910	9.428	8.300	7.412	6.697	1
2		70.711	50.000	38.573	31.381	26.445	22.849	20.113	17.962	16.226	2
3			79.370	61.427	50.000	42.141	36.412	32.052	28.624	25.857	3
4				84.090	68.619	57.859	50.000	44.015	39.308	35.510	4
5					87.055	73.555	63.588	55.984	50.000	45.169	5
6						89.090	77.151	67.948	60.691	54.831	6
7							90.572	79.887	71.376	64.490	7
8								91.700	82.038	74.142	8
9									92.587	83.774	9
10										93.303	10

Rank Order	Sample Size										Rank Order
	11	12	13	14	15	16	17	18	19	20	
1	6.107	5.613	5.192	4.830	4.516	4.240	3.995	3.778	3.582	3.406	1
2	14.796	13.598	12.579	11.702	10.940	10.270	9.678	9.151	8.677	8.251	2
3	23.578	21.669	20.045	18.647	17.432	16.365	15.422	14.581	13.827	13.147	3
4	32.380	29.758	27.528	25.608	23.939	22.474	21.178	20.024	18.988	18.055	4
5	41.189	37.853	35.016	32.575	30.452	28.589	26.940	25.471	24.154	22.967	5
6	50.000	45.951	42.508	39.544	36.967	34.705	32.704	30.921	29.322	27.880	6
7	58.811	54.049	50.000	46.515	43.483	40.823	38.469	36.371	34.491	32.795	7
8	67.620	62.147	57.492	53.485	50.000	46.941	44.234	41.823	39.660	37.710	8
9	76.421	70.242	64.984	60.456	56.517	53.059	50.000	47.274	44.830	42.626	9
10	85.204	78.331	72.472	67.425	63.033	59.177	55.766	52.726	50.000	47.542	10
11	93.893	86.402	79.955	74.392	69.548	65.295	61.531	58.177	55.170	52.458	11
12		94.387	87.421	81.353	76.061	71.411	67.296	63.629	60.340	57.374	12
13			94.808	88.298	82.568	77.525	73.060	69.079	65.509	62.289	13
14				95.169	89.060	83.635	78.821	74.529	70.678	67.205	14
15					95.484	89.730	84.578	79.976	75.846	72.119	15
16						95.760	90.322	85.419	81.011	77.033	16
17							96.005	90.849	86.173	81.945	17
18								96.222	91.322	86.853	18
19									96.418	91.749	19
20										96.594	20

(Continued)

TABLE 5.6 (CONT'D)

Rank Order	Sample Size										Rank Order
	21	22	23	24	25	26	27	28	29	30	
1	3.247	3.101	2.969	2.847	2.734	2.631	2.534	2.445	2.362	2.284	1
2	7.864	7.512	7.191	6.895	6.623	6.372	6.139	5.922	5.720	5.532	2
3	12.531	11.970	11.458	10.987	10.553	10.153	9.781	9.436	9.114	8.814	3
4	17.209	15.734	15.734	15.088	14.492	13.942	13.432	12.958	12.517	12.104	4
5	21.890	20.015	20.015	19.192	18.435	17.735	17.086	16.483	15.922	15.397	5
6	26.574	25.384	24.297	23.299	22.379	21.529	20.742	20.010	19.328	18.691	6
7	31.258	29.859	28.580	27.406	26.324	25.325	24.398	23.537	22.735	21.986	7
8	35.943	34.334	32.863	31.513	30.269	29.120	28.055	27.065	26.143	25.281	8
9	40.629	38.810	37.147	35.621	34.215	32.916	31.712	30.593	29.550	28.576	9
10	45.314	43.286	41.431	39.729	38.161	36.712	35.370	34.121	32.958	31.872	10
11	50.000	47.762	45.716	43.837	42.107	40.509	39.027	37.650	36.367	35.168	11
12	54.686	52.238	50.000	47.946	46.054	44.305	42.685	41.178	39.775	38.464	12
13	59.371	56.714	54.284	52.054	50.000	48.102	46.342	44.707	43.183	41.760	13
14	64.057	61.190	58.568	56.162	53.946	51.898	50.000	48.236	46.592	45.056	14
15	68.742	65.665	62.853	60.271	57.892	55.695	53.658	51.764	50.000	48.352	15
16	73.426	70.141	67.137	64.379	61.839	59.491	57.315	55.293	53.408	51.648	16
17	78.109	74.616	71.420	68.487	65.785	63.287	60.973	58.821	56.817	54.944	17
18	82.791	79.089	75.703	72.594	69.730	67.084	64.630	62.350	60.225	58.240	18
19	87.469	83.561	79.985	76.701	73.676	70.880	68.288	65.878	63.633	61.536	19
20	92.136	88.030	84.266	80.808	77.621	74.675	71.945	69.407	67.041	64.832	20
21	96.753	92.488	88.542	84.912	81.565	78.471	75.602	72.935	70.450	68.128	21
22		96.898	92.809	89.013	85.507	82.265	79.258	76.463	73.857	71.424	22
23			97.031	93.105	89.447	86.058	82.914	79.990	77.265	74.719	23
24				97.153	93.377	89.847	86.568	83.517	80.672	78.014	24
25					97.265	93.628	90.219	87.042	84.078	81.309	25
26						97.369	93.861	90.564	87.483	84.603	26
27							97.465	94.078	90.885	87.896	27
28								97.555	94.280	91.186	28
29									97.638	94.468	29
30										97.716	30

life of the tested springs is not sufficient by itself to establish a judgment on the design, the material, or the producton method which the springs represent. The relationship between the number of applied cycles and the percentage of springs which failed at these cycles can best be analyzed with the help of statistical techniques which will systematically describe the "dispersion" or "spread" or "scatter" of the recorded test results. The extent of the scatter will depend upon the consistency of surface condition, fabrication, and the general quality of the springs which are tested.

Sampling

One of the main purposes of statistical analysis is to draw inferences about the properties of a large group (the "population") from the results of tests on a small group (the "sample"). If the entire population were tested, one would not have to infer anything; one would know how the population reacted to the test. This would be called 100% confidence. If 99% of the population were tested, one would be 100% confident about that 99%. Also, if the sample of 99% were considered representative of the remaining 1%, one would be close to 100% confident of predicting the result if that 1% were tested. If only 5% are tested, one would know about that 5%, but how much could one infer about the remaining 95%? Actually, if certain conditions of sampling are met, one can infer a great deal about the entire population from tests on small samples. What is required is a good, honest sample.

The primary condition for a good sample is that it be taken at random under conditions which ensure that all springs of the population have an equal chance of being chosen. This is obviously impossible in the case where the sample consists of a few handmade springs of a design which has not yet gone into production. Only experience can tell the engineer whether the various properties of the sample which can affect the test result (in regard to material as well as to production methods and controls) will also be present in the production springs (the "population"). This determination is outside the realm of statistics. However, statistical mathematics are based on the inherent assumption that the sample is a true representative of the population.

Distribution

If the entire population were tested under identical test conditions, the results could be shown in graphical form by arranging them in ascending numerical order and plotting the cumulative fraction (or percent) of failures over an abscissa of "life cycles." A sample selected at random from this population can be expected to exhibit a similar distribution of fatigue life; the larger the number of springs in the sample, the closer will be the similarity.

It is possible to calculate the likelihood of similarity for samples of any given size. Tables are available based on such likelihoods; Table 5.6 is an example. It presents the "median rank" of each test result for a sample size between 1 and 30. A rank is assigned to each individual test result corresponding to that portion of the population which it is most likely to represent. The median rank is used as an estimate of the true rank because it is just as likely to be high as low. Table 5.6 lists percent figures.

A good approximation formula for the median rank (which may be used for larger sample sizes than those in Table 5.6) is:

$$100 \times \frac{J - 0.3}{N + 0.4}$$

where:

J = Position (in ascending order) for each test result in the sample
N = Total quantity of springs in the sample

The median rank line constructed from such data predicts that certain percentages of the population will survive specific cycles-to-failure. But any such estimate may err substantially on either the high or low side. The question is: How confident can the engineer be of such an estimate? When he has accumulated a great deal of experience in comparing the results of small samples which represent springs of different materials or different designs or different production methods, he may judge, after contemplating two such median rank lines, that he should give preference to one set of springs over the other because of its apparent superiority in fatigue life. However, when judgment based on experience is not considered adequate for a final decision, then it will be necessary to construct lines of higher confidence.

In many cases a confidence level of 90 or 95 or even 99% will be required, so that there will remain only a 10 or 5 or 1% risk of the estimate being either too high or too low. For a chosen confidence level, the life cycles of a given percentage of the population will be found within a certain "tolerance interval." On the median rank graph this may be represented by a "tolerance band" to either side of the median rank line. Wider bands indi-cate increased doubt about the line truly representing the population. With a given sample size, the bands will be wider for higher confidence levels. With a given confidence level, the bands will be wider for a smaller sample size.

If a 90% confidence level has been chosen, the lower and upper limits to the estimated fatigue life distribution of the population can be shown by constructing "5% rank" and "95% rank" lines (see "Theory and Technique of Variation Research" by Leonard G. Johnson. Elsevier Publishing Co., 1964). The numbers 5% and 95% represent the chance of being either too high or too low in assigning the given ranks to the individual springs in the sample, and they are used to establish the limit within which the true population is expected to lie.

For a concise summary of distribution mathematics see "Engineering Considerations of Stress, Strain, and Strength," by Robert C. Juvinall, McGraw-Hill Book Co., 1967.

Weibull Plot

Several systems of mathematically organizing the test result data have been established. In the past, the normal (or Gaussian) distribution has been most widely used. It is graphically represented by the familiar symmetrical bell-shaped distribution curve which is completely defined by two statistical parameters:

1. The *mean life* — In the test sample it is the sum of all the recorded test result values, divided by the sample size. It then becomes an estimate of the population mean life. A population with normal distribution has the *mean* coinciding with the *median* (which is the middle result when all individual results are arranged in order of magnitude), and also coinciding with the *mode* (which is the cycle value at which the greatest number of failures occurs).

2. The *standard deviation* — It describes the scatter on either side of the mean. For the test sample it is mathematically defined as the square root of the sample variance (which in turn is the sum of the squares of the difference between each recorded test result and the mean, divided by the sample size minus 1). It then becomes an estimate of the population standard deviation.

When the test results are arranged in ascending numerical order, and the cumulative percent of failures is plotted (using median ranks) over an abscissa of life cycles on *normal probability graph paper*, it will be found that a straight line can be fitted to the results as long as the distribution is normal. This becomes an estimate of the population distribution.

While the normal distribution has a number of attractive attributes and has been the subject of many publications, it must be recognized that in spring fatigue testing the results are usually not normal in that they cannot be plotted in the symmetrical bell-shaped distribution curve but in a skewed curve. This has led to other mathematical formulations. In the automotive industry the *Weibull plot* is used because it permits straight-line plotting of the cumulative failure probability versus life cycles on *Weibull probability graph paper*, even when the distribution is skewed.

In the Weibull distribution the relationship between the number of applied cycles and the cumulative percent of failures at these cycles is expressed by a formula which uses three parameters:

1. The *minimum life*, which may or may not be zero. It is denoted by the letter "a."

It is generally assumed that a = zero because that is the condition for which the Weibull formula assures straight-line plotting on Weibull paper.

2. The *Weibull slope*, which is an indicator of the skewness of the distribution. It is called the "shape parameter" and is denoted by the letter "b." It also is a measure of the scatter of the distribution; a low slope value indicates a high degree of scatter, and vice versa.

The slope is the tangent of the angle formed by the distribution line with the abscissa on Weibull probability paper, when the scales are such that the distance representing the factor 100 on the (logarithmic) abscissa scale for life cycles equals the distance from 2.3 to 90.0% on the (log-log) ordinate scale for percentages of failure.

In the Weibull distribution, the mean, the median, and the mode never coincide exactly. But when the Weibull slope is within the range of 3.2-3.5, the differences are small enough to give the Weibull distribution an appearance of symmetry. The Weibull comes nearest to the normal distribution when the Weibull slope equals 3.44 (thus representing an angle of 73.8 deg), because there the mean and the median have identical values.

The more the slope value increases above 3.44, the more the distribution curve will be skewed to the left (with a long tail to the left), where the mean is to the left of (or less than) the median. The more the slope value decreases below 3.44, the more the distribution curve will be skewed to the right (with a long tail to the right), where the mean is to the right of (or more than) the median. Fig. 5.4 shows a graph (for Weibull slopes 1 through 12) which locates the percent of failed springs at the life cycles representing the mean of the population.

3. The *characteristic life*, which is the 63.2% failure point for the population. It is called the "scale parameter" and is denoted by the Greek letter theta (θ).

$$63.2 = 100 \ (1-1/e)$$

where: $e = 2.7183$ (the Napierian base)

Example 7.1 (Fig. 5.5): Eight springs have been fatigue tested under identical conditions. The results are arranged in ascending order of failure cycles and are given rank

order numbers accordingly. In this order they are assigned median ranks from Table 5.6 as follows:

Spring	Order No.	Cycles to Failure	Median Rank
E	1	61 000	8.30
A	2	91 000	20.11
F	3	114 000	32.05
H	4	135 000	44.02
C	5	155 000	55.98
B	6	177 000	67.95
G	7	205 000	79.89
D	8	245 000	91.70

These points are plotted on Weibull graph paper in Fig. 5.5. Drawing a straight line of best fit through the median rank points produces an estimate for the failure rate of the entire population with the parameters b = 2.4 and θ = 170 000.

From Fig. 5.4 it will be seen that for b = 2.4, the percent of failed springs at the mean is 52.7. At that failure level the median rank line in Fig. 5.5 shows 150 000 cycles. At the B-10 life level (that is, at the number of cycles where 10% of the population are estimated to fail), the median rank line shows 66 000 cycles.

Significant Differences

In most cases the fatigue testing of springs will be undertaken for the purpose of comparing different samples, and the probability graph will be expected to convey information on the relative life distribution of the populations represented by those samples. For example, the comparison may involve a sample representing a first design, and another sample representing a second design.

When the two median rank lines for the test data of the two samples are plotted on the same graph, they will readily show if the second design promises some improvement in fatigue life. However, in order to establish if there is a "significant difference" between the two designs, it will be necessary to find quantitative values for the degree of improvement (or degradation) between one design and the other. The questions is this: How confidently can one say the limited test results indicate that the second design assures an improvement in fatigue life for the entire spring population? The answer depends not only on the amount of separation between the two plotted slopes, but also on the size of the two test samples.

Furthermore, the degree of confidence in the superiority of one design over the other need not be constant from one quantile level to another. For example, it is possible to have a significant improvement at the B-50 life (50% failure level) without any improvement at the B-10 life (10% failure level), or vice versa. This is partly due to differences in the Weibull slope, and partly due to the

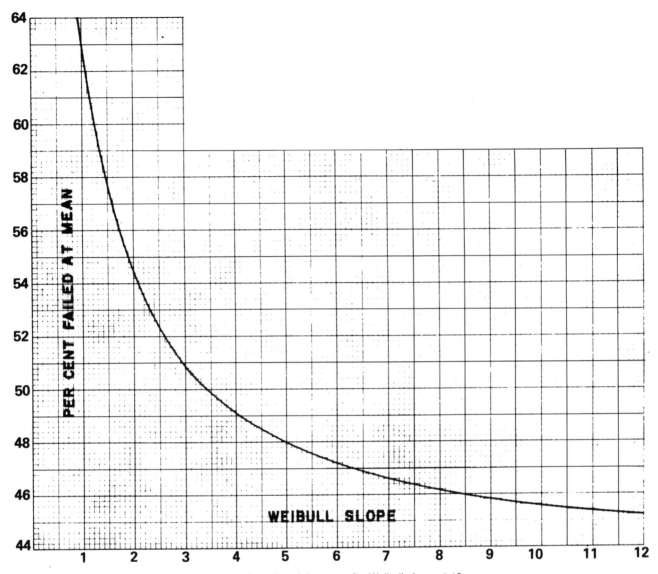

Fig. 5.4—Location of the mean for Weibull slopes 1-12

greater width of the tolerance bands at the lower quantile levels.

Example 7.2 (Fig. 5.5): It has been proposed that the spring design represented by the sample of eight in Example 7.1 be replaced by a new design. Seven springs of the new design have been fatigue tested with results which are shown arranged in ascending order and are assigned median ranks from Table 5.6 as follows:

Spring	Order No.	Cycles to Failure	Median Rank
M	1	132 000	9.43
O	2	195 000	22.85
K	3	233 000	36.41
L	4	275 000	50.00
P	5	315 000	63.59
J	6	365 000	77.15
N	7	440 000	90.57

For this second plot it will be seen from Fig. 5.5 that the Weibull slope b = 2.7 and the characteristic life θ = 310 000; the mean life level is at 51.8% (see Fig. 5.4), therefore, the estimated mean life is 280 000 cycles. The estimated B-10 life is 138 000 cycles.

Since the estimated mean life in Example 7.1 was 150 000 cycles, the mean life ratio on the median rank lines is 280 000/150 000 = 1.87. This represents an estimated improvement of 87%. The "confidence number" corresponding to this mean life ratio (that is, the probability that the true mean life ratio of the population is greater than 1) is found by reference to the mean life nomograph (Fig. 5.6).

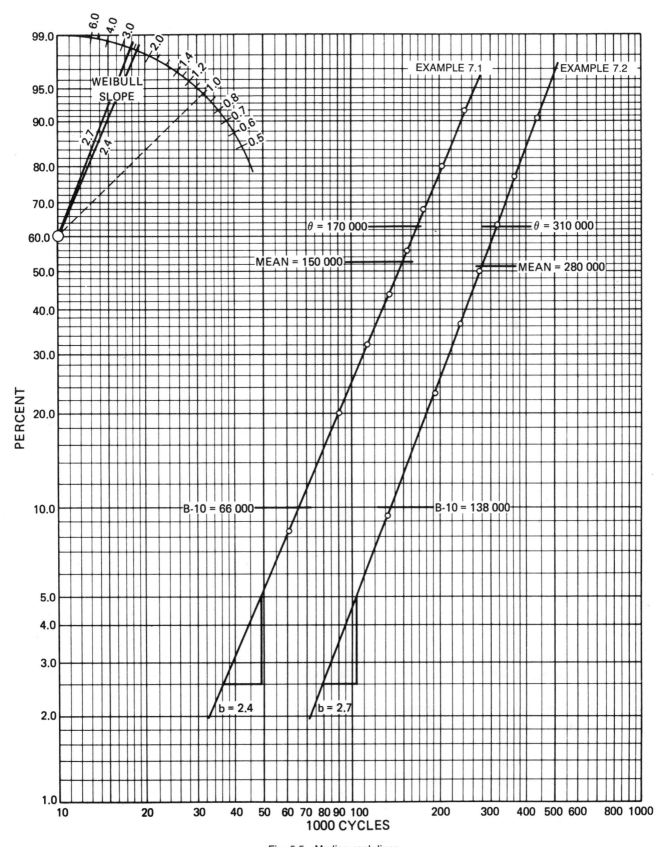

Fig. 5.5—Median rank lines

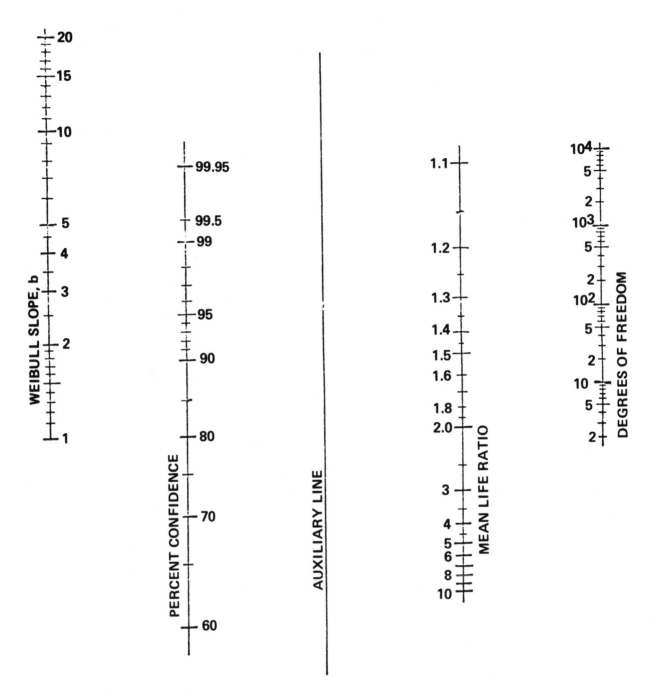

1. Connect total degrees of freedom with Weibull slope and locate intersection point on auxiliary line.

2. Connect life ratio with intersection point and continue to intercept on confidence number.

3. For unequal Weibull slopes perform operation for each slope and average the confidence numbers so obtained.

Fig. 5.6—Confidence nomograph at mean life level

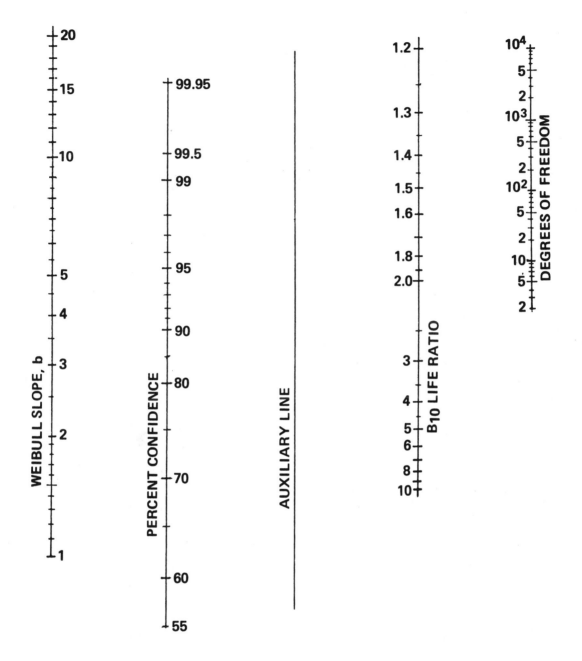

1. Connect total degrees of freedom with Weibull slope and locate intersection point on auxiliary line.

2. Connect life ratio with intersection point and continue to intercept on confidence number.

3. For unequal Weibull slopes perform operation for each slope and average the confidence numbers so obtained.

Fig. 5.7—Confidence nomograph at B-10 life level

2.73

In order to work with the nomograph it is necessary to establish the "degree of freedom" (= freedom of movement of the individual test results about a fixed mean) for the samples representing the two designs. The degree of freedom equals (N - 1) where N is the size of the sample. Thus $N_1 - 1 = 7$ and $N_2 - 1 = 6$, and the product of the two, known as "total degrees of freedom," is 42.

The nomograph furnishes the confidence number 98.8 for b = 2.4 (first design) and 99.4 for b = 2.7 (second design). The average is 99.1, and this means that 99.1 times out of 100 the second design is superior to the first design at the mean life level.

The confidence number corresponding to the B-10 life ratio will be found by reference to the B-10 level nomograph (Fig. 5.7). Since the estimated B-10 life in Example 7.1 was 66 000 cycles, the B-10 life ratio on the median rank line is 138 000/ 66 000 = 2.10. The nomograph furnishes the confidence number 91.0 for the first design and 93.0 for the second design. The average is 92.0, so the second design is superior at the B-10 level 92.0 times out of 100.

Thus, the confidence numbers obtained from the foregoing "significant difference" study indicate a certain superiority of the second design over the first. Quantitative values for the degree of this superiority are obtained by using the information from the nomographs on so-called "confidence interpolation graph paper" or simply "ratio paper" which has a log-log ordinate for percent confidence and an arithmetical abscissa for life ratio (Fig. 5.8).

To obtain quantitative values at the mean life level, the mean life ratio on the median rank lines (in this case 1.87) is plotted on the ratio paper at the 50% confidence level and is connected by a straight line with the confidence number (in this case 99.1) at abscissa 1. The life ratio values at other confidence levels are then found on this line. It will be seen that at the 60% confidence level the ratio is 1.75, so there are 6 out of 10 chances that 75% improvement will occur. At 90% confidence (which is frequently used as a standard) the ratio is 1.32, so there are 9 out of 10 chances that 32% improvement will be realized.

The same procedure will establish quantitative values at the B-10 level. There the percent of improvement will always be comparatively lower for the same degree of confidence. This is due to the greater width of the tolerance bands at the lower quantile levels. In this example, the life ratio at 90% confidence is 1.08, so there are 9 out of 10 chances that 8% improvement will be realized.

Minimum Life Greater than Zero

As stated earlier, Weibull plots are generally constructed with the assumption that the minimum life is

zero. The sample data will then plot a straight line on Weibull probability graph paper.

When the minimum life is greater than zero, the sample points can usually be fitted with a fairly smooth curved line. However, this would mean foregoing one of the major advantages of the Weibull process, which is to analyze the data with the help of a straight line even when the number of test results is small.

A relatively simple technique permits the test results, which indicate a minimum life greater than zero, to be converted to straight-line plotting on the Weibull graph. This requires that the curve drawn through the sample points be extended downward until the abscissa value which it approaches asymptotically can be approximately established as an estimate of the finite minimum life "a." When this is subtracted from each of the plotted data points, it may be possible to fit the new points thus obtained with a straight line.

If it develops that the new points can still be better fitted by a curve than by a straight line, then the estimate of the minimum life was incorrect. If the new line is still concave downward, the minimum life was estimated too small; if the new line is concave upward, the minimum life was estimated too large. A second (and possibly a third) estimate will then be required until the plotted points can be successfully fitted with a straight line. It is well to remember that before any one of the life cycle values found on this straight line is used for comparison with a corresponding value on any other median rank line, it must be increased by the "a" value which was subtracted from the curved median rank line to obtain the straight line.

Example 7.3 (Fig. 5.9): Seven springs have been fatigue tested with results listed below, arranged in ascending order and shown with their assigned median ranks from Table 5.6. The plotting of these values produces a curve which is concave downward. A tentative extension indicates "a" to approximate 50 000 cycles. When the points are replotted with each result reduced by 50 000, they can only be fitted with a curve which is concave upward; therefore, a second attempt is made with a = 45 000, and this brings about a successful straight line fit.

Order No.	Cycles to Failure	Median Rank	Cycle less 50 000	Cycles less 45 000
1	85 000	9.43	35 000	40 000
2	110 000	22.85	60 000	65 000
3	135 000	36.41	85 000	90 000
4	155 000	50.00	105 000	110 000
5	180 000	63.59	130 000	135 000
6	210 000	77.15	160 000	165 000
7	250 000	90.57	200 000	205 000

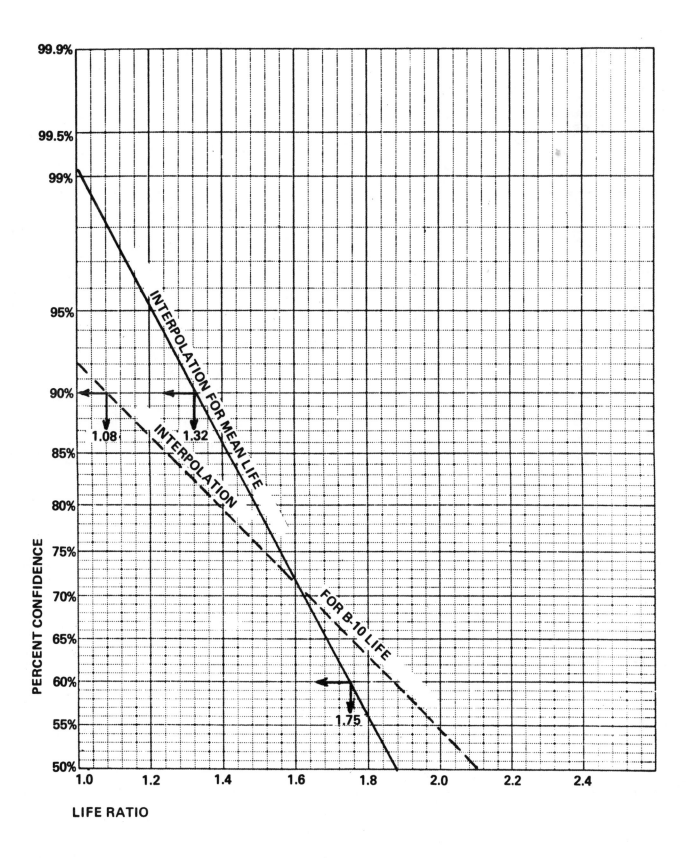

Fig. 5.8—Confidence interpolation

2.75

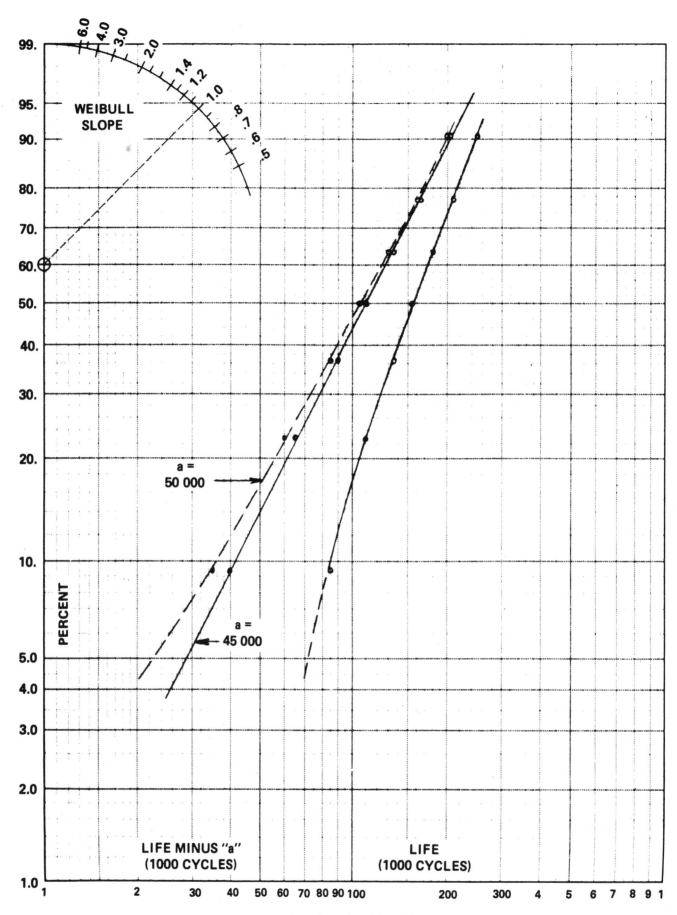

Fig. 5.9 — Weibull plot for springs with minimum life greater than zero

2.76

8. Rectangular Wire Springs

The use of rectangular wire sections (including the special case of square wire) is not recommended, except in cases where space limitations make it necessary. Since these shapes are not produced in large quantities, they are difficult to procure. In addition, the square edge conditions which are prevalent in them tend to limit ductility and may thus promote breakage.

Wire which is rectangular before coiling will upset at the inside of the coil and become trapezoidal in section after coiling. This limits the deflection per coil, as the solid height is predicated in the wire thickness at the inside diameter. An appropriate formula for the upset thickness (t_i) at the inside, compared to the original thickness (t) of the rectangular wire [or (a) of the square wire], is:

$$t_i = t \left[1 + K \frac{(D_o - D_i)}{(D_o + D_i)} \right]$$

where:

D_o = outside diameter of spring
D_i = inside diameter of spring
K = 0.3 for spring temper materials
= 0.4 for annealed materials

Wire which is keystone-shaped before coiling overcomes this difficulty but is costly and hard to obtain, particularly in small quantities.

The principal object in using rectangular section wire is usually to procure maximum load capacity for the spring in a given space. Consequently, such springs are often highly stressed. The problem of insuring reasonably long service life becomes aggravated if the range of stress is high or if elevated temperature conditions prevail. In such cases, shot peening becomes a mandatory requirement.

The design formulae for compression and extension springs coiled from rectangular wire are shown as:

$$P = \frac{K_1 \, G \, b \, t^3 \, F}{D^3 \, N_a}$$

$$R = \frac{K_1 \, G \, b \, t^3}{D^3 \, N_a}$$

$$S = \frac{D \, P}{K_2 \, b \, t^2}$$

The values for the deflection and stress constants are given in Fig. 5.10. They are approximated by the following formulae:

$$K_1 = \text{deflection constant} = \frac{0.430 C_R - 0.217}{C_R + 0.200}$$

$$K_2 = \text{stress constant} = \frac{0.685 C_R + 0.355}{C_R + 1.500}$$

where:

C_R = b/t
b = long side of rectangular cross-section wire
t = short side of rectangular cross-section wire

The previously cited formulae are uncorrected for stress increase due to curvature. They are suitable for the basis of design for springs which are preset and then used for static loading. Coil curvature in rectangular wire springs alters both rate and maximum stress and needs to be considered where fatigue loading is present. Factors for determining corrected rates and stresses were developed by G. Liesecke and appear in graph form in spring design textbooks. The reader is referred to these sources for these finer points in rectangular wire spring design. See book entitled "Mechanical Springs," by A. M. Wahl.

9. Square Wire Springs

Uncorrected stress — The uncorrected stress S in a square wire spring is obtained by assuming the spring to act essentially as a straight wire under torsion. This gives:

$$S = \frac{2.4 \, P \, D}{a^3}$$

Deflection — The deflection F for a square wire helical spring is given by:

$$F = \frac{5.59 \, P \, D^3 \, N_a}{G a^4}$$

$$\text{Spring rate} = \frac{P}{F} = \frac{G a^4}{5.59 \, D^3 \, N_a}$$

Corrected stress — The corrected stress S_c which includes effects of curvature and direct shear and which should be used to calculate stress range for fatigue loading, is given by:

$$S_c = K's$$

$$K' = 1 + \frac{1.2}{C} + \frac{0.56}{C^2} + \frac{0.5}{C^3}$$

In this case, the spring index is $C = D/a$ and K' represents the curvature correction factor.

C_R	1.0	1.2	1.5	2.0	2.5	3	4	5	6	8	10
K_1	0.178	0.214	0.252	0.292	0.318	0.335	0.358	0.372	0.381	0.393	0.400
K_2	0.416	0.436	0.461	0.493	0.517	0.536	0.563	0.582	0.595	0.614	0.627

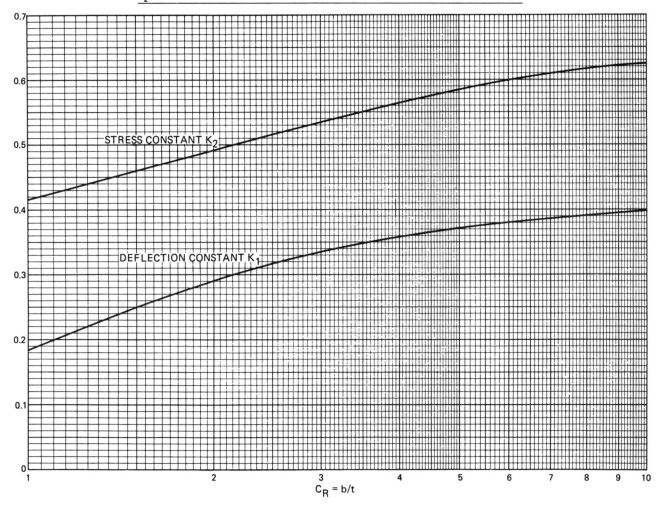

Fig. 5.10 — Constants for springs of rectangular wire

10. Buckling of Compression Springs

A helical compression spring which is made excessively long compared with its diameter can buckle in the same manner that elastic columns buckle under compressive loads. Regardless of the size or stiffness of the spring, the critical deflection at which buckling occurs depends on the free length, the slenderness ratio L_θ/D, and the method of constraining the ends. Fig. 5.11 gives curves for the ratio of critical deflection to free length for springs supported with hinged and fixed ends.

In springs with hinged ends, the free length must be measured as the distance between hinges when the spring is free, not between the ends of the spring. This applies only when the distance between the hinges is not much different than the distance between the spring ends. If the distance is much greater or much less, the relations do not apply. These data are derived from calculations and are substantiated by experiments on springs with the conventional closed and ground ends and with reasonable concentric loading. Eccentric loading and poor end support will cause buckling at slightly lower deflections. Interference between coils will delay buckling and in some cases will suppress it entirely.

If a spring is compressed gradually beyond the buckling point, lateral deflections will build up while the axial rate decreases rapidly and then becomes negative. Sudden collapse may not occur at the buckling point, and the spring may be used right up to the buckling limit. Where the slenderness ratio is very large, loading without guides may cause complete collapse with the spring flying out transversely to its axis. If buckling cannot be avoided by spring design, provision must be made for proper guiding; care must be taken to prevent binding on the guides. The lateral pressure on the guides increases with lateral deflections of the spring; therefore, the guide

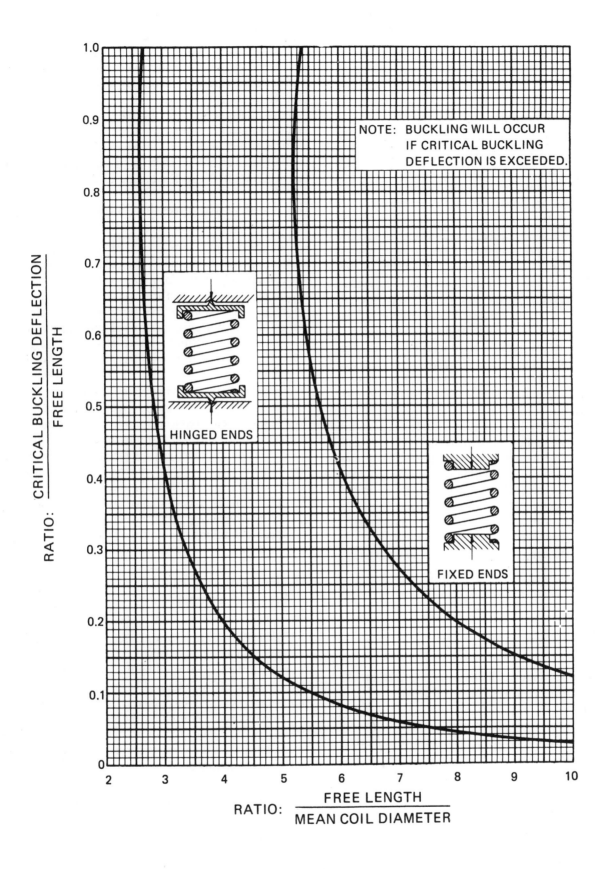

NOTE: BUCKLING WILL OCCUR IF CRITICAL BUCKLING DEFLECTION IS EXCEEDED.

HINGED ENDS

FIXED ENDS

RATIO: $\dfrac{\text{CRITICAL BUCKLING DEFLECTION}}{\text{FREE LENGTH}}$

RATIO: $\dfrac{\text{FREE LENGTH}}{\text{MEAN COIL DIAMETER}}$

Fig. 5.11 — Buckling criteria for compression springs

clearances should be as small as is compatible with diameter tolerances and eccentricity of springs and spring seats.

11. Diameter Changes

The spring designer who must use springs in a confined space should be aware of the natural diameter changes which occur in springs when they are compressed. These diameter changes are due to two combined effects. One is an increase in the diameter of a helical spring upon compression when the coils, which were originally inclined, assume a position more nearly at right angles to the spring axis. The second effect is the result of the normal winding or unwinding of the spring during extension or compression. In the case of the conventional round wire helical spring, this effect is one of unwinding during compression. Associated with this unwinding (or slight decrease in the number of coils) is an increase in the spring diameter. For springs using wire of other cross sections, this last effect may be zero or it may even be negative, resulting in a decrease in spring diameter during compression.

With a conventional round wire helical spring under compression, the combined diameter increase coupled with the normal permissible dimensional tolerances may be large enough to cause jamming of a spring required to work within a tube or sleeve. Therefore, it is desirable to have a sufficiently accurate formula to determine the increase of the spring diameter during compression.

When spring ends are constrained from unwinding during compression, the relative increase of diameter is:

$$\frac{\Delta D}{D} = 0.05 \frac{p^2 - d^2}{D^2}$$

where:

ΔD = increase of mean coil diameter due to compression from free length to solid length

When one or both ends of the spring are free to unwind freely without friction, the relative increase of the spring diameter is nearly twice as great and is given by the expression:

$$\frac{\Delta D}{D} = 0.10 \left[\frac{p^2 - 0.8\, p\, d - 0.2\, d^2}{D^2} \right]$$

For a normally supported compression spring, the diameter increase will be somewhere in the range between these two extreme cases, depending upon the friction conditions at the spring seats.

For extension springs, the same formulae apply. A minus sign appears for ΔD, since an extension spring decreases in diameter during extension. The type of end hooks must also be considered, as it may or may not allow for winding of the spring during extension.

12. Conical Compression Springs

Conical compression springs made with a uniform taper from end to end may be calculated with the use of the compression spring formulae by using the average or mean coil diameter. The formulae will not hold after the spring is deflected more than that required to close the largest coil of the spring. Conical springs, like volute springs, are manufactured with variable coil spacings. The pitch of the larger coils may be increased to maintain linearity closer to the solid length. Conical springs with a large taper from end to end are very much like volute springs in that the coils can telescope inside each other to give a markedly reduced solid length. In other cases, however, where the taper is small, the conical feature has little effect on the solid length which then becomes similar to that of the straight helical compression spring. For volute springs, the reader is referred to the SAE Volute Spring Manual. Barrel-shaped or hourglass-shaped springs may be calculated as two conical springs mounted in series, each taking one-half the total deflection under a given load.

13. Relaxation or Load Loss in Helical Springs at Elevated Temperatures

If a helical spring is compressed by a given amount between parallel plates at elevated temperature, it will be found that the load exerted by the spring will gradually relax or drop off with time. The amount of this relaxation, or set, generally increases as the stress and/or temperature increases. Normally, the set is greater for long periods of time than for short ones. Fig. 5.12 shows spring relaxation for various spring wires. For information on heat setting see Chapter 1, Section 4.

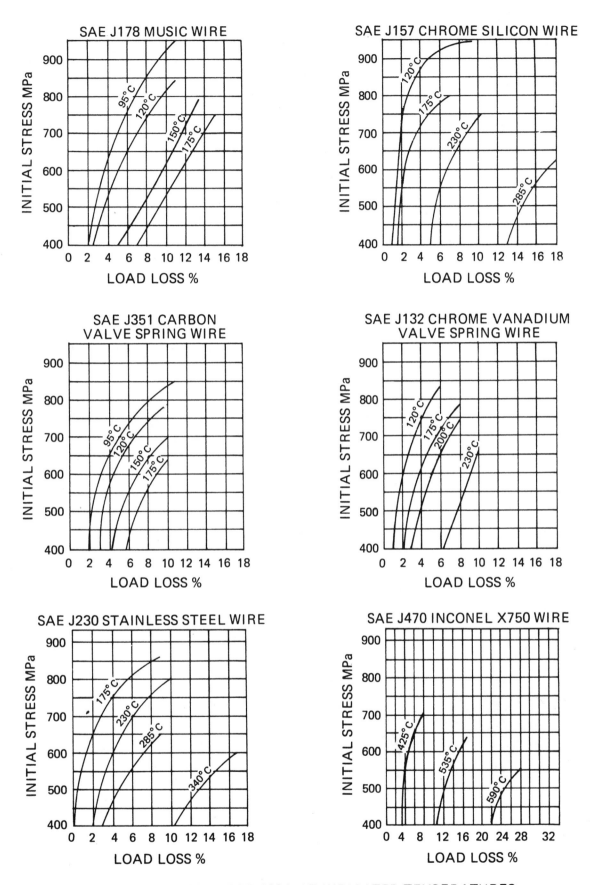

SPRINGS TESTED FOR 100 h AT INDICATED TEMPERATURES

Fig. 5.12 — Spring relaxation for various spring wires

Part 3

Design and Manufacture of Torsion Bar Springs

SAE HS 796

SPRING COMMITTEE

H. M. Reigner (Sponsor), Eaton Corp., Engineering & Research Center

J. F. Kelly (Chairman), Detroit Steel Products, Div. of Marmon Group

K. Campbell (Vice-Chairman), Rockwell International Corp., Suspension Components Div.

J. A. Alfes, Pontiac Motor Div., General Motors Corp.

T. A. Bank, Firestone Industrial Products Co.

J. J. Bozyk, Chrysler Corp.—Product Planning & Development

G. W. Folland, Rockwell International Corp., Suspension Components Div.

L. A. Habrle, Engineering Consultant

R. E. Hanslip, Toledo Spring Co.

D. J. Hayes, United States Steel Corp.

E. H. Judd, Associated Spring—Barnes Group, Inc.

W. Mayers, Peterson American Corp.

M. W. Mericle, Caterpillar Tractor Co., Materials Div.

G. W. Myrick, XM1 Tank System

E. C. Oldfield, Burton Auto Spring Corp.

W. Platko, Chevrolet Motor Div., General Motors Corp.

G. L. Radamaker, Eaton Corp., Suspension Div.

F. T. Rowland, Registered Professional Engineer

H. L. Schmedt, Caterpillar Tractor Co.

G. Schremmer, Schnorr-Neise Disc Spring Corp.

K. E. Siler, Ford Motor Co., Chassis Engineering

J. E. Silvis, Winamac Steel Products Div.—Norris Industries

B. Sterne, Bernhard Sterne Associates

W. M. Wood, Associated Spring—Barnes Group, Inc.

TORSION BAR SPRING SUBCOMMITTEE

K. Campbell (Chairman), Rockwell International Corp., Suspension Components Div.

R. Siorek (Vice-Chairman), U.S. Army Tank Auto Command

W. Allison, Engineering Consultant

J. Bozyk, Chrysler Corp.

G. Dentel, Chrysler Corp.

R. E. Hanslip, Toledo Spring Co.

J. Marsland, Chrysler Corp.

W. Platko, General Motors Corp., Chevrolet Motor Div.

D. W. Schumann, Ford Motor Co., Light Truck Div.

A. F. Skover, Machine Products Co., Inc.

B. Sterne, Bernhard Sterne Associates

D. Tuttle, General Motors Corp., Oldsmobile Div.

W. Young, White Motor Corp.

TABLE OF CONTENTS

Chapter 1

Introduction

Torsion bar springs are used in a wide range of installations, from precision instruments through balance springs, to automotive and military tank suspension springs. The designs vary widely in efficiency of material utilization (stored energy per unit volume of spring material), in complexity, and in production costs. This Manual presents information particularly pertinent to high efficiency springs as used on suspensions for surface vehicles. It is applicable, however, to many types of torsion bar springs for a wide variety of uses.

The Manual has been revised to reflect up-to-date designs and manufacturing processes. It also incorporates the SI metric units in all definitions, design details, and specifications. In the International System of Units (SI), force is not defined by the action of gravity. Accordingly, in this SI (Metric) edition of the Manual, the kilogram (kg) is restricted to the unit of mass [in place of the pound-mass, or pound avdp], and the newton (N) is the unit of force [in place of the pound-force]. The millimeter (mm) has been chosen as the unit of length.

Chapter 2
Design Calculations

1. Symbols Used in Formulae (for SI Units see Appendix)

P = Force at end of lever, perpendicular to reference line (Fig. 2.3), N

P_{static} = Static load (used for P in suspension computations), N

α = Angle between reference line and lever when P is applied; positive as shown in Fig. 2.3, rad

α_{static} = Angle α when P_{static} is applied; positive as shown in Fig. 2.1, rad

β = Angle between reference line and lever at zero load position; positive as shown in Fig. 2.1, rad

θ = Windup angle when P is applied ($= \alpha + \beta$), rad

θ_{static} = Windup angle when P_{static} is applied ($= \alpha_{static} + \beta$) (Fig. 2.1), rad

θ_{jounce} = Windup angle when the maximum operating

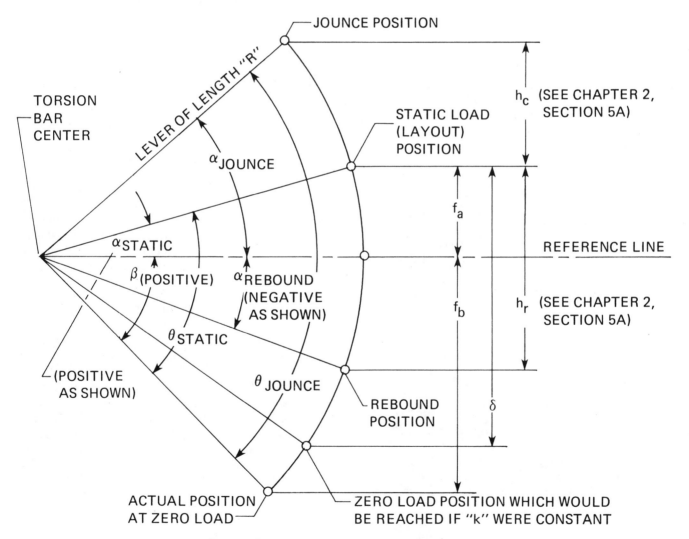

Fig. 2.1—Relation of various load positions at end of lever

load is applied, with the lever in jounce position (= $\alpha_{jounce} + \beta$) (Fig. 2.1), rad

d = Diameter of round bar (outside diameter of tubular bar), mm

d_i = Inside diameter of tubular bar, mm

d_{max} = Diameter of tapered bar at large end, mm

d_{min} = Diameter of tapered bar at small end, mm

t = Thickness of rectangular bar or square bar, mm

w = Width of rectangular bar (long side of cross section), mm

L = Active length of bar (defined in Chapter 3, Section 3), mm

R = Length of lever (Fig. 2.3), mm

f = Deflection at end of lever from reference line when P is applied, measured parallel to direction of P (= R sin α); positive as shown in Fig. 2.3, mm

f_a = Deflection at end of lever from reference line when P_{static} is applied, measured parallel to direction of P_{static} (= R sin α_{static}); positive as shown in Fig. 2.3, mm

f_b = Deflection at end of lever from zero load position to reference line, measured parallel to direction of P (= R sin β), mm

$f_a + f_b$ = Total static deflection at end of lever, from zero load position to static load position, measured parallel to direction of P_{static}, mm

δ = Effective static deflection at end of lever, equal to load divided by spring rate prevailing at that load, mm

T = Torque applied to bar (= P R cos α), N · mm

k = Spring rate at end of lever (variable), measured parallel to direction of P, N/mm

k_T = Torsional spring rate of bar (= T/θ), N · mm/rad

γ = Shear strain

G = Shear modulus (see Chapter 5, Section 1), MPa

S_s = Shear stress, MPa

η_2 = Saint Venant's stress coefficient (Fig. 2.2)

η_3 = Saint Venant's stiffness coefficient (Fig. 2.2)

C_1 = Load factor ⎫ defined in
C_2 = Rate factor ⎬ Chapter 2,
C_3 = Static deflection factor ⎭ Section 4

N_L = Number of laminae in laminated bar

2. Round Bar in Torsion

Table 2.1 shows the relation which exists for the following conditions:

a) Bar is straight.

b) Uniform cross section of solid cylindrical, tubular cylindrical, or solid tapered circular configuration.

c) If tubular, inside and outside diameters are concentric.

d) Torsionally loaded only.

e) Tapered bars have constant taper, with d_x (at distance x from end with diameter d_{min}) = d_{min} + (d_{max} − d_{min}) · x/L

When the three types are to have identical torsional rate k_T and identical stress rate S_s/θ, the following relations exist between them:

1. The ratio L/d (L/d_{min} for the solid tapered bar) is the same for all three.

2. The volume (mass) of the solid cylindrical bar can be reduced by changing to one of the other two types:

a) The space requirements for the tubular cylindrical bar are larger than for the solid cylindrical bar because both d and L increase by a factor

$$\sqrt[3]{1/(1 - (d_i/d)^4)}$$

However, these spacial increases are more than balanced for volume by the growing interior space as d_i (and therefore the factor) are reduced.

b) In the solid tapered bar, as the ratio d_{max}/d_{min} increases, d_{max} increases against d in the solid cylindrical bar, but d_{min} decreases by a larger percentage, and therefore L also decreases. The net result is that the volume of the solid tapered bar becomes ever smaller as the ratio d_{max}/d_{min} is increased.

3. Rectangular Bar in Torsion

A. Straight Bar of Solid Rectangular Cross Section

For a straight bar of solid rectangular cross section, loaded in torsion only, the following relations exist:

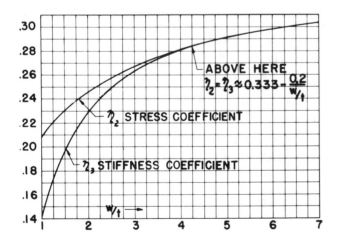

Fig. 2.2—Saint Venant's coefficients for rectangular bar in torsion

Windup angle (radian):

$$\theta = \frac{T\,L}{\eta_3\,t^3\,w\,G} = \frac{\eta_2\,S_s\,L}{\eta_3\,t\,G}$$

Torsional rate (N · mm/rad):

$$k_T = \frac{T}{\theta} = \frac{\eta_3\,t^3\,w\,G}{L}$$

Stress (MPa):

$$S_s = \frac{T}{\eta_2\,t^2\,w} = \frac{\eta_3\,\theta\,t\,G}{\eta_2\,L}$$

Stress rate (MPa/rad):

$$\frac{S_s}{\theta} = \frac{\eta_3\,t\,G}{\eta_2\,L}$$

B. Straight Laminated Bar

For a straight laminated bar (each of the N_L laminae having equal rectangular cross section, same material, same length), loaded in torsion only, the relations are:

$$k_T = \frac{\eta_3\,N_L\,t^3\,w\,G}{L}$$

$$\frac{S_s}{\theta} = \frac{\eta_3\,N_L\,t\,G}{\eta_2 L}$$

When the bar has N_{L1} laminae of t_1 thickness, w_1 width, T_1 torque, $\eta_{2,1}$ and $\eta_{3,1}$ coefficients; also N_{L2} laminae of t_2, w_2, T_2, $\eta_{2,2}$ and $\eta_{3,2}$; also N_{L3} laminae . . . . . :

Windup angle:

$$\theta = \frac{T_1\,L}{\eta_{3,1}\,t_1^3\,w_1\,G} = \frac{\eta_{2,1}\,S_s\,L}{\eta_{3,1}\,t_1\,G}$$

$$= \frac{T_2\,L}{\eta_{3,2}\,t_2^3\,w_2\,G} = \frac{\eta_{2,2}\,S_s\,L}{\eta_{3,2}\,t_2\,G}$$

andso on

Torsional rate:

$$k_T = \frac{T}{\theta} = N_{L1}k_{T1} + N_{L2}k_{T2} + \ldots$$

$$= \frac{N_{L1}\,\eta_{3,1}\,t_1^3\,w_1\,G + N_{L2}\,\eta_{3,2}\,t_2^3\,w_2\,G + \ldots}{L}$$

Stress:

$$S_s = \frac{\eta_{3,1}\,\theta\,t_1\,G}{\eta_{2,1}\,L} = \frac{\eta_{3,2}\,\theta\,t_2\,G}{\eta_{2,2}\,L}$$

and so on

Stress rate:

$$\frac{S_s}{\theta} = \frac{\eta_{3,1}\,t_1\,G}{\eta_{2,1}\,L} = \frac{\eta_{3,2}\,t_2\,G}{\eta_{2,2}\,L}$$

and so on

TABLE 2.1—FORMULAE FOR ROUND BARS IN TORSION

Quantity	Symbol	Type of Round Bar			Unit
		Solid Cylindrical	Tubular Cylindrical	Solid Tapered	
Windup Angle	θ	$\dfrac{32\,T\,L}{\pi\,d^4\,G}$ $\dfrac{2\,S_s\,L}{d\,G}$	$\dfrac{32\,T\,L}{\pi\,d^4\,G\,[1-(d_i^4/d^4)]}$ $\dfrac{2\,S_s\,L}{d\,G}$	$\dfrac{32}{\pi\,G} \cdot \dfrac{T\,L}{\left[\dfrac{3\,D^3\,d^3}{D^2+Dd+d^2}\right]}$	radian
Torsional Rate	k_T	$\dfrac{T}{\theta}$ $\dfrac{\pi\,d^4\,G}{32\,L}$	$\dfrac{T}{\theta}$ $\dfrac{\pi\,d^4\,G\,[1-(d_i^4/d^4)]}{32\,L}$	$\dfrac{T}{\theta}$ $\dfrac{\pi\,G}{32\,L} \cdot \dfrac{3\,D^3\,d^3}{D^2+Dd+d^2}$	N · mm/rad
Stress	S_s	$\dfrac{16\,T}{\pi\,d^3}$ $\dfrac{\theta\,d\,G}{2\,L}$	$\dfrac{16\,T}{\pi\,d^3\,[1-(d_i^4/d^4)]}$ $\dfrac{\theta\,d\,G}{2\,L}$	$\dfrac{16\,T}{\pi\,d_x^3}$ $\dfrac{\theta\,G}{2\,L\,d_x^3} \cdot \dfrac{3\,D^3 d^3}{D^2+Dd+d^2}$	MPa
Stress Rate	$\dfrac{S_s}{\theta}$	$\dfrac{d\,G}{2\,L}$	$\dfrac{d\,G}{2\,L}$	Variable	MPa/rad

In actual practice, the dimensions for a laminated torsion bar can be established by the following procedure, considering the case where the complete bar is to have a rate k_T and this cross section:

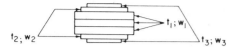

From given data the maximum windup angle (θ_{jounce}) will be computed at which all laminae may be subjected to the maximum allowable stress ($S_{s(jounce)}$), with all laminae having the same active length (L), the maximum allowable thickness for each lamina is:

$$t_{max} = \frac{\eta_2}{\eta_3} \times \frac{S_{s(jounce)} \; L}{\theta_{jounce} \; G}$$

Any lamina thicker than t_{max} will be overstressed.

All laminae are wound up to the same θ_{jounce} and would carry the same maximum stress if the thickness of all laminae could be held to t_{max}. This would result in the most efficient spring or in a spring of minimum weight. In most cases this is not practical and the thickness of the laminae must be adjusted to give the required rate. It should, however, be kept in mind that the stresses should be held as close as possible to the allowable maximum stress in order to keep the spring weight to a minimum.

The torsional rate of the complete bar is:

$$k_T = N_{L1}k_{T1} + N_{L2}k_{T2} + \ldots . N_{Ln}k_{Tn}$$

In general, this result will not coincide with the desired value calculated from

$$k_T = \frac{P \; R \cos \alpha}{\alpha + \beta}$$

See Chapter 2, Section 4.

If the rate as checked by load-deflection test turns out too low, another lamina will have to be added. If it turns out too high, thinner laminae must be used, or the widths of the laminae must be reduced.

It is important to remember that no laminae should exceed t_{max}, but all laminae should be kept as close as possible to this thickness. The stress is not affected by a change in width (w) as long as the ratio $\eta_3/\eta_2 = 1$; this holds true for most practical applications where (w/t) > 3.5.

Consequently, changing w is a convenient way of rate adjustment if the widths of the laminae are not required to be held to existing commercial sizes.

Changing the active length (L) of the bar, provided the specific design allows this, is also a means of rate adjustment; however, L also affects the stress, and the whole computation will have to be repeated.

4. Torsion Bar Spring and Lever

Fig. 2.3 shows a combination of torsion bar spring and lever which is frequently used in suspensions. In this diagram the deflection f and the angles α and β are measured from a reference line which is perpendicular to the applied load and passes through the center of the torsion bar. They are counted positive when their relations to the reference line are as shown in Fig. 2.3.

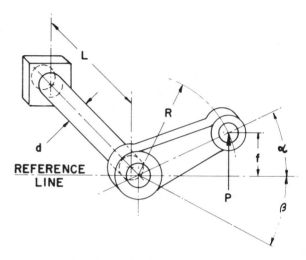

Fig. 2.3—Torsion bar spring and lever

The load deflection characteristics of this mechanism are not linear but are given by the following:

$$P = \frac{T}{R \cos \alpha}$$

$$k_T = \frac{T}{\theta}$$

$$\theta = \alpha + \beta$$

$$k_T = \frac{T}{a + \beta}$$

$$P = \frac{k_T (\alpha + \beta)}{R \cos \alpha} = \frac{k_T}{R} C_1 \left[C_1 = \frac{\alpha + \beta}{\cos \alpha} = \frac{\theta}{\cos \alpha} \right]$$

$$\frac{f}{R} = \sin \alpha$$

The function C_1 is plotted in Fig. 2.4 against the ratio f/R or sin α as abscissa, so that the curves represent load-deflection diagrams for the end of the lever.

The rate will be a minimum when dk/df is equal to zero, or when

$$C_1 = \frac{-3 \sin \alpha}{2 \sin^2 \alpha + 1}$$

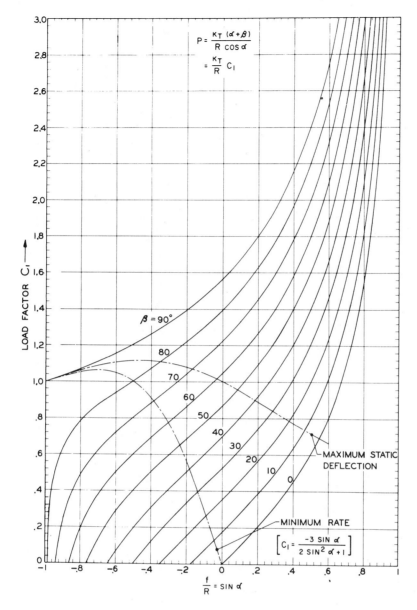

$$P = \frac{k_T\,(\alpha + \beta)}{R\,\cos\alpha}$$

$$= \frac{k_T}{R}\,C_1$$

Fig. 2.4—Load factor versus deflection

It will be noted from the chart that C_1 becomes equal to zero when $\beta = -\alpha$.

The vertical rate at the end of the lever is given by $k = (dP/df)$. Using the value of P above and the relation $f = R\sin\alpha$

$$P = \frac{k_T}{R} \times \frac{(\alpha + \beta)}{\cos\alpha}$$

$$\frac{dP}{d\alpha} = \frac{k_T}{R}\left[(\alpha + \beta)\frac{\tan\alpha}{\cos\alpha} + \frac{1}{\cos\alpha}\right]$$

$$\frac{dP}{d\alpha} = \frac{k_T}{R}\left[\frac{1 - (\alpha + \beta)\tan\alpha}{\cos\alpha}\right]$$

$$f = R\sin\alpha$$

$$\frac{df}{d\alpha} = R\cos\alpha$$

$$k = \frac{dP}{df} = \frac{dP}{d\alpha} \cdot \frac{d\alpha}{df}$$

$$= \frac{k_T}{R^2} \times \frac{1 + (\alpha + \beta)\tan\alpha}{\cos^2\alpha}$$

$$k = \frac{k_T}{R^2} \cdot C_2$$

$$C_2 = \frac{1 + (\alpha + \beta)\tan\alpha}{\cos^2\alpha}$$

Fig. 2.5 shows the function C_2 plotted against the ratio

3.7

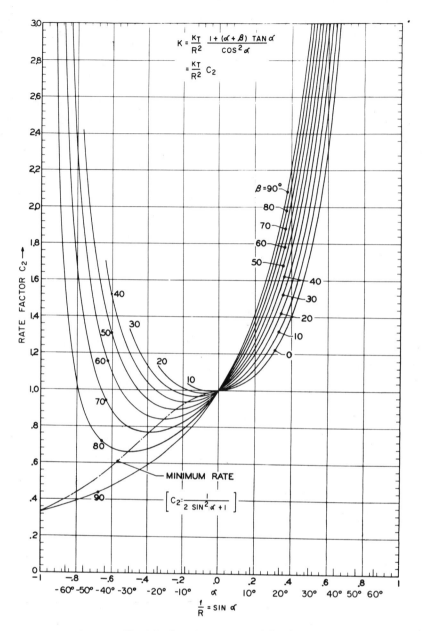

$$K = \frac{K_T}{R^2} \frac{1 + (\alpha + \beta)\ \tan \alpha}{\cos^2 \alpha}$$

$$= \frac{K_T}{R^2} C_2$$

$\beta = 90°$

MINIMUM RATE

$$\left[C_2 = \frac{1}{2\ \sin^2 \alpha + 1} \right]$$

$$\frac{f}{R} = \sin \alpha$$

Fig. 2.5—Rate factor versus deflection

f/R or sin α as abscissa. It shows that the rate of such a torsion bar and lever is by no means constant, and that the minimum rate occurs at a position of lever center-line below the horizontal. When f/R is zero then for all values of β, $C_2 = 1.0$.

The rate will be a minimum when dk/df is equal to zero, or when $C_2 = 1/(2 \sin^2 \alpha + 1)$ only for negative values of sin α.

The static deflection at any point is defined as $\delta = P/k$. Using the above values this becomes

$$\delta = R \frac{\cos \alpha}{\dfrac{1}{\alpha + \beta} + \tan \alpha}$$

$$\delta = R\ C_3$$

$$C_3 = \frac{\cos \alpha}{\dfrac{1}{\alpha + \beta} + \tan \alpha}$$

Fig. 2.6 shows the function C_3 plotted against the ratio f/R or sin α as abscissa, so that the curves represent a comparison of static vertical deflection with the vertical deflection from the reference line. By definition, the abscissa values in Fig. 2.6 actually represent sin α_{static} or f_a/R.

The rate will be a minimum when dk/df is equal to zero, or when $C_3 = -3 \sin \alpha$.

3.8

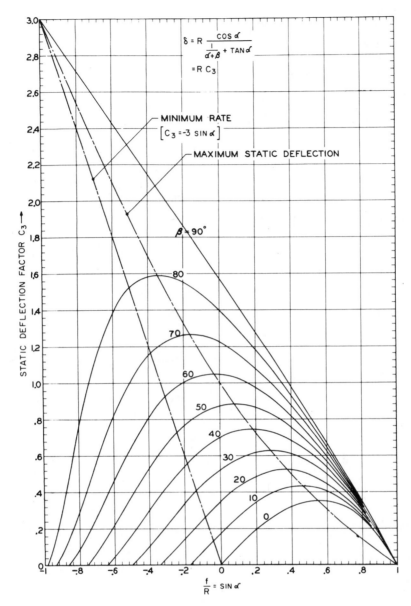

$$\delta = R \frac{\cos \alpha}{\frac{1}{\alpha+\beta} + \tan \alpha}$$
$$= R\, C_3$$

MINIMUM RATE
$[C_3 = -3 \sin \alpha]$

MAXIMUM STATIC DEFLECTION

$\beta = 90°$

Fig. 2.6—Static deflection factor versus deflection

It is to be noted that the curves for minimum rate, maximum static deflection, and $\beta = 90$ deg all intersect at $f/R = -1.0$ or $C_3 = 3.0$.

Note: All angles in previous equations are expressed in radians. In Figs. 2.4–2.6, the angle β is given in degrees, but for calculation of loads and rates from the constants C_1 and C_2 the torsional rate must be expressed in the dimensions N · mm/radian. Where necessary, angles are converted from degrees to radians by the relation 1 radian = 57.296 deg as shown in the examples.

5. Sample Computation

A. Requirements for Round Bar

Required: Determination of diameter (d) and active length (L) on a torsion bar spring of round cross section for a passenger car suspension for the given values:

Static load:	P_{static} = 4000 N
Rate at static load:	k = 16 N/mm
Length of lever:	R = 400 mm
Position of lever at static load:	α_{static} = 7 deg
	= +0.122 rad

Deflection, static load to jounce
position: $h_c = 100$ mm
Deflection, static load to rebound
position (with rebound stop): $h_r = 125$ mm
Stress at static load: $S_{s(static)} \leq 670$ MPa
Stress at jounce position: $S_{s(jounce)} \leq 900$ MPa

For recommended stress values see Chapter 2, Section 6.

B. Computation for Round Bar

The effective static defection at static load $\delta = 4000/16 = 250$ mm.
Therefore

$$C_3 = \frac{\delta}{R} = \frac{250}{400} = 0.625$$

Position of end of lever above horizontal at static load (all positions shown in Fig. 2.1):

$$f_a = R \sin \alpha_{static} = +0.122 \cdot 400 = +48.8 \text{ mm}$$

From Fig. 2.6 or the formula

$$C_3 = 0.625 = \frac{\cos \alpha_{static}}{1/(\alpha_{static} + \beta) + \tan \alpha_{static}}$$

$$= \frac{0.993}{1/(0.122 + \beta) + 0.123}$$

$$\frac{1}{0.122 + \beta} = \frac{0.993}{0.625} - 0.123 = 1.466$$

$$\beta = \frac{1}{1.466} - 0.122 = 0.560 \text{ radians} = 32 \text{ deg}$$

From Fig. 2.5 with $f_a/R = +0.122$ and $\beta = 32$ deg; $C_2 = 1.10$

Therefore, $k_T = \dfrac{k R^2}{C_2} = \dfrac{16 \cdot 400^2}{1.10} = 2\,327\,000$ N · mm/rad

Rechecking with Fig. 2.4 or formula we obtain

$$P_{static} = \frac{k_T(\alpha_{static} + \beta)}{R \cos \alpha_{static}}$$

$$= \frac{2\,327\,000 (0.122 + 0.560)}{400 \cdot 0.993} = 4000 \text{ N}$$

The lever angle at jounce position is obtained from

$$\sin \alpha_{jounce} = \frac{f_a + h_c}{R} = \frac{48.8 + 100}{400} = 0.372$$

$$\alpha_{jounce} = +21.8 \text{ deg} = +0.38 \text{ radian}$$

The lever angle at the rebound position is obtained from

$$\sin \alpha_{reb} = \frac{f_a - h_r}{R} = \frac{48.8 - 125}{400} = -0.191$$

$$\alpha_{reb} = -11.0 \text{ deg} = -0.19 \text{ radian}$$

The torsional windup angle is $\beta + \alpha_{jounce} = 32$ deg $+ 21.8$ deg $= 53.8$ deg $= 0.94$ radians.
The maximum torque $T_{jounce} = k_T (\beta + \alpha_{jounce}) = 2\,327\,000 (0.56 + 0.38) = 2\,187\,000$ N · mm
Then,

$$d^3 = \frac{16 \, T_{jounce}}{\pi \, S_{s(jounce)}} = \frac{16 \cdot 2\,187\,000}{3.1416 \cdot 900} = 12\,376$$

$$d = 23.1$$

Use d = 23 mm

$$S_{s(static)} = \frac{16 \, P \, R}{\pi \, d^3} = \frac{16 \cdot 4000 \cdot 400}{3.1416 \cdot 12\,167} = 670 \text{ MPa}$$

To produce the required rate, the active length must be

$$L = \frac{\pi d^4 G}{32 k_T} = \frac{3.1416 \cdot 280\,000 \cdot 76\,000}{32 \cdot 2\,327\,000}$$

$$= 898$$

Use L = 900 mm
We shall sketch a line corresponding to $\beta = 32$ deg in Fig. 2.5 and mark the limits

$$f/R = -0.53 \text{ at free position}$$
$$f/R = -0.19 \text{ at rebound}$$
$$f_a/R = -0.12 \text{ at static}$$
$$f/R = -0.38 \text{ at jounce}$$

The curve shows that the rate, going from static to jounce, builds up rapidly from $C_2 = 1.10$ to a value of $C_2 = 1.60$, or a rate change of 45%. The rebound rate first decreases and then builds up again. At the rebound position $C_2 = 0.96$ indicating a decrease of 13% from the rate at static load. At zero load $C_2 = 1.40$. This arrangement, with the radius arm at a small angle above the horizontal at the static load position, gives an increasing rate from this position to the jounce position which is desirable. The rate increase near the free position limits the deflection of the arm and thus facilitates assembly.

This spring requires presetting. (See Chapter 2, Table 2.2.)

For method of selecting preset angle, see Chapter 5.

C. Computation for Laminated Bar

If the active length of the round bar is found to be considerably in excess of the available space, then consideration may be given to a laminated bar. This will require more cross-sectional space, but much less length. Use of rectangular cross section material in the "as rolled" condition may dictate a slight reduction in maximum allowable stress.

A laminated bar with square cross section ($N_L t = w$) can be fitted into a square hole anchor. A bar with stepped width laminae requires a more complicated anchor configuration, but it has the advantage of reducing the compressive stresses between bar and anchor, since each step in the anchor takes only the torque of one or two laminae; also it improves the environmental clearance conditions as its cross section approaches a circle.

For this sample computation it will be assumed that a 35% reduction in active length is required (from 900 to 585 mm), that the stress at jounce position is to be limited to 860 MPa, and that w/t will exceed 4.0 so that $\eta_2 = \eta_3$.

The requirements for load, angular position, and deflection are to remain unchanged; therefore, the torsional rate is to be maintained at $k_T = 2\,327\,000$ N · mm/rad. Then

$$t_{max} = \frac{\eta_2}{\eta_3} \frac{S_{s_{max}} L}{\theta\, G} = 1 \frac{860 \cdot 585}{0.940 \cdot 76\,000} = 7.04 \text{ mm}$$

D. Computation for Laminated Bar of Square Cross Section

For the square cross section ($N_L = w$), and assuming $\eta_3 \approx 0.3$:

$$k_T = N \frac{\eta_3\, t^3\, w\, G}{L} = \frac{\eta_3\, t^2\, w^2\, G}{L}$$

$$tw = \sqrt{\frac{L\, k_T}{\eta_3\, G}} = \sqrt{\frac{585 \cdot 2\,327\,000}{0.3 \cdot 76\,000}}$$
$$= \sqrt{59\,706} = 244$$

Using $t\, w = N\, t^2 = 244$ and $t_{max} = 7.04$:

$$N_{L(min)} = \frac{244}{7.04^2} = 4.92$$

Let $N = 5$; then:

$$t^2 = 244/5 = 48.8$$
$$t = 7.0$$
$$w = 5 \cdot 7.0 = 35.0$$
$$\eta_3 = 0.33 - (0.2/5) = 0.333 - 0.040 = 0.293$$

Check

$$k_T = \frac{5 \cdot 0.293 \cdot 343 \cdot 35 \cdot 76\,000}{585}$$
$$= 2\,290\,000$$

(which is within 1½% of the required rate, thus satisfactory)

$$S_{s_{max}} = 860 \cdot \frac{7.0}{7.04} = 855 \text{ MPa}$$

(This solution is satisfactory.)

With density of steel = 7850 kg/m³:

$$\text{Mass} = 7.85 \cdot 10^{-6} \cdot 585 \cdot 35^2$$
$$= 5.6 \text{ kg}$$

Required Clearance = diagonal of the square

$$1.414 \cdot 35 = 49.5 \text{ mm}$$

6. Operating Stresses

A. Suspension Springs

Suspension springs are generally loaded in one direction only and, as can be seen from Table 2.2, they are usually shot peened and preset. The bars are subjected to both processes because this enables the springs to operate at the high stresses required of them. Without shot peening and presetting, the bars would encounter early fatigue failures and would suffer excessive settling. Both shot peening and presetting are further discussed in other sections of this manual. It can also be noted that research and development is pushing the upper edges of Table 2.2 upward. As materials and processes are improved, added classes can be expected. Extension of applications to reduce weight and size will follow as experience in the higher stress classes is gained.

Classes 1250–1100—Operation of torsion bars at these stress levels requires close attention to material and production quality as the critical flaw size becomes smaller in inverse proportion to the operating stress level. With proper precautions in design and manufacturing, the expected limited spring life can be extended to acceptable levels. Only minimal applications have been logged in these classes, primarily in advanced military vehicles. Extension to commercial applications will bear close cost-benefit analyses because of the costs involved in the design and manufacture of material in these stress ranges. With proper handling, very extended lives have been displayed by early production material.

TABLE 2.2—ROUND TORSION BAR CLASSES

Class No.	Operating Shear Stress, Max (MPa)	Settling % of Max Deflection	Load Direction		Shot Peened	Preset	SAE Steel (Typical)		Typical Application
			Single Only	Reverse			Diameter Up to 40 mm	Diameter Up to 65 mm	
1250	1250	10 Max	Yes	—	Yes	Yes	T20811(MOD)		Heavy vehicle suspensions—military and off highway
1200	1200	10 Max	Yes	—	Yes	Yes			
1150	1150	10 Max	Yes	—	Yes	Yes	H43400 (ESR)[b] H43500 (ESR)[b]		
1100	1100	10 Max	Yes	—	Yes	Yes			
1000	1000	10 Max	Yes	—	Yes	Yes	H43400 H51600 H61500 H15621	H41500 H43400 H86550 H86600	
900	900	2–4	Yes	—	Yes	Yes			Passenger car suspensions
800	800	2–4	Yes	—	Yes	Yes			Truck suspensions
						—	G10650 to G10900		Stabilizer bars
700	700	2–4	—	Permissible	As required	—			
550	550	2 Max	—	Permissible	As required	—	G51600 G61500	—	Hatch covers, lids, doors, machine components
550S[a]	550	2 Max	—	Permissible	As required	—	G10450 to G10900	—	

[a] Torsion bars in class 550S are normally produced from shallow hardening carbon steel grades; they cannot be expected to have the depth of hardenability which can be achieved with alloy steels.
[b] (ESR)—Electro Slag Remelt.

Class 1000—Bars operating at these stress levels have now been in production for over 20 years. Operation at these stress levels implies limited spring life, but with proper attention to design and manufacturing details, acceptable extended life can be obtained. Particular emphasis on minimizing operating stress range can do much to assure extended life. By referring to the conservative "Life Test" diagram, (Chapter 6, Fig. 6.1) it will be observed that a shot peened and preset bar operating between a maximum stress of 950 MPa and a minimum stress of 170 MPa is estimated to have a minimum fatigue life of about 60 000 full cycles. On the other hand, a bar operating under a stress range reduced from 780 to 600 MPa (between a maximum stress of 950 MPa and a minimum stress of 340 MPa) will have a minimum life of about 140 000 cycles.

Class 900—This might be called a typical passenger car torsion bar spring. A bar required to have a life in the laboratory in excess of 100 000 cycles, with a maximum stress of about 900 MPa and a minimum of about 275 MPa has proved very satisfactory in the field. From the diagram, a minimum life in the laboratory of about 120 000 cycles would be obtained.

Class 800—This is a typical truck suspension torsion bar spring. To achieve the longer life required of a truck spring and to take into account the harder usage, the maximum stress is limited to 800 MPa. According to the chart, a truck spring designed to a maximum stress of 800 MPa and a minimum of 200 MPa should have a minimum life of 250 000 cycles in the laboratory.

In the discussion of fatigue life, it was noted in each case that the expected life was that to be obtained in the laboratory. In the laboratory, the bar is usually cycled between the metal-to-metal jounce and rebound positions, but in actual service, very seldom is the bar called on to withstand these extreme cycles. A formal road test program or a record of failures, or lack of failures, should be made to attempt to find a correlation between actual service life and laboratory life. After a correlation has been established between actual service and laboratory fatigue life, the laboratory testing becomes a quality control tool.

Classes 1000, 900, and 800—Rectangular and Laminated Bars—Since most bars of rectangular cross section are fabricated from "as rolled" stock, their inferior surface finish, as well as the secondary stresses that occur in such bars, when twisted, must be taken into account. A reduction of the maximum operating stresses (Table 2.2) by 5% is, therefore, recommended when using Class 1000, 900, and 800 steels.

B. Classes 700 and 550—Other Torsion Bar Springs

Counterbalance Springs—Counterbalance springs run the gamut of stress ranges. Fortunately, they do not usually have the service life required of them that suspension springs do; for example, 10 000 cycles of a tilt cab bar represents about 30 years life at the rate of one cycle per day. Generally, counterbalance springs are under load most of the time and are only relaxed when the mechanism they are counterbalancing is operated. Because of this condition, the maximum stress is usually limited to about 700 MPa. Counterbalance torsion bars sometimes, because of space limitations, have to operate at stresses over 700 MPa and it is a practical solution to preset the spring.

Stabilizer Bars—Stabilizer bars are generally designed to give a certain roll resistance, and the stress is of secondary consideration. Because a stabilizer bar is a part of the suspension and is subjected to reverse loading, it should be shot peened if some areas are stressed in excess of 700 MPa, but it need only be shot peened in the highly stressed areas.

Chapter 3

Design of End Fastening

1. End Configuration

A. Serrated End Connection

Of all known arrangements, serrated ends permit the smallest end diameter. Satisfactory static strength may be obtained even if the diameter of the serrated end is small. However, experience indicates that the minor diameter of the smallest serrated end should not be less than 1.15 times the diameter of the bar.

Refer to ANSI B92.1 (latest revision) for spline and serration design and inspection.

The serrations themselves must be designed with two features in mind:

Stress Concentration at the Serration Root—At the point where the anchor first contacts the serration (next to the transition section) the bar still is under the full windup torque. Even with the enlarged end diameter, the stress at the bottom of the serration root may be nearly as high as that in the body of the bar. Added to it is a stress pattern due to the localized pressures on the serration flanks, so that the total stress here may be very high.

This stress concentration can be minimized by providing the largest root radius compatible with the anchor serration contour. The standard serration pressure angle (at the pitch diameter) is 45 deg which is large enough to permit effective shot peening. Surface finish of the serration contour is usually not specified because the methods of forming produce an acceptable finish.

Compressive Stress on the Flanks of the Serration—A design that is favorable for the first condition (large root radius) reduces the flank area, and therefore increases the pressure on the flanks. If this pressure is too high, an excessive local permanent deformation will occur somewhere near the contact area, initiating a fatigue fracture. Experience has shown that a compressive stress equal to approximately 140% of ultimate tensile stress is permissible when it is calculated for full length contact on approximately 25% of total serrations in accordance with the hertz equations for stress due to pressure between curved elastic bodies.

Formation of serrations may be done before or after heat treatment depending on manufacturing procedure. However, it should be noted that the serrations will be subject to a final acceptance gage. The possibility of serration distortion exists when heat treatment follows the serration forming operation.

Serrations shall be parallel to the centerline of the bar in accordance with tolerances specified on the individual part drawing.

A slight change of serration size is produced by shot peening operations. Assembly problems sometimes result when a serrated bar end is machined to the maximum size, as defined on a drawing, and gaging is conducted directly after the machining operation with shot peening following later. To insure absolutely proper assembly of serrated torsion bars into their anchors, it is necessary to conduct composite "go gage" checks of the serrations after all shot peening is completed.

External and internal serrations when new will not show full contact. During the first few load applications, the high spots will take a local compressive set and the required contact area will be established. On parts which have been fabricated with reasonable accuracy, the permanent set required to produce this contact is not large enough to cause damage. Both external and internal serrations deform elastically during load application, and the contact will therefore be limited in length. Experience has shown that it is satisfactory to make the length of the serrations equal to 0.4 times the major diameter.

B. Hexagonal End Connection

Hexagonal ends for round torsion bars can be produced without machining. They offer advantages in high volume production where upsetting machines in various degrees of automation can be utilized.

The usual practice is to centerless grind the bar stock, cut to a predetermined length, upset the ends to a hexagon shape, and finish with a coining operation to improve the flatness of the sides. There is no need for further machining or grinding. Part numbers, arrows or letters indicating the direction of rotation, and other coded data are usually put in during the upsetting operation. The transition section, which is being formed by the plastic flow of the material, is left in the "as forged" condition. The slightly rougher surface of the transition section is not detrimental as long as no pits or inclusions are present.

The ratio of width across flats to bar diameter must

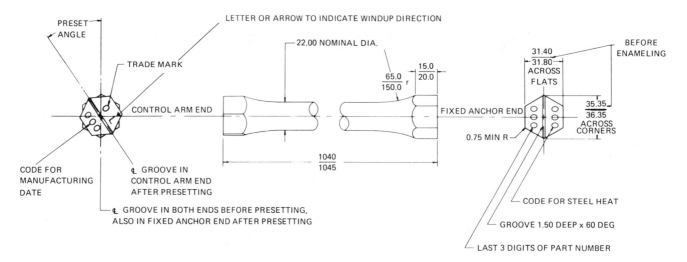

Fig. 3.1—Typical design of hexagonal end connection

be sufficiently large to preclude high stress in the hex section which would cause end failures. Ratios of 1.4, 1.3, and 1.2 for classes 1000, 900, and 800, respectively, have proved satisfactory in laboratory and field testing. For classes below 800, a minimum ratio of 1.20 is recommended.

A typical design is shown in Fig. 3.1. The dimension across the hexagon flats has an overall tolerance of 0.40 mm. A minimum corner radius of 0.75 mm is specified, to go with a 1.00 mm maximum corner radius in the anchor. To insure against the possibility of the corners being too round, it is necessary to specify a dimension across the corners of the torsion bar hexagon. Otherwise, a cam action by the hexagon in the anchor is conceivable which may cause a bursting of the anchor.

2. Anchor Member

A. Serrated Anchor

Stresses in the torsion bar anchor or hub will not be as high as they are in the bar, although high local stresses exist. The internal serrations must also be designed with generous fillets in order to avoid high stress concentrations. Shot peening is not necessary.

The high pressure angle required for a durable serration design will produce large radial components of the flank pressures, and, therefore, a tendency to burst the anchor, as if it were loaded by internal hydraulic pressure. Adequate material must be provided behind the teeth to withstand this internal pressure.

The internal serrations should be longer than the external serrations and so positioned that they will overlap the external serrations at both ends, regardless of tolerances and assembly variations.

The hardness of the anchor can be considerably lower than that of the bar. In many installations it has been held below Rockwell C 30. If broaching is scheduled to follow heat treating, the hardness should not exceed Rockwell C 36 in consideration of tool life. (See ANSI B92.1.)

B. Hexagonal Anchor

As in the serrated anchor, local high stress areas exist in the hexagonal anchor. They are located at the corners of the internal hexagon, and unless sufficient wall thickness is maintained, the anchor will split through the corners 180 deg apart. This will be particularly true if the torsion bar has overgenerous radii at the corners. The bar will then act as a cam and cause a wedging action.

The internal portion of the hexagon anchor can be formed by broaching or extruding, but the parallelism of the sides must be held within 0.004 mm/mm of length. A surface finish of approximately 1.9 μm or less should be maintained.

The hardness of the anchor should be substantially below the hardness of the torsion bar. Anchor hardness (of internal faces) of Rockwell C 25-30 has been found satisfactory. This lower hardness permits the relatively rough finish on the bar (generally in the as-forged condition) to "seat" into the anchor faces. Otherwise, the bar tends to bear only on the high spots. This introduces serious stress concentrations in the bar which have been found to lead to premature failure. However, the hardness cannot be substantially less than Rockwell C 25 or settling of the bar end into the anchor face will change the relative position of the anchor and torsion bar appreciably.

3.16

The length of the internal hexagon should be such that the torsion bar is in full contact with the anchor. The transition portion must be protected from corrosion if the anchor extends beyond the bar so as to form a pocket for the retension of corrosive elements.

A maximum clearance of 0.50 mm across the flats has been found to provide the necessary clearance between the bar and the anchor. The minimum clearance is determined by the relative position of the anchors and the desired ease of assembly.

3. Transition Section

The transition from bar diameter to end diameter should be gradual in order to keep stress concentrations to a minimum. A taper of 30 deg included angle and a fillet radius of 1.5 times bar diameter have been found to be satisfactory. Some users have specified an arc instead of a uniform taper in blending the end diameter with the bar diameter. A radius of four times the bar diameter has been found to be satisfactory.

With splined torsion bars where thin-walled anchors are utilized, and especially where a tubular anchor structure is used to reverse the direction between a solid torsion bar and a tubular torsion bar, additional transition configuration controls are required which are not shown in Fig. 3.2. Specifically, the sharp edge produced at the juncture of the transition radius or taper and the end diameter must be radiused. Experience to date has shown a radius of three times the spline depth is adequate.

The active length "L" of the torsion bar spring is not immediately apparent on the spring drawing which generally notes the overall length (L_{oa}) and the lengths of the two end sections (L_{end}). Adjacent to each end section there is a "transition section" of length l which may be considered to consist of one *inactive* portion close to the end section of diameter D, and one *active* portion close to the active bar diameter (d). This portion has the same flexibility as a bar of diameter d and an "equivalent length" l_e. See the sketches in Fig. 3.2.

The type of transition section (uniform taper or arc, Fig. 3.2) affects the flexibility of the bar.

When the transition section has the form of a uniform taper where the length l can be readily computed or measured, the equivalent length l_e equals the product of l and a factor Q which is found in Fig. 3.3 at the intersection between the appropriate D/d value on the abscissa and the Q curve (see right hand ordinate scale).

When the transition section has the form of a circular arc with radius r, the length l cannot be readily established on the drawing, but it can be computed as the product of the bar diameter d and the factor l/d which in turn is found on the left hand ordinate scale in Fig. 3.3 at the intersection between the appropriate D/d value on the abscissa and the appropriate r/d curve (with interpolation between two curves if necessary). Then the equivalent length l_e equals the product of l and a factor V which is found in Fig. 3.3 at the intersection between D/d and the V curve (see right hand ordinate scale).

The active torsion bar spring length is:

$$
\begin{aligned}
L &= L_{oa} - 2\,L_{end} - 2\,l + 2\,l_e \\
&= L_{oa} - 2\,L_{end} - 2\,(l - l_e) \\
&= L_{oa} - 2\,L_{end} - 2\,l\,(1 - Q) \text{ or} \\
&= L_{oa} - 2\,L_{end} - 2\,l\,(1 - V)
\end{aligned}
$$

The derivation of the Q and V factors follows.

The windup angle of a short element of length dx and diameter y under an applied torque T is

$$
d\theta = \frac{32\,T\,dx}{\pi\,y^4\,G}
$$

From this, the windup angle of a uniformly tapered section can be calculated by integration as

$$
\theta_{taper} = \frac{32\,T\,l}{\pi\,d^4\,G} \times \frac{1}{3}\left[\frac{1}{D/d} + \left(\frac{1}{D/d}\right)^2 + \left(\frac{1}{D/d}\right)^3\right]
$$

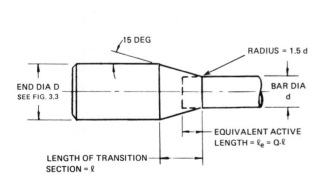

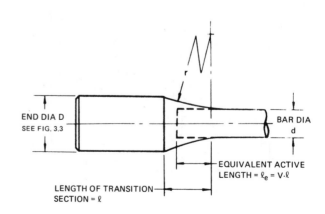

Fig. 3.2—Transition with uniform taper (left); and with circular arc (right)

For the transition section of length ℓ between the active bar of diameter d and the inactive end of diameter D (D = root diameter for the serrated end; D = inscribed diameter for hexagonal end), the equivalent active length ℓ_e at diameter d is V·ℓ when transition is an arc of radius r, Q·ℓ when transition is a uniform taper.

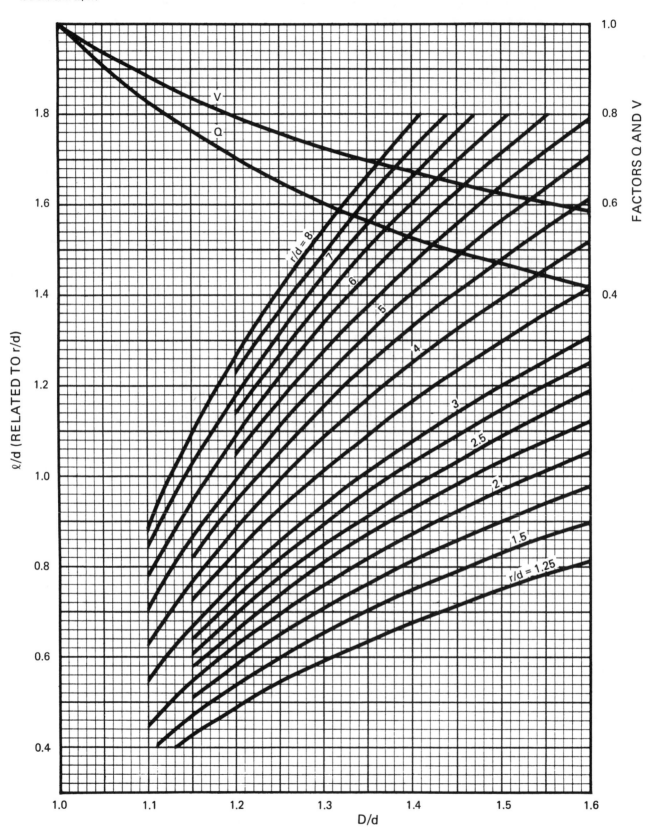

Fig. 3.3—Torsion bar springs, transition data

where:

θ_{taper} = Windup angle of a uniformly tapered section

    d = Diameter of small end, equal to diameter of bar
    D = Diameter of large end
    ℓ = Length of taper

This is correct for any taper angle. Comparing this equation with the equation for windup angle in Chapter 2, we see that the tapered portion has the same flexibility as a bar of diameter d and a length ℓ_e

$$\frac{\ell_e}{\ell} = \frac{1}{3}\left[\frac{1}{D/d} + \left(\frac{1}{D/d}\right)^2 + \left(\frac{1}{D/d}\right)^3\right] = Q \text{ (Fig. 3.3)}$$

where:

ℓ_e = Equivalent active length of transition section at bar diameter d

From the formula for windup angle d θ, the windup angle of the transition section formed as a circular arc can be found by integration as

$$\theta_{arc} = \frac{32\,T\,\ell}{\pi\,G\,d^4} \times \frac{1}{48\left(\frac{D}{d}\right)^3}\left[8 + 10\frac{D}{d} + 15\left(\frac{D}{d}\right)^2\right.$$

$$\left. + 15\frac{\left(\frac{D}{d}\right)^3}{\sqrt{\frac{D}{d} - 1}}\text{ arc tan }\sqrt{\frac{D}{d} - 1}\right]$$

Comparing this equation with the equation for windup angle in Chapter 2, we see that the transition section

has the same flexibility as a bar of diameter d and a length ℓ_e

$$\frac{\ell_e}{\ell} = \frac{1}{48\left(\frac{D}{d}\right)^3}\left[8 + 10\frac{D}{d} + 15\left(\frac{D}{d}\right)^2\right.$$

$$\left. + 15\frac{\left(\frac{D}{d}\right)^3}{\sqrt{\frac{D}{d} - 1}}\text{ arc tan }\sqrt{\frac{D}{d} - 1}\right] = V \quad \text{(Fig. 3.3)}$$

It should be noted that this equation is developed for a parabolic arc rather than a circular arc because of ease of calculation. This license introduces an insignificant error.

We can consider the value ℓ_e the equivalent active length of the transition section. Fig. 3.3 gives the value of this equivalent in a convenient form. The total active length of the bar will be L = Length between transition sections plus equivalent active length of both transition sections.

A transition section formed by a circular arc may be specified by the use of the following formulae (Fig. 3.3).

$$\frac{r}{d} = \frac{\left(\frac{\ell}{d}\right)^2}{\left(\frac{D}{d} - 1\right)} + \frac{\frac{D}{d} - 1}{4}$$

$$\frac{\ell}{d} = \sqrt{\frac{r}{d}\left(\frac{D}{d} - 1\right) - \left(\frac{\frac{D}{d} - 1}{2}\right)^2}$$

For a given bar diameter and end diameter, the length of the transition section will be directly specified by the radius of the arc.

Chapter 4

Control of Assembly Position

1. Methods to Insure Correct Vehicle Height

A. Infinitesimal Adjustment by Screw

This system is desirable where vehicle height is important and where a definite preload is necessary to prevent reversals of load. The screw adjustment is being used to overcome tolerance stackups of mating parts, variances in fit between external and internal parts, variances in load to be carried, and also, to overcome settling problems in service.

Fig. 4.1 shows two typical torsion bar adjusting mechanisms using hexagonal torsion bar anchor ends to transmit torque from the torsion bar to the anchor housing. A screw operating a lever on the anchor housing converts the torque into a force and lever arrangement. Spherical or cylindrical seats must be used to toggle the bolt in order to eliminate any possibility of bolt bending.

During assembly, the parts are positioned near the desired free position, the torsion bar is assembled to the anchor housing which is attached to the reactionary member, and the torsion bar is wound up to the desired angle.

Variances of the adjusting mechanism range from bolts being used in tension, to bolts being used in compression and a number of other combinations.

Provisions for protecting the assembly from rust are commonly used.

B. Adjustment by Vernier Steps

This system is desirable for experimental installations where the static load or standing height may need to be changed. It is subject to errors and, therefore, not desirable in large production.

The two ends of the bar are made with different numbers of serrations; for example, 55 serrations on one end, 57 serrations on the other. Each step of adjustment is then equal to $(360/n_1) - (360/n_2)$ deg, where n_1 and n_2 are the numbers of serrations at the two ends.

For convenience in assembly, the end with the smaller number of serrations should also be smaller in outer diameter, so that it will pass through the bore of the large internal serration.

In assembly, the lever must then be located in the desired free position (zero load position) and the bar turned until a position is found at which both serrated ends will enter.

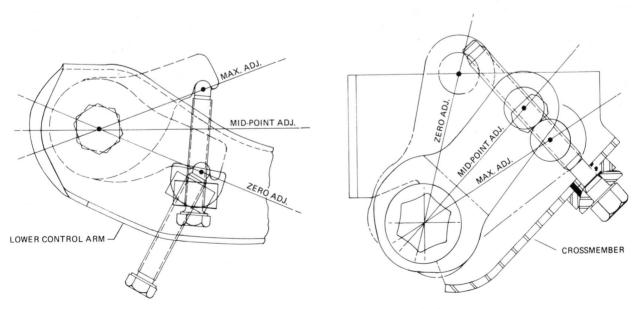

Fig. 4.1—Two typical torsion bar adjusting mechanisms

C. Equal Serrations at Both Ends

In this system, the smallest step of adjustment equals 360/n deg, where n is the number of serrations. The assembly tolerances will usually be smaller than 360/n, so that adjustment during assembly is not possible.

The internal serrations in lever and anchor must be broached in relation to fixed reference lines and the bar must be so machined and preset that the serrations at both ends are in proper relation. The permanent set after the first windup must be measured, and the angle of succeeding windups adjusted correspondingly. This has been done in large production runs without difficulty.

During assembly, the lever is positioned near the desired free position, and the bar is inserted, forcing the lever to the correct position. This type of assembly is used for volume production where vehicle height consideration and settling problems are of relatively minor importance.

D. Blocked Serrations at Both Ends

This system can be applied to (Section C) in order to insure correct assembly. One tooth is removed from each external serration and one space is omitted in each internal serration (by removing one tooth from the broach, see Fig. 4.2). These blocking serrations are kept in proper position in anchor, lever, and bar, so that only the correct assembly position is possible.

Where bars with clockwise and counterclockwise windup are used, the blocking serrations can be so arranged that bars will fit only in the correct station.

E. Variable Height Systems

These systems fall into two broad categories; mechanically actuated and hydraulically actuated. It is usual to apply one of the systems outlined in Sections B, C, and/

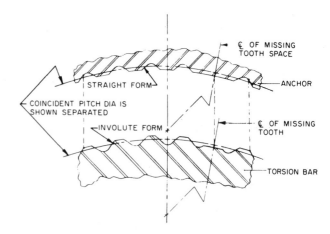

Fig. 4.2—Application of missing tooth and missing tooth space for indexing

or D to assure proper system assembly and component alignment. Beyond that, proper vehicle elevation and/ or attitude may be manually or automatically controlled or a combination of these.

Power supply and control system configurations also present endless combinations and can involve many mechanical and hydraulic subsystems outside of the scope of this Manual.

2. Marking of Windup Direction

Even where blocked serrations are used, bars should be marked for direction of windup to avoid mistakes in fabrication and assembly. This should also be done on bars which are not preset, but will be stressed at or near the recommended operating stresses. Such bars will take a small permanent set and must be reinstalled in a station with the same windup direction if they have been removed for servicing.

Chapter 5
Material and Processing

1. General Requirements

Selection of material and proper specification and control of processing are vital factors in the production of torsion bars. Applications involving high stresses necessitate strict hardenability and quality requirements for the material and suitable control of decarburization, heat treating, shot peening, and presetting, plus quality assurance provisions, even to the extent of fatigue testing of samples of each production lot. From this ultimate category, torsion bars vary all the way to parts using heat treated carbon steels at low stresses involving no special requirements or processing. The following brief discussions and tables are presented as basic information to assist the designer in detailed specifications. They will also serve to direct attention to certain aspects of processing which are of critical importance.

A. Material

The hardenability of the specific material selected for a torsion bar must not only guarantee that an adequate surface hardness can be achieved, but must also insure that the proper center hardness to achieve the optimum residual stress distribution—as determined by experience—will be obtained. Since the surface of a spring is of major importance, the quality of steel ordered should reflect experience gained under service conditions.

Bars which operate at a high stress level may require purchase of Magnaflux Quality steel (Aircraft Quality if an aircraft application is specified), which details special steel testing practices such as magnetic particle inspection. Also, rigid inspection standards for finished springs should be specified as the application requires. For bars designed to operate at somewhat lower stresses, it is recommended that a standard of surface quality be established by negotiation between the manufacturer and the steel producer, and that surface inspection standards be made as definite as possible to avoid misunderstandings.

Centerless grinding of hot rolled steel may be utilized for two primary purposes. The grinding insures the tolerances required on the bar diameter; also the amount of steel ground off may be sufficient to improve surface conditions.

Table 5.1 suggests hardenability specifications for common sizes of alloy bars. To achieve a 90% minimum martensitic structure, it is necessary to select a steel that will meet these minimum hardenability criteria. Typical hardenability by grade can be determined from Hardenability Bands for Carbon and Alloy H Steels, SAE Handbook, J1268. The depth of hardness for a given steel may be improved by utilizing a strong oil quench with violent agitation. The extent to which the required improvement can be achieved is usually determined by experiment.

Grade selection of carbon steels is usually based on carbon and manganese content, with standard SAE 1065 (G 10650) through SAE 1095 (G 10950) analyses generally chosen; these are usually fine grained killed steels. Standard Hardenability Bands for Carbon H Steels, SAE Handbook J 1268.

Table 5.2 gives recommendations for magnetic particle inspection specifications. Category A is in the form currently used for heavy duty class 1000 applications. See Table 2.2 in Chapter 2, Section 6A.

Category B represents the limited stated requirements

TABLE 5.1—RECOMMENDED MINIMUM HARDENABILITY

Approximate Dia. of Body, mm	Minimum Hardenability, Jominy Distance in mm at 55 Rockwell C
25	9.5
38	13.0
51	16.0
63	19.0

TABLE 5.2—RECOMMENDED MAGNETIC PARTICLE INSPECTION REQUIREMENTS

Category	Requirement (Applicable to Hardened Springs)
A	No evidence of circumferential indications, longitudinal cracks, heavy seams, continuous light indications, heat checks, laminations, or flakes. Light longitudinal indications within a circumferential width of 8 mm shall not exceed a longitudinal length of 100 mm individually and/or collectively, or as agreed between steel supplier and user.
B	No evidence of circumferential indications, longitudinal cracks, or heavy seams.
C	As agreed between supplier and manufacturer.

for highly stressed bars when extensive negotiations have taken place. Similarly the lower stressed bars designated as category C generally do not have written drawing requirements and the supplier agreements are less comprehensive in scope.

B. Value of Shear Modulus

The value of the shear modulus "G" recommended for use in design and stress calculations is 76.0×10^3 MPa for steel. While somewhat lower values have been used in the past, recent tests confirm the recommended figure, which has become accepted as standard for hardenable carbon and low allow steels in many modern reference works and handbooks. Since tests are partly dependent on shape, the values of G thus obtained show some fluctuation, but for torsion bars above 16 mm diameter and flat bars requiring hot forming, the suggested figure of 76.0×10^3 is applicable. 79.3×10^3 MPa is recommended as the value of the shear modulus for round torsion bars below 16 mm diameter. The shear modulus is also known as the modulus of rigidity and torsional modulus of elasticity.

Identical values of G should be used for torsion bars with and without presetting. A substantial number of windup tests show that the rate of the torsion bar—and therefore G—is noticeably reduced as an immediate consequence of presetting (while all dimensional values remain unchanged in the presetting operation). However, the greater part of this loss is restored within relatively few days, and the remaining difference between the original and final G values is less than the normal variation between products of successive heats of steel, or between successive tests on the products of any one heat of steel.

C. Upsetting and Machining

For torsion bars with machined ends (such as serrations) and machined surfaces in the A and B categories, it is recommended that the material either be die forged or upset from bar stock. The die should be designed to result in uniform material removal in machining over the body and ends of the bars. A microscopic study of the longitudinal cross section of the upset or forged ends should show a grain flow structure following the contour of the bar.

When bars are upset, it is usually considered advisable to normalize them. If experience has indicated that the heating for forging or normalizing can lead to cracking, especially with high hardenability materials, the cooling should be controlled. This may be accomplished either by slowing down the cooling rate near the exit end of the furnace or by covering the heated portion. Since the surface must be of superior quality, it is necessary not only to provide adequate stock for removal, but to achieve a quality surface in machining. It is generally recommended that highly stressed parts be ground rather than merely turned.

No circumferential grind or tool marks should be evident. Regardless of machining methods, which are optional to the manufacturer, there must be no stress risers in the surface. These generally result from cutting or grinding blends, tool tears, or similar faulty machining conditions.

Table 5.3 suggests machining allowances based on the diameter of the raw stock. For torsion bars of other than round section, it is not a general practice to machine or grind the surfaces, but it is desirable to limit decarburization and surface imperfections by agreement between steel supplier and manufacturer. For certain applications, such as deck lid springs, even the normal hot rolled surface may suffice. Because of the cost of surface removal, each application should be surveyed individually for stock removal requirements. Also, for very high volume applications, special surface quality negotiations with suppliers may result in less allowance than would be otherwise necessary.

TABLE 5.3—RECOMMENDED MINIMUM STOCK REMOVAL

Dia. of Body, mm	Diametral Allowance, mm	
	Classes 700—900	Classes 1000—1250
Over — To Include		
0.0 — 12.5	0.75	1.50
12.5 — 19.0	1.06	2.24
19.0 — 25.0	1.40	3.00
25.0 — 37.5	1.70	3.75
37.5 — 50.0	2.24	5.00
50.0 — 63.0	2.80	6.30
63.0 — 90.0	3.55	8.00

Most automotive torsion bars have hexagon forged ends that are neither ground nor machined. Experience has shown that such torsion bars manufactured of SAE 15B62H (H15621) and SAE 5160H (H51600) steels require no annealing after upsetting and forging prior to heat treating.

D. Heat Treatment

In heating for hardening, bars should be adequately soaked prior to quenching. Highly stressed bars should be heated in a controlled atmosphere furnace, in a neutral salt bath, or by electrical methods to minimize pitting, scaling, and surface decarburization. This is mandatory if either the bar body or serrations are finished before heat treatment. Quenching should be done in oil at a temperature of 40—80°C with rather vigorous agitation. Bars should be removed from the quenching medium while still warm and transferred immediately to the tempering furnace to avoid cracking. The actual process must be checked metallurgically to insure that the hardening transformation is complete. To avoid the necessity of straightening, bars are frequently heated and quenched while hanging in a vertical position or quenched in a machine which spins them during the cooling. Periodic checks should be made of the as-quenched hardness in cross sections of the bars as well as of the tempered parts. Adequate metallurgical tests should also be performed to check for retained austenite, decarburization, and proper microstructure. The presence of decarburization, retained austenite, or nonmartensitic structure will result in decreased fatigue life.

Hardness of highly stressed bars (Class 800 and above) is generally specified as Rockwell C 47-51. Steels of superior surface and internal quality as processed can be used at values up to about Rockwell C 54. Ordinary quality steels intended for lower stress (Class 550) applications may be as low as Rockwell C 40. It is desirable to specify as high a hardness as possible consistent with quality so as to minimize permanent set in operation, but not so high that excessive breakage is encountered.

Hardness testing on the surface of torsion bars must be avoided because any indention becomes a stress riser. Sample coupons or sacrifical bars may be included in the heat treat lot for testing purposes.

E. Straightening

For most applications, the straightness of a torsion bar is not critical, and requirements would be on the order of perhaps 3 mm Total Indicator Reading (TIR) in 300 mm of length. It is essential, however, that steps be taken to avoid cracking should straightening be undertaken. In the case of hardened alloy steel springs, it is recommended that straightening be performed while the bars are still hot from the tempering operation.

F. Shot Peening

Springs should be shot peened to the maximum obtainable intensity consistent with cost limitations. In the instance of the most highly stressed bars, a minimum intensity of 10 C on the body and 7 C on the serrated ends is desirable. Because of the necessity of peening the root radii of the serrations, a smaller diameter shot must be used, than would normally be economical and consistent with the body peening requirement. As a general rule, the shot diameter for reaching into the serration root

radius should be approximately one half the radius. One common practice is to peen the spring with a mixture of 60% large and 40% small diameter shot (by weight) using S550 and S280 nominal diameter steel shot, for example. Intensity measurements in accordance with SAE J443 should be made at regular intervals. The common requirement is for 90% minimum visual surface coverage. It is the intent to provide complete "visual" coverage or even, in most cases, coverage in excess of 100%, obtained by repeated slow passes through the peening equipment.

Reduced peening intensity even to the extent of 12 A minimum is frequently specified on automotive applications because of the added cost of obtaining the very high intensities required for critical military usage. If experience so indicates, limited or no peening on the relatively lightly stressed ends of some commercial bars may be acceptable or peening may be limited to stressed areas such as bends.

Recent work in the 1000—1250 Classes of high stress level series torsion bars has indicated the desirability of ascertaining the quality of the shot peened surface through residual surface stress measurements. Use of high speed, x-ray diffraction equipment allows the collection of a statistically significant number of distributed readings from a production sample item. Desirable residual stress levels can be ascertained during the development program of the particular item as it is dependent on stress level imposed and expected cyclic life history.

G. Presetting

The preferred method of presetting bars is to twist them through a specified angle, release and twist at least two more times to the same end position. The twist angle should correspond to about 0.022 radians strain and the resultant set should not exceed 0.008 radians strain as discussed in Chapter 5, Section 2. Set in excess of the latter amount indicates the physical properties of the bar are subnormal. In large volume operations with hexagon ends, it is often more satisfactory to establish an angle for each lot which will give the desired set condition, and twist one time to the value established.

It is generally considered desirable to peen bars prior to presetting, and most processing schedules show that sequence. Nevertheless, the two operations have been successfully used in reverse order. Since the aim for peening and presetting is the same—to secure favorable residual stresses—the sequence can be influenced by peening intensity and bar diameter as well as the possible obliterating effect of one operation on the other. Practical considerations concerning processing requirements and preset angle tolerances may frequently be the deciding factors for the choice of sequence.

H. Corrosion Protection

Since corrosion results in pits and other surface imperfections leading to reduced fatigue life, it may be desirable that precautions be taken in this regard depending on the material and application. Corrosion protection should be accomplished by adequate cleaning followed immediately by applying an acceptable primer. Materials and procedures as defined in Federal Specification TT-C-490 for cleaning and pretreatment methods, TT-P-636 for alkyd coatings, and TT-P-1757 for zinc chromate primer, have been found satisfactory. Commercial practice has been to apply one or two coats of inexpensive baked enamel or epoxy paint following the cleaning operation. Serrations and other exposed areas should be coated with a good grade of No. 2 grease or equivalent. When additional protection to the surface of the bar is required, it is recommended that a double layer of plastic tape be applied. Both vinyl and polyethylene materials have been used. Other materials may be used depending upon the nature of the environment. Rubber sleeves and plastisol coatings have been used, but care must be taken that deterioration of the coating will not permit hidden corrosion.

I. Special Testing

As is evident from the previous discussions, a wide variety in detailed processing can be employed to produce satisfactory parts. To provide the most economical torsion bars of suitable quality, it is necessary to define requirements accurately and completely and then substantiate the selected processing by adequate fatigue testing.

To establish the confidence limits inherent in a given material and process, or to prove the efficacy of process control, it is necessary to perform fatigue tests on the finished parts. Detrimental residual stresses from improper heat treatment or inadequate peening will readily show up in reduced fatigue life, as will poor surface, improper presetting, or other defects. It is common practice to test at a higher loading than expected in service to provide an expedited test, as discussed elsewhere in this Manual. Since the fatigue test is the ultimate test of spring quality, it provides the basic data for evaluation of processing changes and variables.

Corrosion testing to evaluate coatings and protective covers at the attaching end may be carried out in conjunction with fatigue testing or as a separate procedure.

The following test procedures have been established by SAE, all of which are published in the SAE Handbook:

SAE J406, Methods of Determining Hardenability of Steels

SAE J417, Hardness Tests and Hardness Number Conversions

SAE J419, Methods of Measuring Decarburization

SAE J420, Magnetic Particle Inspection

SAE J421, Cleanliness Rating of Steels by the Magnetic Particle Method

SAE J443, Procedures for Using Standard Shot Peening Test Strip

SAE J448, Surface Texture

It will be noted however, that no standard fatigue testing procedure is given and it is believed important to impart a note of caution when establishing such a procedure. By use of the information contained in this Manual, the engineer can design a test which is meaningful and capable of accomplishment.

2. Presetting

A. Load Deflection Curve

Presetting is an operation which allows the designer to use, in his calculation, operating stresses higher than the initial elastic limit of the material. It increases the load capacity of the torsion bar in the direction of preset; but reduces it in the opposite direction. The operating torque must always be in the direction of the preset torque and should never be reversed. Torsion bars used for clockwise and counterclockwise operation should not be preset. An exception may be made for bars highly stressed in one direction only. A preset in this direction will benefit the life of the bar even if this results in an actual stress increase in the opposite (low stress) windup direction. However, care must be taken that this actual stress does not exceed the safe limit.

Theory and practice of the presetting operation have been well established for torsion bars of circular cross section, and the following paragraphs deal with such bars. While the presetting operation can and should also be applied to rectangular torsion bars, the limits of presetting windup and the amounts of allowable set strain on these bars have not been as thoroughly explored.

Fig. 5.1 shows the load deflection diagram obtained during the presetting of a torsion bar spring of approximately 58 mm diameter and 1930 mm active length. The loading begins at point 0 and progressing to point I shows a linear increase in torque with windup angle. The slope of the line 0-I indicates a torsional rate of 765×10^3 N · mm/deg. After point I, partial yielding occurs and the rate of torque buildup decreases up to point II which represents the maximum specified presetting torque. Upon removal of the load, the bar returns from point II to point III elastically and along a straight line. The distance 0-III on the abscissa represents a permanent set of approximately 30 deg. The slope of line II-III now indicates a torsional rate of 740×10^3 N · mm/deg. This drop of torsional rate immediately after presetting appears in most

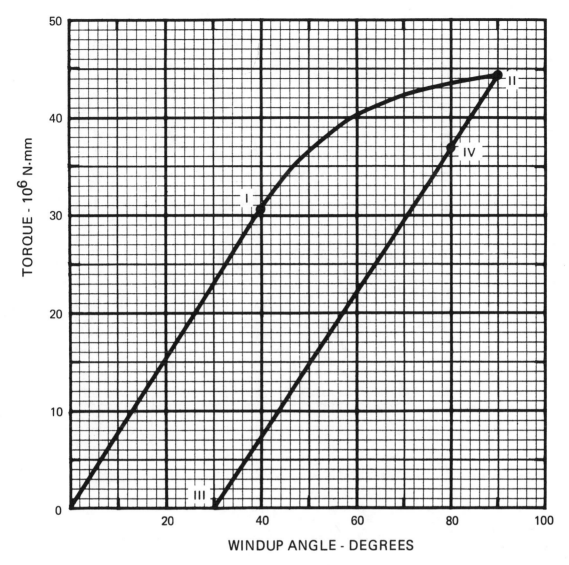

Fig. 5.1—Load-deflection diagram during presetting

cases but it is of a temporary nature, in other words, the apparent drop in shear modulus will persist for a short time only. (See Section 1 of this chapter.) At point II, all the material in the bar was either in the yield range or below. Since the yield point is now at point II, any subsequent deflection from point III, for example to point IV, will be elastic. Then, if the operating torque is lower than the presetting torque, the maximum load torque in service will correspond to a point such as point IV, and the yield point will not again be reached.

Yielding does not occur instantly; therefore, if point IV is to be very close to point II, the bar should be held at point II for several seconds or it should be loaded to that point several times.

B. Stress Distribution

The mechanism by which the bar produces the extra load carrying capacity is illustrated by Fig. 5.2. Here

the curve 0-I-II is identical with that in Fig. 5.1, but the coordinates have been changed to stress versus strain.

Strain γ, shown along the abscissa, is determined from $\gamma = \theta d/2\,L$.

Nominal stress S_s, shown along the ordinate, is derived from $S_s = 16\,T/\pi\,d^3$.

This stress has no physical significance above point I, because it is calculated by a formula which is true only when Hooke's law holds—that is, below the elastic limit. But the actual stress at the surface can be derived from curve 0-I-II by a method shown by G. B. Upton[1] as follows:

Through a point A on line I-II draw a vertical AD. Lay a tangent to curve I-II through point A, which intersects the Y axis at point B. Drop vertically from point A by a distance equal to 1/4 OB to point C. The actual

[1] G. B. Upton, "Materials of Construction." Timoshenko, "Strength of Materials," Part 2, should also be consulted.

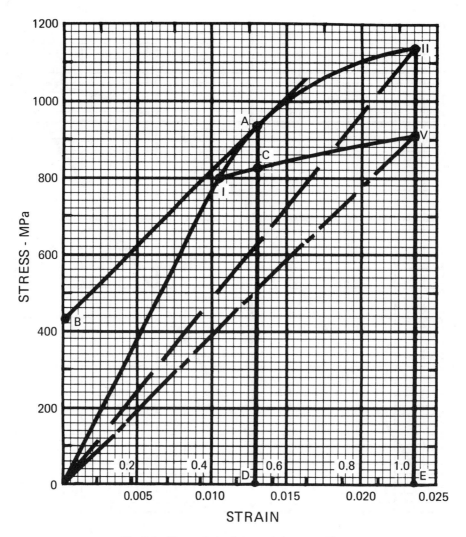

Fig. 5.2—Stress-strain diagram during presetting

stress at the surface of the bar is equal to DC when the nominal stress equals DA and the strain equals OD.

Curve I-V is drawn by connecting a number of points found in this way and is the actual stress-strain curve for a point at the surface of the bar.

We see that the material has strain hardened from 800 MPa at point I to 910 MPa at point V, which explains part of the increased load capacity. But even more is due to a change in stress distribution.

If the material of the bar is homogeneous, any other element must have the same stress-strain curve as a particle at the surface. Also shown along the abscissa is a linear scale, with the maximum strain OE equal to unity (upper scale). It is generally considered proved that during twisting of a round bar, the cross sections remain plain and undistorted, so that the strain at any point is proportional to radius. Therefore, at full preset, when the surface is at strain OE, a point at 0.6 radius is under a strain of 0.6 times OE and its stress is shown by the point of curve I-V above that abscissa.

The curve 0-I-V then shows the actual stresses under

full preset windup plotted against radius as abscissa. Linear stress distribution would give stresses along the straight line 0-V, and the excess of stresses above this line accounts for the increase in stored energy.

The ordinate E-II shows the nominal stress at the surface when the bar is under full preset windup. When the bar is released, this nominal stress becomes zero, and the actual stress will be reduced by stress E-II. A trapped stress equal to E-II minus E-V will then remain at the surface. Similar reasoning applied to points below the surface shows that the trapped stresses at any radius are indicated by the difference between curve 0-I-V and the straight line 0-II. Thus the beneficial trapped stresses for this example do not go any deeper below the surface than where the diameter is 75% of the bar diameter which is the point of intersection of 0-I-V and O-II.

The trapped stresses have been measured experimentally and found to agree with this diagram.[2]

[2] H. O. Fuchs and R. L. Mattson, "Measurement of Residual Stress in Torsion Bar Springs." Proceedings of the Society for Experimental Stress Analysis, Vol. 4, No. 1.

C. Selection of Preset Strain

In Fig. 5.3 the curve 0-I-II of Fig. 5.2 is replotted on different coordinates. The abscissa is the same as in Fig. 5.2 except it is now called "applied strain γ_{ap}."

Plotted as ordinate is "elastic recovery strain γ_{er}" which is equal to the nominal stress (curve 0-I-II in Fig. 5.2) divided by the shear modulus. The difference between applied strain and elastic recovery strain is the permanent set strain γ_{set}.

For the range 0-I, the elastic recovery strain is equal to the applied strain, since we are operating elastically or below yield point I.

For range I-II, the elastic recovery strain is less than the applied strain or, in other words, the bar has taken a permanent set. The difference between applied strain and recovery strain is the permanent set strain $\gamma_{set} = \gamma_{ap} - \gamma_{er}$.

The ordinates of any point along the diagonal O-I-VI are, by construction, equal to the applied strain as explained for the range 0-I. Consequently, the vertical distance between curve 0-I-II and diagonal 0-I-VI represents the set strain γ_{set}.

The nominal stress, identical with that shown in Fig. 5.2, has also been plotted along the ordinate.

In this form, the diagram applies to any straight bar of circular cross section, made of the same material and heat treated the same way, as the original test bar. For

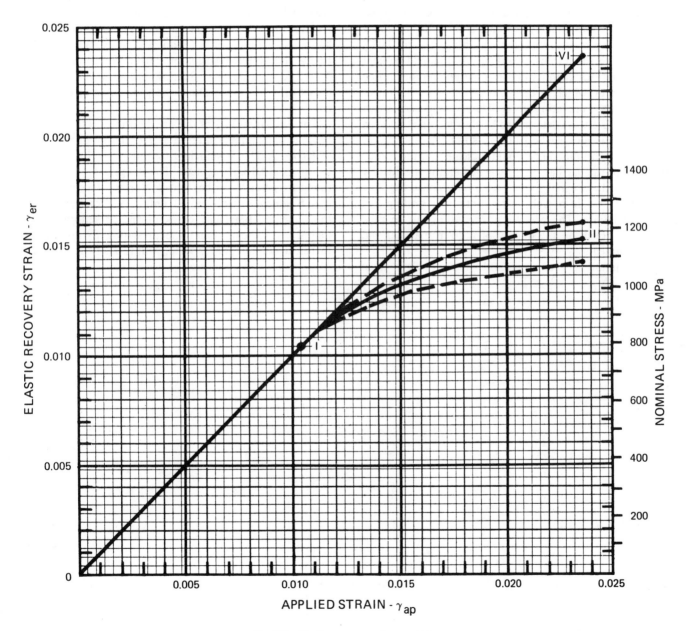

Fig. 5.3—Elastic recovery of a round bar

3.29

any other material or heat treatment, a new load deflection diagram similar to Fig. 5.1 must be established by a simple torsion test. From this data, the corresponding elastic recovery diagram of Fig. 5.3 can then be constructed. Any point along curve I-II shows the elastic recovery strain or nominal stress which is produced by presetting up to the corresponding applied strain. The diagram may therefore be used to determine the proper presetting for a new design. Assuming that Fig. 5.3 represents the elastic recovery characteristics of our material and heat treatment, we proceed to select the proper presetting strain.

Due to production tolerances in chemical composition and heat treatment, the elastic recovery will vary in a range limited by the two dashed lines. If, for example, we want a bar to withstand an operating stress of 1000 MPa nominal, we should preset to a strain $\gamma_{ap} = 0.022$ so that the minimum expected nominal presetting stress S_s becomes approximately 1070 MPa or about 7% above the operating stress. With this applied strain of 0.022, the set strain should not exceed 0.008. It should be kept in mind that this quantity depends entirely on the chosen applied strain, which in turn is determined by the maximum expected operating stress. The set strain, as shown by Fig. 5.3 is always determined by the difference between applied strain and elastic recovery strain. Any set strain higher than this difference would indicate that the torque capacity of the bar has been exceeded and it is evidently inferior to the original test bar used to establish Fig. 5.3. Such a bar would be subject to continuous settling and eventual failure. The presetting, therefore, can be used as a final check on the quality of a given material and heat treatment.

The relation between strain and windup angle is given by

$$\theta = 2 \, L \, \gamma/d \text{ (radians)}$$

or

$$\theta = 114.6 \, L \, \gamma/d \text{ (degrees)}$$

If presetting is used on through hardened bars, there is no reason to stop the process short of an applied strain of about $\gamma_{ap} = 0.022$, even if the operating stress is less than 1000 MPa. At this strain, the presetting penetrates to about one-half diameter so that three-quarters of the material is loaded to yield stress. Presetting to a larger applied strain will increase the torque capacity very little more, but the chances of damaging the material are increased. Therefore, an applied strain of $\gamma_{ap} = 0.022$ is recommended for torsion bar springs made of through hardened material of a hardness near Rockwell C 50. Using this applied strain of 0.022, the windup angle $\theta = 2.52 \, L/d$ degrees. Shallow hardened bars may be preset only to slightly above the yield stress of the material.

Torsion bar springs of the size described in this chapter have at maximum windup a stored energy of the order of 16 000 J. This is sufficient to impart a velocity of 900 km/h to a 0.5 kg fragment. The danger to personnel is therefore great if the bar breaks during presetting, even if a fragment is accelerated only by a fraction of the total enengy. Bars may crack during presetting due to inclusions or seams, and the presetting fixture should therefore contain guards which will protect the operator.

Chapter 6

Fatigue Life

Fatigue testing is an accelerated method of examining springs for design adequacy and for quality control purposes. Fatigue life is expressed by the number of deflection cycles a spring will withstand without failure. It can be estimated by the use of the fatigue strength diagram shown in Fig. 6.1. Reexamination of the design will be in order if the fatigue tests result in failures which are confined to one section of the spring.

The procedure to establish suitable parameters for a fatigue test frequently requires the exercise of judgment when the spring application is known to involve random stress amplitudes. As an example, a suspension spring will undergo a large number of cycles of small amplitude near the design load position without failure. Under greater amplitudes, the number of cycles without failure will be reduced, since the maximum stress as well as the stress range are increased, and both are determining factors in the fatigue life of a spring.

Fig. 6.1 shows a relationship between the initial and maximum stress in the test cycle, and the estimated minimum number of cycles to failure, for a solid round alloy steel torsion bar, heat treated to Rockwell C 47-51, shot peened, preset, then subjected to unidirectional loading only, as are most preset bars.

The chart covers fatigue tests under this loading condition, with the specified stresses maintained throughout the test. It provides a ready means of estimating minimum fatigue life for any selected method of testing or a means

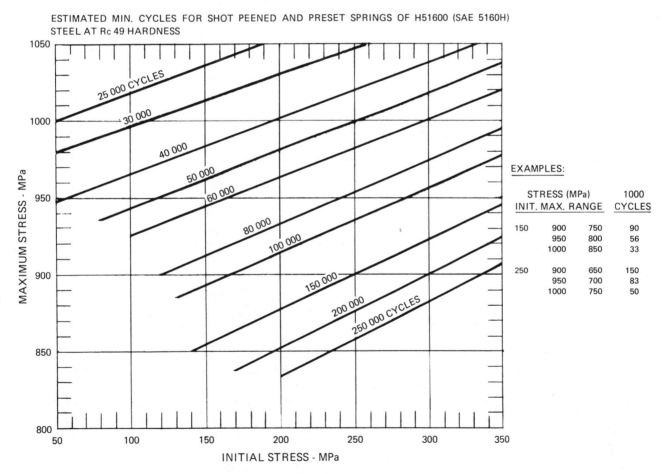

ESTIMATED MIN. CYCLES FOR SHOT PEENED AND PRESET SPRINGS OF H51600 (SAE 5160H) STEEL AT Rc 49 HARDNESS

EXAMPLES:

| STRESS (MPa) | | | 1000 |
INIT.	MAX.	RANGE	CYCLES
150	900	750	90
	950	800	56
	1000	850	33
250	900	650	150
	950	700	83
	1000	750	50

Fig. 6.1—Torsion bar spring life tests

TABLE 6.1—MEDIAN RANKS (PERCENT) FOR SAMPLE SIZES 1 TO 30

Sample Size

Rank Order	1	2	3	4	5	6	7	8	9	10	Rank Order
1	50.000	29.289	20.630	15.910	12.945	10.910	9.428	8.300	7.412	6.697	1
2		70.711	50.000	38.573	31.381	26.445	22.849	20.113	17.962	16.226	2
3			79.370	61.427	50.000	42.141	36.412	32.052	28.624	25.857	3
4				84.090	68.619	57.859	50.000	44.015	39.308	35.510	4
5					87.055	73.555	63.588	55.984	50.000	45.169	5
6						89.090	77.151	67.948	60.691	54.831	6
7							90.572	79.887	71.376	64.490	7
8								91.700	82.038	74.142	8
9									92.587	83.774	9
10										93.303	10

Sample Size

Rank Order	11	12	13	14	15	16	17	18	19	20	Rank Order
1	6.107	5.613	5.192	4.830	4.516	4.240	3.995	3.778	3.582	3.406	1
2	14.796	13.598	12.579	11.702	10.940	10.270	9.678	9.151	8.677	8.251	2
3	23.578	21.669	20.045	18.647	17.432	16.365	15.422	14.581	13.827	13.147	3
4	32.380	29.758	27.528	25.608	23.939	22.474	21.178	20.024	18.988	18.055	4
5	41.189	37.853	35.016	32.575	30.452	28.589	26.940	25.471	24.154	22.967	5
6	50.000	45.951	42.508	39.544	36.967	34.705	32.704	30.921	29.322	27.880	6
7	58.811	54.049	50.000	46.515	43.483	40.823	38.469	36.371	34.491	32.795	7
8	67.620	62.147	57.492	53.485	50.000	46.941	44.234	41.823	39.660	37.710	8
9	76.421	70.242	64.984	60.456	56.517	53.059	50.000	47.274	44.830	42.626	9
10	85.204	78.331	72.472	67.425	63.033	59.177	55.766	52.726	50.000	47.542	10
11	93.893	86.402	79.955	74.392	69.548	65.295	61.531	58.177	55.170	52.458	11
12		94.387	87.421	81.353	76.061	71.411	67.296	63.629	60.340	57.374	12
13			94.808	88.298	82.568	77.525	73.060	69.079	65.509	62.289	13
14				95.169	89.060	83.635	78.821	74.529	70.678	67.205	14
15					95.484	89.730	84.578	79.976	75.846	72.119	15
16						95.760	90.322	85.419	81.011	77.033	16
17							96.005	90.849	86.173	81.945	17
18								96.222	91.322	86.853	18
19									96.418	91.749	19
20										96.594	20

of selecting a desired stress range for the fatigue test. It is recommended that this fatigue setup produce at least an average of 50 000 and preferably an average of 100 000 cycles.

There are several considerations why the test setup should be preferred to be such that it will result in a preponderance of cycle figures which are neither too short nor too long. At higher stresses (shorter lives), the scatter of the cycle lives is theoretically reduced, so that fewer test samples will produce a given degree of precision in the estimated life of the entire population. However, lower stresses (longer lives) give more realistic results, since they duplicate more nearly the actual service conditions and prevent spring settling during the test. Also, comparisons between different groups of springs will be more distinct at lower stresses, since different S-N curves tend to diverge the more they approach the fatigue limit (or limiting value of the stress at which 50% of the population would survive a very large number of cycles, say $n = 10\ 000\ 000$).

In order to establish the fatigue life cycles which are acceptable in any spring design, it is desirable to have road durability tests run over a prescribed course so that fatigue life test data and actual road durability results may be correlated.

It must be understood that the fatigue life cycles for any group of springs will vary considerably even under closely controlled test conditions. Moreover, the average life of the tested springs is not sufficient by itself to establish a judgment either on the design or on the material or on the production method which they represent. The relationship between the number of applied cycles and the percentage of springs which failed at these cycles can best be analyzed with the help of statistical techniques which will systematically describe the "dispersion" or "spread" or "scatter" of the recorded test results. The extent of the scatter will depend upon the consistency of surface condition, fabrication, and the general quality of the springs which are tested.

The following paragraphs have been more fully treated

TABLE 6.1—(CONT'D)

Rank Order	Sample Size										Rank Order
	21	22	23	24	25	26	27	28	29	30	
1	3.247	3.101	2.969	2.847	2.734	2.631	2.534	2.445	2.362	2.284	1
2	7.864	7.512	7.191	6.895	6.623	6.372	6.139	5.922	5.720	5.532	2
3	12.531	11.970	11.458	10.987	10.553	10.153	9.781	9.436	9.114	8.814	3
4	17.209	15.734	15.734	15.088	14.492	13.942	13.432	12.958	12.517	12.104	4
5	21.890	20.015	20.015	19.192	18.435	17.735	17.086	16.483	15.922	15.397	5
6	26.574	25.384	24.297	23.299	22.379	21.529	20.742	20.010	19.328	18.691	6
7	31.258	29.859	28.580	27.406	26.324	25.325	24.398	23.537	22.735	21.986	7
8	35.943	34.334	32.863	31.513	30.269	29.120	28.055	27.065	26.143	25.281	8
9	40.629	38.810	37.147	35.621	34.215	32.916	31.712	30.593	29.550	28.576	9
10	45.314	43.286	41.431	39.729	38.161	36.712	35.370	34.121	32.958	31.872	10
11	50.000	47.762	45.716	43.837	42.107	40.509	39.027	37.650	36.367	35.168	11
12	54.686	52.238	50.000	47.946	46.054	44.305	42.685	41.178	39.775	38.464	12
13	59.371	56.714	54.284	52.054	50.000	48.102	46.342	44.707	43.183	41.760	13
14	64.057	61.190	58.568	56.162	53.946	51.898	50.000	48.236	46.592	45.056	14
15	68.742	65.665	62.853	60.271	57.892	55.695	53.658	51.764	50.000	48.352	15
16	73.426	70.141	67.137	64.379	61.839	59.491	57.315	55.293	53.408	51.648	16
17	78.109	74.616	71.420	68.487	65.785	63.287	60.973	58.821	56.817	54.944	17
18	82.791	79.089	75.703	72.594	69.730	67.084	64.630	62.350	60.225	58.240	18
19	87.469	83.561	79.985	76.701	73.676	70.880	68.288	65.878	63.633	61.536	19
20	92.136	88.030	84.266	80.808	77.621	74.675	71.945	69.407	67.041	64.832	20
21	96.753	92.488	88.542	84.912	81.565	78.471	75.602	72.935	70.450	68.128	21
22		96.898	92.809	89.013	85.507	82.265	79.258	76.463	73.857	71.424	22
23			97.031	92.809	89.447	86.058	82.914	79.990	77.265	74.719	23
24				97.153	93.105	89.847	86.568	83.517	80.672	78.014	24
25					97.265	93.628	90.219	87.042	84.078	81.309	25
26						97.369	93.861	90.564	87.483	84.603	26
27							97.465	94.078	90.885	87.896	27
28								97.555	94.280	91.186	28
29									97.638	94.468	29
30										97.716	30

in Section 3 of Chapter 8 of the Manual on Design and Application of Leaf Springs, SAE HS J788 APR80 and the interested reader is referred to that Manual.

One of the main purposes of statistical analysis is to draw inferences about the properties of a large group (the "population") from the results of tests on a small group (the "sample"). It is, in fact, possible to infer a great deal from the small sample when it has been properly selected. Only experience can tell the engineer whether the various properties of the sample which can affect the test result (in regard to material, to production methods, to controls) will also be present in the production springs (the "population"). This determination is outside the realm of statistics. However, statistical mathematics are based on the inherent assumption that the sample is a true representative of the population.

If the entire population were tested under identical test conditions, the results could be shown in graphical form by arranging them in ascending numerical order and plotting the cumulative fraction (or percent) of failures over an abscissa of "life cycles." A sample selected at random from this population can be expected to exhibit a similar distribution of fatigue life; the larger the number of springs in the sample, the closer will be the similarity.

It is possible to calculate the likelihood of similarity for samples of any given size. Tables are available based on such likelihoods; as an example, Table 6.1 presents the "median rank" of each test result for sample sizes between 1 and 30. A rank is assigned to each individual test result corresponding to that portion of the population which it is most likely to represent. Table 6.1 lists percent figures.

The median rank line constructed from such data predicts that certain percentages of the population will survive specific cycles-to-failure. But any such estimate may err substantially on either the high or low side. How confident can the engineer be of such an estimate? When judgment based on experience is not considered adequate for a final decision on this question, it will be necessary to construct lines of higher confidence.

In many cases, a confidence level of 90 or 95 or even 99% will be required, so that there will remain only a 10 or 5 or 1% risk of the estimate being either too high or too low. For a chosen confidence level, the life cycles of a given percentage of the population will be found within a certain "tolerance interval." On the median rank graph this may be represented by a "tolerance band" to either side of the median rank line. Wider bands indicate increased doubt about the line truly representing the population. With a given sample size, the bands will be wider for higher confidence levels. With a given confidence level, the bands will be wider for a smaller sample size.

Several systems of mathematically organizing the test result data have been established. In the automotive industry, the Weibull plot has found wide acceptance because it permits straight-line plotting of the cumulative failure probability versus life cycles on Weibull probability graph paper.

In the Weibull distribution, the relationship between the number of applied cycles and the cumulative percent of failures at these cycles is expressed by a formula which uses three parameters:

a) The minumum life, which may or may not be zero. It is denoted by the letter "a." When a = zero, the sample data will plot a straight line on Weibull paper. Otherwise, the points will plot a curved line, but this can be converted to straight-line plotting by a relatively simple technique which is detailed in the Leaf Spring Manual SAE HS J788 APR80).

b) The Weibull slope, which is an indicator of the skewness of the distribution. It is called the "shape parameter" and is denoted by the letter "b." It also is a measure of the scatter of the distribution; a low slope value indicates a high degree of scatter, and vice versa. The Weibull comes nearest to the normal (or Gaussian) distribution when b = 3.44.

The Leaf Spring Manual shows a graph (for Weibull slopes 1 through 12) which locates the percent of failed springs at the life cycles representing the mean of the population.

c) The characteristic life, which is the 63.2% failure point for the population. It is called the "scale parameter" and is denoted by θ:

$$63.2 = 100\,(1 - 1/e)$$

where:

$$e = 2.7183 \text{ (the Napierian base)}$$

Example 1: Eight springs have been fatigue tested under identical conditions. The results are arranged in ascending order of the failure cycles and are given rank order numbers accordingly. In this order, they are assigned median ranks from Table 6.1 as shown in Table 6.2.

These points are plotted on Weibull graph paper in Fig. 6.2. Drawing a straight line of best fit through the median rank points produces an estimate for the failure rate of the entire population with the parameters b = 2.4 and θ = 170 000.

For b = 2.4, the percent of failed springs at the mean is 52.7 according to the Leaf Spring Manual. At that failure level the median rank line in Fig. 6.2 shows

TABLE 6.2

Spring	Order No.	Cycles to Failure	Median Rank
E	1	61 000	8.30
A	2	91 000	20.11
F	3	114 000	32.05
H	4	135 000	44.02
C	5	155 000	55.98
B	6	177 000	67.95
G	7	205 000	79.89
D	8	245 000	91.70

150 000 cycles. At the B-10 level (that is, at the number of cycles where 10% of the population is estimated to fail) the median rank line shows 66 000 cycles.

In most cases, the fatigue testing of springs will be undertaken for the purpose of comparing different samples, and the probability graph will be expected to convey information on the relative life distribution of the populations represented by these samples. For example, the comparison may involve a sample representing a first design, and another sample representing a second design.

When the two median rank lines for the test data of the two samples are plotted on the same graph, they will readily show if the second design promises some improvement in fatigue life. However, in order to establish if there is a "significant difference" between the two designs, it will be necessary to find quantitative values for the degree of improvement (or degradation) between one design and the other. The question is this: "How confidently can one say the limited test results indicate that the second design assures an improvement in fatigue life for the entire spring population?" The answer depends not only on the amount of separation between the two plotted slopes, but also on the size of the two test samples.

The degree of confidence in the superiority of one design over the other need not be constant from one quantile level to another. For example, there may be a significant improvement at the B-50 life (50% failure level) without any improvement at the B-10 life (10% failure level). This is partly due to differences in the Weibull slopes, and partly due to the greater width of the tolerance bands at the lower quantile levels.

Example 2: It has been proposed that the spring design represented by the sample of eight in Example 1 be replaced by a new design. Seven springs of the new design have been fatigue tested with results which are shown arranged in ascending order and assigned median ranks from Table 6.1 as shown in Table 6.3.

For this second plot, it will be seen from Fig. 6.2 that the Weibull slope b = 2.7 and the characteristic life θ = 310 000; the mean life level is at 51.8%, therefore the estimated mean life is 280 000 cycles. The estimated B-10 life is 138 000 cycles.

The Leaf Spring Manual explains the method by which

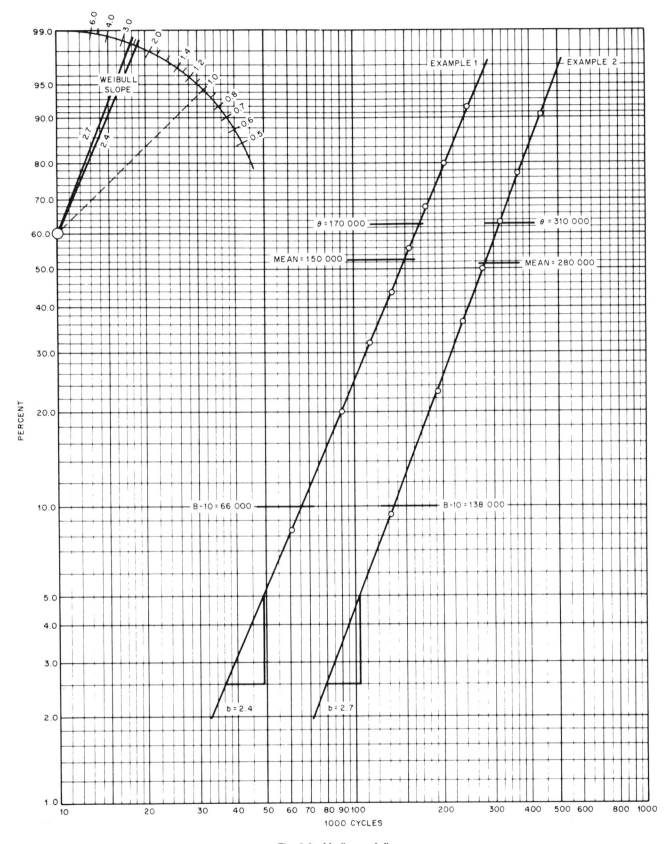

Fig. 6.2—Median rank lines

3.35

TABLE 6.3

Spring	Order No.	Cycles to Failure	Median Rank
M	1	132 000	9.43
O	2	195 000	22.85
K	3	233 000	36.41
L	4	275 000	50.00
P	5	315 000	63.59
J	6	365 000	77.15
N	7	440 000	90.57

the amount of improvement over the original design (Example 1) can be reduced to percentage figures at certain degrees of confidence. The result is in this case that with 90% confidence a 32% improvement can be expected at the mean life level, an 8% improvement at the B-10 level.

Chapter 7

Torsion Bar Spring Applications

1. Square Torsion Bar for Boat Trailer (Fig. 7.1)

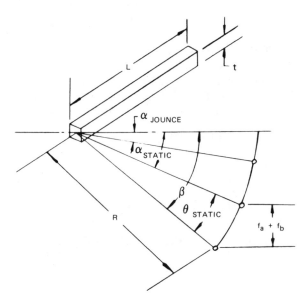

Fig. 7.1—Square torsion bar, boat trailer

Specifications

Installation:	Boat trailer suspension: two transverse bars, one on each side of vehicle; see Fig. 7.1 for schematic layout
Length of lever:	$R = 450$ mm
Static load:	$P_{static} = 2000$ N
Static deflection:	$f_a + f_b = 100$ mm
Position of lever at static load:	$\alpha_{static} = -22.5$ deg
Angle of lever at jounce position (limited by component interferences):	$\alpha_{jounce} = -12.5$ deg
Maximum torsion bar length:	$L = 550$ mm
Maximum torsion bar size:	$t^2 = 25 \times 25 = 625$ mm²

Material: UNS G41500 (SAE 4150)
Hardness: 415-461 Bhn
Shear modulus: 76.0×10^3 MPa

Determined from Calculations

$L = 529$ mm
$t = 20$ mm
$S_{s(static)} = 500$ MPa
$S_{s(jounce)} = 840$ MPa
$k_T = 56.6 \times 10^3 \dfrac{N \cdot mm}{deg}$
Total deflection to jounce position = 175 mm
$P_{jounce} = 3182$ N

Calculation

Determine static torque:

$$T_{static} = P_{static} \times R \times \cos \alpha_{static}$$
$$= 2000 \times 450 \times 0.9239$$
$$= 831.5 \times 10^3 \ N \cdot mm$$
$$f_a + f_b = R \sin \alpha + R \sin \beta = 100 \ mm$$

Therefore $\sin \beta = \dfrac{100 + 450 \times \sin 22.5 \ deg}{450}$

$$= 0.6049$$

$$\beta = 37.2 \ deg$$

Total deflection to
jounce position $= R \sin \beta + R \sin \alpha_{jounce}$
$$= 450 \times 0.6049 - 450 \times 0.2164$$
$$= 175 \ mm$$

Torsional spring rate:

$$k_T = \frac{T_{static}}{\theta_{static}} = \frac{T_{static}}{\beta + \alpha_{static}} = \frac{831.5 \times 10^3}{14.7}$$

$$= 56.6 \times 10^3 \frac{N \cdot mm}{deg}$$

$$= 3240 \times 10^3 \ N \cdot mm/rad$$

From $k_T = \dfrac{T_{jounce}}{\beta + \alpha_{jounce}}$ we obtain:

$$T_{jounce} = 56\,600\,(37.2 - 12.5)$$

$$= 1398 \times 10^3\,N \cdot mm$$

$$T_{jounce} = P_{jounce} \times R \times \cos \alpha_{jounce}$$

$$P_{jounce} = \frac{1398 \times 10^3}{450 \times 0.9763} = 3182\,N$$

From Fig. 2.2: $\eta_2 = 0.208$ $\eta_3 = 0.141$
Assume $t = 20.0$ mm

$$L = \frac{\eta_3\,t^4\,G}{k_T} = \frac{0.141 \times 20^4 \times 76 \times 10^3}{3240 \times 10^3}$$

$$= 529\,mm$$

$$S_{s(static)} = \frac{T_{static}}{\eta_2\,t^3} = \frac{831.5 \times 10^3}{0.208 \times 20^3} = 500\,MPa$$

$$S_{s(jounce)} = \frac{k_T\,(\beta + \alpha_{jounce})}{\eta_2\,t^3}$$

$$= \frac{56\,600\,(37.2 - 12.5)}{0.208 \times 20^3}$$

$$= \frac{1398 \times 10^3}{1.664 \times 10^3} = 840\,MPa$$

Note: This high stress is considered acceptable in this application as experience has shown that the jounce position is attained infrequently during operation.

2. Hexagonal Torsion Bar for Truck Tilt Cab (Fig. 7.2)

Specifications

Installation: Hexagonal torsion bar for cab-over-engine truck cab counter-balance; torque requirement and maximum inscribed diameter given; secured to chassis on one end, other end torqued.

Torque required (T): 2050×10^3 N · mm under tilt α
1880×10^3 N · mm under tilt β
(see Fig. 7.2)

Windup angles: $\alpha = 23.5$ deg
$\beta = 21.5$ deg

Torsional rate desired:

$$k_T = \frac{T}{\theta} = \frac{(2050 + 1880)10^3}{23.5 + 21.5}$$

$$= 87.3 \times 10^3 \frac{N \cdot mm}{deg}$$

$$= 5000 \times 10^3 \frac{N \cdot mm}{rad}$$

Maximum bar size allowable: 28.0 mm across hexflats

Material: UNS G51600 or G41400 (SAE 5160 or 4140)

Hardness: 415-461 Bhn

Maximum allowable stress in either direction (S_s): 725 MPa

Shear modulus (G): 76.0×10^3 MPa

Data Determined from Calculations

Bar size selected:
L = 683 mm active length
S_s = 698 MPa under 2050×10^3 N · mm torque

Actual torsional rate (k_T) = $87.3 \times 10^3 \dfrac{N \cdot mm}{deg}$

Calculation

Using the formulae for stress and for windup angle with an hexagonal bar from Timoshenko:[1]

$$S_s = \frac{T}{0.188\,d'^3}$$

where d' = diameter of inscribed circle
Therefore:

$$d'^3 = \frac{T}{0.188\,S_s}$$

$$= \frac{2050 \times 10^3}{0.188 \times 725}$$

$$= 15.04 \times 10^3$$

$$d' = 24.7\,mm$$

Select 25 mm hexagonal bar ($d' = 25$ mm)

$$\theta = \frac{T\,L}{0.115\,d'^4\,G}$$

[1] Timoshenko, "Strength of Materials." Part 2, Chapter VI.

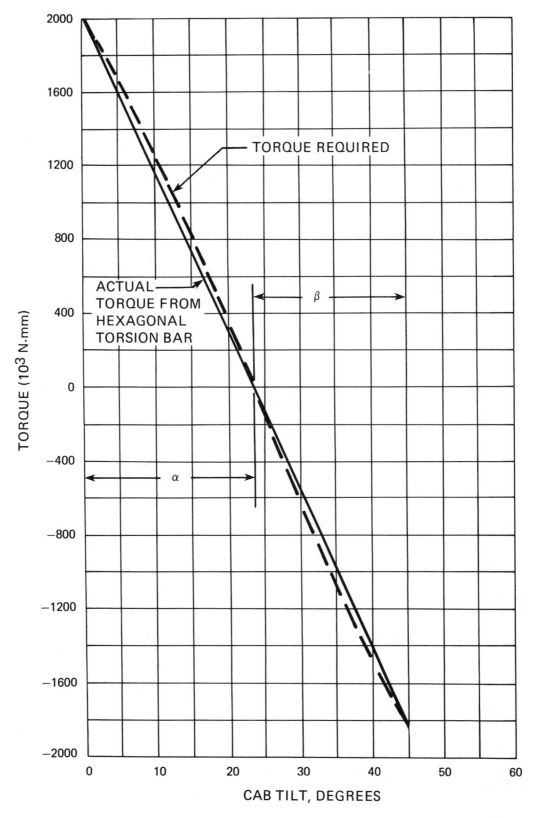

Fig. 7.2—Hexagonal torsion bar—truck tilt cab

3.39

Therefore:

$$L = \frac{0.115 \, d'^4 \, G}{k_T}$$

$$= \frac{0.115 \times 25^4 \times 76.0 \times 10^3}{5000 \times 10^3}$$

$$= 683 \text{ mm}$$

Then:

$$S_s = \frac{2050 \times 10^3}{0.188 \times 25^3} = 698 \text{ MPa}$$

3. Laminated Torsion Bar for Truck Tilt Cab (Fig. 7.3)

Specifications

Installation:	Torsion bar for cab-over-engine truck cab counterbalance; torque requirement and bar length given; torsion bar secured to chassis at centerline, torqued at both ends.
Torque required (T) per side:	1835×10^3 N · mm under tilt α
	770×10^3 N · mm under tilt β
	(see Fig. 7.3)
Windup angles:	$\alpha = 39.5$ deg $= 0.6894$ rad
	$\beta = 15.5$ deg $= 0.2705$ rad
Torsional rate desired (k_T) per side:	$\dfrac{(1835 + 770)10^3}{39.5 + 15.5}$
	$= 47.36 \times 10^3$ N · mm/deg
	$= 2714 \times 10^3$ N · mm/rad
Length per side (L):	700 mm
Material:	G51600 (SAE 5160), leaf spring stock
Hardness:	415-461 Bhn
Maximum allowable stress in either direction (S_s):	725 MPa
Shear modulus (G):	76.0×10^3 MPa

Data Determined from Calculations

Width (w) of all laminae	$= 50$ mm
Number of laminae (N_L)	$= 2$

Thickness of laminae (t)	$= 9.50$ mm
Maximum working stress (S_s)	$= 711$ MPa
Actual torsion rate (k_T)	$= 47.43 \times 10^3$ N · mm/deg per side or 94.86×10^3 N · mm/deg total

Calculation

$$k_T = N_L \cdot \frac{\eta_3 \, t^3 \, w \, G}{L} \text{ (per side)}$$

$$N_L \cdot \eta_3 \, t^3 \, w = \frac{k_T \times L}{G} = \frac{2718 \times 10^3 \times 700}{76.0 \times 10^3}$$

$$= 25\,000 \text{ (required)}$$

$$S_s = \frac{\eta_3 \, \alpha \, t \, G}{\eta_2 \, L}$$

Assuming $\eta_2 = \eta_3$ (Fig. 2.2)

$$t_{max} = \frac{725 \times 700}{76.0 \times 10^3 \times 0.6894} = 9.69 \text{ mm}$$

Assuming w = 50 and t = 9.50 (nearest standard leaf spring size below 9.69)

$$w/t = 50/9.50 = 5.26$$

$$\eta_3 = 0.292$$

$$N_L = \frac{25\,000}{\eta_3 w \, t^3}$$

$$N_L = \frac{25\,000}{0.292 \times 50 \times 9.5^3} = 1.997$$

$$\text{Use } N_L = 2$$

Then

$$k_T = \frac{2 \times 0.292 \times 9.5^3 \times 50 \times 76 \times 10^3}{700}$$

$$= 2718 \times 10^3 \text{ N · mm/rad} = 47.43 \times 10^3 \text{ N · mm/deg (Satisfactory because it is within 0.15\% of the specified rate)}$$

$$S_s = \frac{\eta_3 \, \alpha \, t \, G}{\eta_2 \, L} \quad (\eta_3/\eta_2 = 1; \text{ Fig. 2.2})$$

$$= \frac{0.6894 \times 9.5 \times 76.0 \times 10^3}{700} = 711 \text{ MPa}$$

4. Torsion Bar with Integral Torque Arm

The example which follows is similar to torsion bar springs used in the front suspension of some compact and mid-size passenger cars. Two bars are usually so ar-

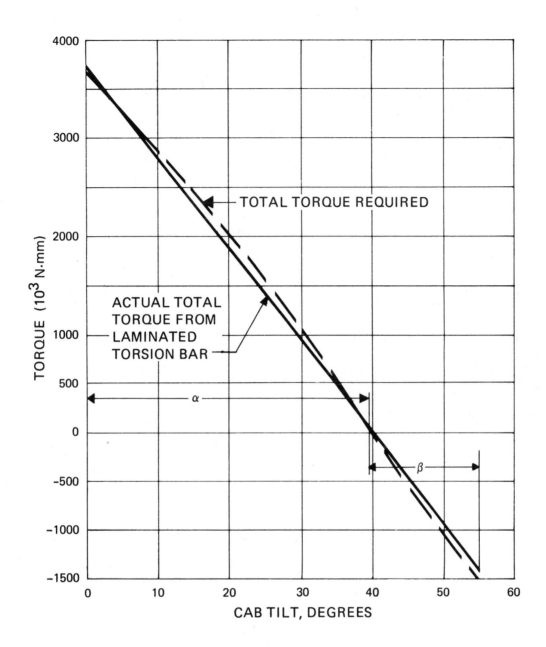

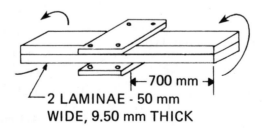

Fig. 7.3—Laminated torsion bar—truck tilt cab

ranged that their torsionally stressed portions are positioned transverse or perpendicular to the centerline of the vehicle. The torque arm portions are substantially parallel to the centerline of the vehicle. The example is simplified in that only two diameters are used in the torsional part of the spring. In practice, both ends are extruded to at least three different diameters to improve the efficiency of the spring.

Bar deflection:

δ_A = Static deflection of bar
P_A = Static load on bar (approximately perpendicular to arm in design position)
k_A = Rate of bar from load P_A

$$\delta_A = \frac{P_A}{k_A} \text{_____} \tag{eq. 1}$$

$$\delta = P\left[\frac{1}{3\,E\,I_1}\left[\!\left[V^3 + (a+b)^3 - a^3\right.\right.\right.$$
$$+ \frac{1}{(1-q_1)}\left\{(L - f - eq_1)^3 - (a+b)^3\right.$$
$$+ q_2\left[\overset{zero}{(L - q_1(e+f))^3} - (L - f - eq_1)^3\right]\!\left.\!\right\}\!\left.\!\right]\!\right]$$
$$+ \frac{(V^2 - a^2)}{G\,J_1}\left[b + e + q_2(f - n)\right]\right]$$

$$\delta = P\left[\frac{1}{3\,E\,I_1}\left[V^3 - a^3 + (a+b)^3 \cdot \frac{q_1}{q_1^{-1}}\right.\right.$$
$$+ (L - f - eq_1)^3 \cdot \frac{q_2^{-1}}{q_1^{-1}}\right]$$
$$+ \frac{(V^2 - a^2)}{G\,J_1} \cdot [b + e + q_2(f - n)]\right]$$
$$\text{_____} \tag{eq. 2}$$

Where:

$$L = a + b + e + f \text{\_\_\_\_\_} \tag{eq. 3}$$

$$q_1 = \frac{L}{e + f} \text{\_\_\_\_\_} \tag{eq. 4}$$

$$q_2 = \frac{I_1}{I_2} = \frac{J_1}{J_2} = \left(\frac{d_1}{d_2}\right)^4 \text{\_\_\_\_\_} \tag{eq. 5}$$

n = Correction for end transition
(see Chapter 3, Section 3)
Load on bar at jounce
$$P_J = P_A + \delta_J k_A \text{\_\_\_\_\_} \tag{eq. 6}$$
δ_J = Jounce travel when α_J (angle of arm travel considered to be small)
Windup angle

$$\alpha_W = \frac{P_A}{k_A} \times \frac{1}{(V^2 - a^2)^{1/2}} \times \frac{180}{\pi} \text{\_\_\_\_\_} \tag{eq. 7}$$

$$S = \frac{16\,P}{\pi\,d_1^3}\left[\left\{\frac{N}{\cos\theta} + (R - N\tan\theta)\sin\theta\right\}\right.$$
$$+ \left\{\left[\frac{N}{\cos\theta} + (R - N\tan\theta)\times\sin\theta\right]^2\right.$$
$$+ \left.\left.[R(1 - \cos\theta) + N\sin\theta]^2\right\}^{1/2}\right]$$
$$\text{\_\_\_\_\_} \tag{eq. 9}$$

Stress at cushion point "C"
$M = P(a + b)$
$T = P(Z^2 - a^2)^{1/2}$

$$S = \frac{16\,P}{\pi\,d_1^3}\left\{(a + b) + [(a + b)^2\right.$$
$$\left.+ (Z^2 - a^2)]^{1/2}\right\} \text{\_\_\_\_\_} \tag{eq. 10}$$

Stress at step point "D"
$M = P\{a + b + e(1 - q_1)\}$
$T = P(Z^2 - a^2)^{1/2}$

$$S = K_s \times \frac{16\,P}{\pi\,d_2^3}\left\{\left[a + b + e(1 - q_1)\right]\right.$$
$$+ \left[[a + b + e(1 - q_1)]^2\right.$$
$$+ \left.\left.(Z^2 - a^2)\right]^{1/2}\right\} \text{\_\_\_\_\_} \tag{eq. 11}$$

K_s = Stress concentration factor at step ≈ 1.1

Design of bar as shown in Fig. 7.4:
Material: SAE 15B62H (UNS H15621)
Hardness: 444-495 Bhn
Preset: 7% over max load
Shotpeened: 2 passes at 90% coverage
Dimensions: (Ref. Fig. 7.4)
$\quad a = 122$ mm
$\quad b = 108$ mm
$\quad e = 90$ mm
$\quad f = 595$ mm
$\quad L = 915$ mm
$\quad Z = 420$ mm
$\quad R = 75$ mm
$\quad d_1 = 32$ mm
$\quad d_2 = 28.5$ mm
$\quad \delta_J = 63.5$ mm
$\quad E = 207 \times 10^3$ MPa
$\quad G = 79.3 \times 10^3$ MPa

$$q_1 = \frac{915}{90 + 595} = 1.33577$$

$$q_2 = \left(\frac{32}{28.5}\right)^4 = 1.60$$

$I_1 = 51\,471$ mm^4
$J_1 = 102\,943$ mm^4
$n = 0.5\,D\,(min) + (1 - V)\ell$
$\quad = 0.5 \cdot 1.2\,d_2 + 0.2 \cdot 0.9\,d_2$
$\quad = d_2(0.60 + 0.18)$
$\quad = 28.5 \cdot 0.78 = 22$ mm
with $D/d = 1.20$ (min)
and $r/d = 4$
$\quad \ell/d = 0.9$
$\quad V = 0.8$

Determine rate of bar from eq. 2:

$$\frac{1}{k_A} = \frac{1}{3 \cdot 207 \cdot 10^3 \cdot 51\,471} \cdot \left[420^3 - 122^3 \right.$$

$$+ (122 + 108)^3 \cdot \frac{1.33577}{0.33577} + (915 - 595$$

$$\left. - 90 \cdot 1.33577)^3 \cdot \frac{0.6}{0.33577} \right]$$

$$+ \frac{(420^3 - 122^3)}{79.3 \cdot 10^3 \cdot 102\,943} \cdot [108 + 90$$

$$+ 1.60 \cdot (595 - 22)]$$

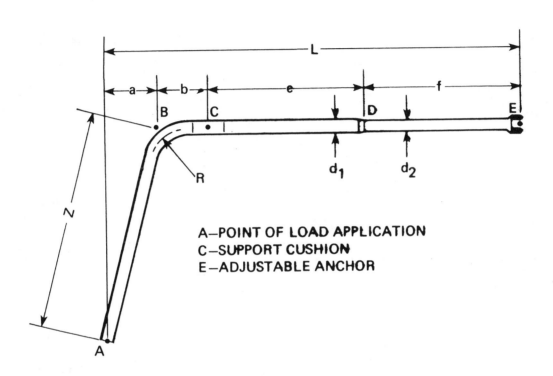

A—POINT OF LOAD APPLICATION
C—SUPPORT CUSHION
E—ADJUSTABLE ANCHOR

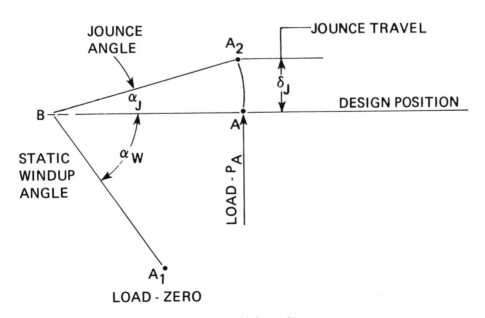

Fig. 7.4—Torsion bar with integral torque arm

$$= \frac{1}{31\,963.5 \cdot 10^6} \cdot \left[\!\!\left[420^3 - 122^3 \right.\right.$$

$$\left.\left. + \frac{230^3 \cdot 1.33577 + 199.7807^3 \cdot 0.6}{0.33577} \right]\!\!\right]$$

$$+ \frac{161\,516}{8163.38 \cdot 10^6} \cdot 1114.8$$

$$= \frac{(74.088 - 1.815848)\cdot 10^6 + \dfrac{(16.25231 + 4.78423)\cdot 10^6}{0.33577}}{31\,963.5\cdot 10^6}$$

$$+ 19.78543 \cdot 10^{-6} \cdot 1114.8$$

$$= \frac{(72.27215 + 62.65164) \cdot 10^6}{31\,963.5 \cdot 10^6} + 0.022057$$

$$= \frac{134.92379}{31\,963.5} + 0.022057$$

$$= 0.004221 + 0.022057$$

$$= 0.026278$$

$$k_A = 38.05 \text{ N/mm}$$

Determine load capacity of bar by rearranging eq. 11, and using stresses for point "D"

at jounce load: $S_J = 1240$ MPa
at design position load: $S_A = 930$ MPa

$$P_J = \frac{1240 \cdot \pi \cdot 28.5^3}{1.1 \cdot 16}$$

$$\cdot \frac{1}{\dfrac{[122 + 108 + 90(1 - 1.33577)]}{+\left[[122 + 108 + 90(1 - 1.133577)]^2\right.} } $$

$$\left. + (420^2 - 122^2)\right]^{1/2}$$

$$= \frac{5\,123\,815}{199.7807 + [199.7807^2 + 161\,516]^{1/2}}$$

$$= \frac{5\,123\,815}{199.7807 + 448.8077} = 7900 \text{ N}$$

$$P_A = 7900 \cdot 930/1240 = 5925 \text{ N}$$

Allowable Travel: $\delta_J = \dfrac{1}{k_A} (P_J - P_A)$

$$= \frac{1}{38.05} (7900 - 5925)$$

$$= 51.9 \text{ mm}$$

Windup angle: $\alpha_W = \dfrac{5925}{38.05} \cdot \dfrac{1}{(420^2 - 122^2)^{1/2}}$

$$\frac{180}{\pi} = 22.2 \text{ deg}$$

Stress in Bend of Bar Point "B"

The maximum value of principal stress was found to be at $\theta = 22$ deg into bend (See Fig. 7.5).

$$\psi = \cos^{-1}\left(\frac{122}{420}\right) = 73.1 \text{ deg}$$

$$N = 420 - 75 \tan (73.1/2) = 364 \text{ mm}$$

Using eq. 9
Stress at jounce

$$S_J = \frac{16 \times 7900}{\pi \times 32^3} \left[\!\!\left[\left\{ \frac{364}{\cos 22 \text{ deg}} + (75 - 364 \right.\right.\right.$$

$$\times \tan 22 \text{ deg}) \times \sin 22 \text{ deg} \Big\} + \Big\{ \Big[\frac{364}{\cos 22 \text{ deg}}$$

$$+ (75 - 364 \times \tan 22 \text{ deg}) \times \sin 22 \text{ deg} \Big]^2$$

$$+ [75 (1 - \cos 22 \text{ deg}) + 364$$

$$\left.\left. \times \sin 22 \text{ deg}]^2 \right\}^{1/2} \right]$$

$$= 1.2279 \left[\!\! \left[365.59 \right.\right.$$

$$\left.\left. + [365.59^2 + (5.461 + 136.357)^2]^{1/2} \right]\!\!\right]$$

$$= 930 \text{ MPa}$$

Stress at Bushing Point "C"

Using eq. 10
Stress at jounce

$$S_J = \frac{16 \times 7900}{\pi \times 32^3} \{(122 + 108) + [(122 + 108)^2$$

$$+ (420^2 - 122^2)]^{1/2}\}$$

$$= 851 \text{ MPa}$$

The allowable stress at points "B" and "C" depends on the margin of safety required in this section of the bar.

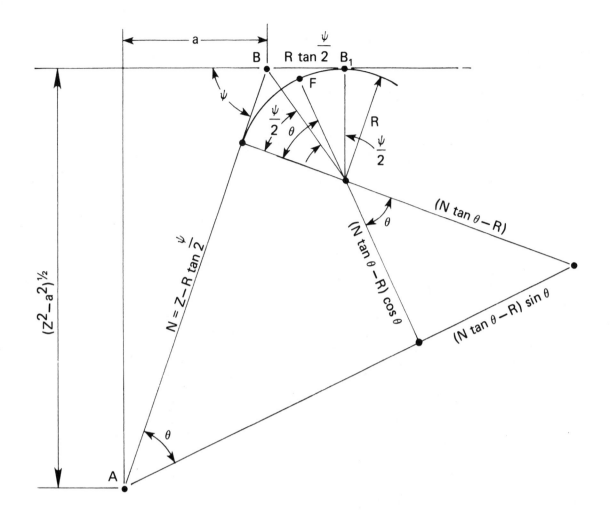

STRESS AT POINT "F" ON ARC OF BEND

MOMENT ON RADIAL SECTION AT

BENDING: $M = P\left[\dfrac{N}{\cos\theta}\ (N\tan\theta - R)\sin\theta\right]$

TORSION: $T = P\left[R + (N\tan\theta - R)\cos\theta\right]$

WHERE: $N = Z - R\tan\dfrac{\psi}{2}$ —————— (eq. 8)

$\psi = \cos^{-1}\left(\dfrac{a}{Z}\right)$

COMBINING TO OBTAIN MAXIMUM PRINCIPAL STRESS
(IN ANY POSITION)

$$S_{max} = \frac{S_b}{2} + \left[\left(\frac{S_b}{2}\right)^2 + S_s^2\right]^{\frac{1}{2}} = \frac{16}{\pi d_1^{\ 3}}\left[M + (M^2 + T^2)^{\frac{1}{2}}\right]$$

Fig. 7.5—Stress in bend

Chapter 8

Stabilizer Bars

1. Applications

Many vehicles utilize stabilizer (anti-roll) bars to increase roll rate for satisfactory handling characteristics. Stabilizer bars are laterally mounted torsional springs which resist vertical displacement of the wheels relative to one another. Vertical suspension rates are not increased when both wheels are deflected simultaneously, however, stiffness is increased for one wheel bump. Passenger car suspensions, tuned to give a soft ride with low rate springs, use stabilizer bars to reduce vehicle roll with only a minor deterioration of ride. Motor homes and pick-up trucks with slide in campers employ stabilizer bars to control body roll caused by a high center of gravity. Stabilizer bars are generally installed in both front and rear suspensions or in front suspension only. Use of a stabilizer bar on the rear suspension only can sometimes have an adverse effect on vehicle handling. Such installations should be tested under severe cornering conditions to ensure the desired handling characteristics.

2. Design Elements

Stabilizer bars are generally one piece with bends to provide arms and meet package constraints. Stabilizer bars consisting of a torsion bar and separate arms are used on race cars and experimental vehicles. Bar diameter is easily changed to determine optimum roll rates.

Material and processing requirements of stabilizer bars are identical to regular torsion bar springs with the exception that stabilizer bars are not preset. Refer to Chapter 2 Table 2.2 and Chapter 5 for details.

Attachments to frame and suspension components are usually rubber insulated. Care must be taken in locating attachment points to avoid interference with suspension motion unless a specific constraint is required. Typical attachments are:

- Bar and bayonet type link (Fig. 8.1)—Rubber insulators and washers are generally common with shock absorbers
- Bar with rubber insulated eye (Fig. 8.2)
- Bar with eye type link (Fig. 8.3)
- Bar with trapped rubber and washers (Fig. 8.4)—

Provides lateral in addition to vertical resistance. May be used as a drag strut on SLA and MacPherson suspensions or a torque arm on live axle suspensions
- Bar with rubber insulator block (Fig. 8.5)
- Bar with silent block bush eye (Fig. 8.6)—Similar to Fig. 8.2—Used to improve durability life

3. Roll Rate Calculations

The total suspension roll rate is due in part to the suspension springs and in part to the stabilizer. When connected to a rigid axle suspension, the roll rate of the stabilizer bar equals its effective roll rate at the wheels. When applied to an independent suspension, however, the bar roll rate is greater than its effective roll rate at the wheels. The required bar roll rate can be obtained by multiplying the desired effective roll rate at the wheels by the square of the ratio of wheel travel to bar travel at the bar end, and by the square of the ratio of bar length to wheel tread:

$$k_R = k_{RW} \times \left(\frac{f_w}{f_A}\right)^2 \times \left(\frac{L}{T}\right)^2$$

where:

k_R = Bar roll rate, N · mm/rad
k_{RW} = Effective roll rate at wheels, N · mm/rad
f_w = Wheel travel
f_A = Deflection at bar end "A"
L = Bar length
T = Wheel tread

When rubber supports are used, as is usually the case, they will add considerable deflection and thereby reduce the calculated roll rate of the bar. The effect of rubber supports, when the rubber rates are known, can be determined from the loads at the points of support C and bar end A. Any connecting links with rubber joints will add to the reduction of roll resistance in the critical stages of initial roll motion. Usually the roll rate of the steel bar is reduced 15–30% by rubber deflections.

The overall configuration of the stabilizer bar, and the points of attachment to the suspension and vehicle struc-

STABILIZER BAR ATTACHMENTS

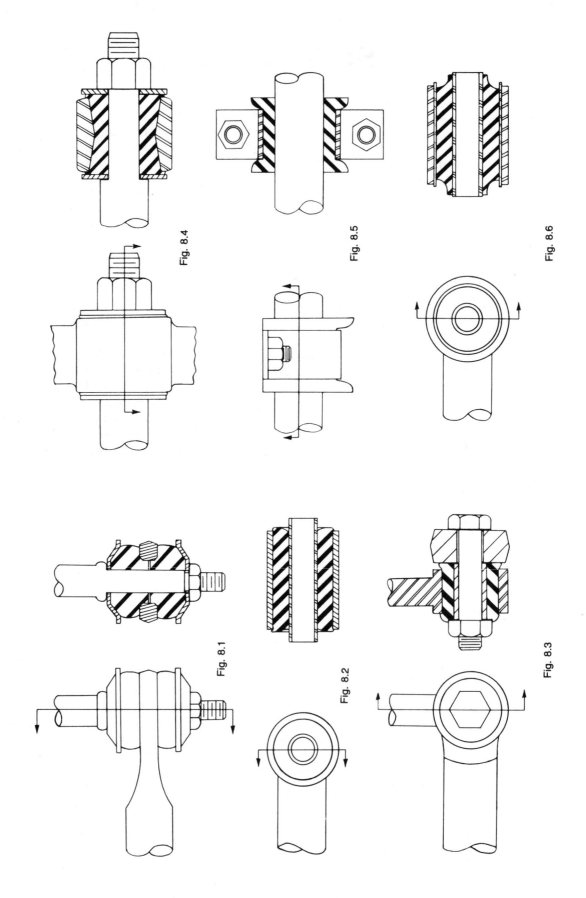

Fig. 8.4

Fig. 8.5

Fig. 8.6

Fig. 8.1

Fig. 8.2

Fig. 8.3

ture are usually determined by layout from available space in the vehicle. With the configuration defined, a bar diameter can be found which will provide the desired roll rate.

Deflection of the bar ends "A" may be determined by the method of unit loads:

$$f_A = \int \frac{M\,m}{E\,I}\,ds + \int \frac{T\,t}{G\,J}\,ds$$

Where M and T represent the bending and torsional moment equations respectively in terms of the distance s from the end of the bar to any section, and m and t are the bending and torsional moment equations due to a force of one Newton acting at the bar end "A" where the deflection is to be found. The integration indicated must be performed over each portion of the beam for which either M, m, T, or t are expressed by a different equation.

The following formulae were derived by this method and may be used to determine the diameter or rate of bars approximating the configuration shown in Fig. 8.7. Rates calculated below are reduced for bars having bends between point C and the centerline of the vehicle.

DEFLECTION OF "A"

$$f_A = \frac{P}{3EI}\left[\ell_1{}^3 - a^3 + \frac{L}{2}(a + b)^2 + 4\ell_2{}^2(b + c) \right]$$

BAR ROLL RATE

$$k_R = \frac{PL^2}{2f_A}$$

$$= \frac{3EIL^2}{2\left[\ell_1{}^3 - a^3 + \dfrac{L}{2}(a + b)^2 + 4\ell_2{}^2(b + c) \right]}$$

BAR DIAMETER

$$d = \sqrt[4]{\left[\frac{128}{3\,\pi} \cdot \frac{k_R}{L^2E} \right]} \cdot$$

$$\left[\ell_1{}^3 - a^3 + \frac{L}{2}(a + b)^2 + 4\ell_2{}^2(b + c) \right]$$

where:

f_A = Deflection at bar end "A", mm
P = Applied load at bar end, N
E = 200 000 MPa
I = $\dfrac{\pi\,d^4}{64}$ mm^4
k_R = Bar roll rate, N · mm/rad

4. Stress

The maximum stress normally occurs at a point on the inside surface of the curved section B. Its magnitude depends upon the inside radius of curvature R_i and the Wahl factor K used in calculating the maximum stress in coil springs.

Since the allowable stress should not exceed 700 MPa (Table 2.2) for a fully hardened bar, R_i is determined as follows:

$$S_s \leq \frac{16\,P\,\ell_2\,K}{\pi\,d^3} \leq 700 \text{ MPa}$$

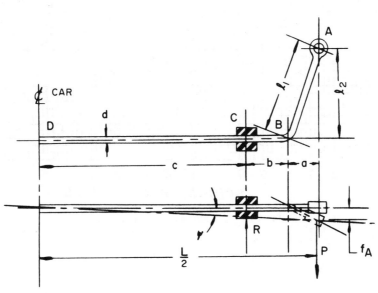

Fig. 8.7—Stabilizer bar

3.49

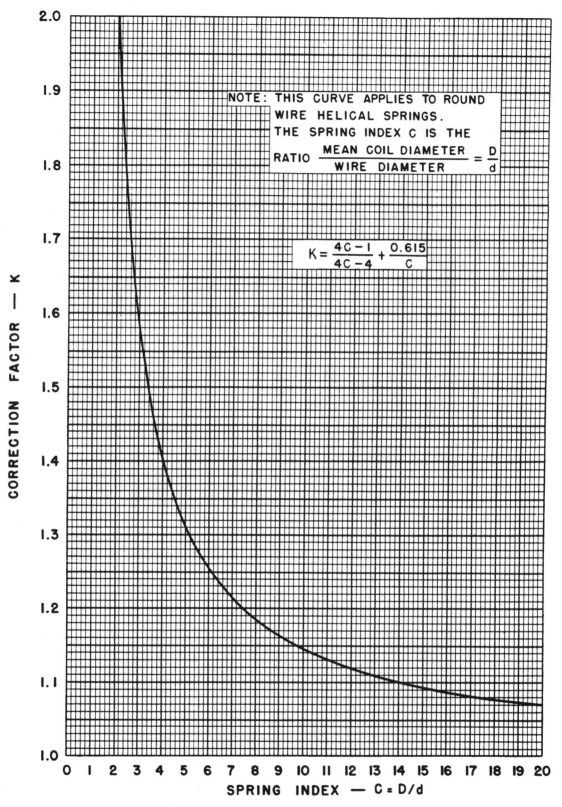

NOTE: THIS CURVE APPLIES TO ROUND WIRE HELICAL SPRINGS. THE SPRING INDEX C IS THE RATIO $\dfrac{\text{MEAN COIL DIAMETER}}{\text{WIRE DIAMETER}} = \dfrac{D}{d}$

$$K = \frac{4C - 1}{4C - 4} + \frac{0.615}{C}$$

CORRECTION FACTOR — K

SPRING INDEX — C = D/d

Fig. 8.8—Wahl stress correction factor

3.50

Therefore $K \leq \dfrac{137\, d^3}{P\, \ell_2}$ = Wahl factor (Fig. 8.8)

from which D/d may be determined, and

$$R_i = \frac{D - d}{2}$$

since $D = 2\, R_i + d$

It is recommended that R_i should at least equal and preferably exceed 1.25d.

The principal shear stress at section B produced by torsion and bending is less than the maximum torsional shear stress on the inside surface of the bend due to the stress concentration factor. On bars with length a + b approaching $2\,\ell_2$, check principal stress (bending and torsion) at attachment point C for critical stress.

Maximum bending stress $S_b = 32\, P\, \ell_1 / \pi d^3$ is usually not critical in a well designed stabilizer bar but should never exceed the maximum yield point in tension (1250 MPa).

5. Design Example
Specifications

Stabilizer bar connected to solid axle.
Required effective roll rate at the wheels 280 000 N · mm/deg
Roll rate loss due to rubber = 30%

L = 1100 mm	ℓ_1 = 250 mm	a = 90 mm
f_A = 75 mm	ℓ_2 = 230 mm	b = 70 mm
		c = 390 mm

Calculation:

$$k_R \text{ (bar roll rate)} = \frac{280\,000}{0.7} = 400\,000 \text{ N · mm/deg}$$

$$= 400\,000 \times 57.3$$

$$= 22.9 \times 10^6 \text{ N · mm/rad}$$

$$\text{maximum } P = \frac{2\, k_R\, f_A}{L^2} = \frac{2 \times 22.9 \times 10^6 \times 75}{1100^2}$$

$$= 2840 \text{ N}$$

$$d = \sqrt[4]{\frac{\left[\dfrac{13.58 \times 22.9 \times 10^6}{1100^2 \times 200\,000} \right] \cdot}{[250^3 - 90^3 + (160^2 \times 550) + (4 \times 230^2 \times 460)]}}$$

$$= \sqrt[4]{\frac{3.113 \times 10^8 \times 1.263 \times 10^8}{2.42 \times 10^{11}}}$$

$$= \sqrt[4]{1.63 \times 10^5} = 20 \text{ mm}$$

For S_s = 700 MPa

Wahl factor $K = \dfrac{137\, d^3}{P\, \ell_2} = \dfrac{137 \times 20^3}{2840 \times 230} = 1.68$

$$D/d \text{ (from Fig. 8.8)} = 2.6$$

Then D = 2.6 × 20 = 52 and

$$R_i = \frac{D - d}{2} = \frac{52 - 20}{2} = 16 \text{ mm}$$

In line with the recommendation that R_i should at least equal 1.25d, it is apparent that R_i should be increased and thus S_s be decreased.

$$\text{maximum bending stress } S_b = \frac{32\, P\, \ell_1}{d^3}$$

$$= \frac{32 \times 2840 \times 250}{\pi \times 20^3}$$

$$= 904 \text{ MPa}$$

To Convert from SI Unit to U.S. Customary Unit, Divide by the <u>Factor</u>
To Convert from U.S. Customary Unit to SI Unit, Multiply by the <u>Factor</u>

Quantity	SI Unit		Factor (§ = Exact)	U.S. Customary Unit	
Length	kilometer	km	1.609 344§	mile	
	meter	m	0.304 8§	foot	ft
	millimeter	mm	25.4§	inch	in
Area	square millimeter	mm²	645.16§	square inch	in²
Volume	cubic millimeter	mm³	16 387.064§	cubic inch	in³
	cubic millimeter	mm³	3 785 412.0	gallon	gal (U.S.)
	liter	L	3.785 412	gallon	gal (U.S.)
Area Moment of Inertia	millimeter to the fourth power	mm⁴	416 231.425 6§	inch to the fourth power	in⁴
Mass	kilogram	kg	0.453 592 37	pound-mass	lb$_m$
Force (or Load)	newton	N	4.448 221 6[a]	pound-force	lb$_f$
Elastic Energy, Work	joule ⎰ J(N · m) ⎱ (= kN · mm)		0.112 984 8	pound inch	lb$_f$ · in
Bending Moment, Torque	newton millimeter ⎰ N · mm ⎱ (= mN · m)		112.984 8	pound inch	lb$_f$ · in
Spring Rate (Linear)	newton per mm N/mm (= kN/m)		0.175 126 8	pound per inch	lb$_f$/in
Torsional Spring Rate	newton millimeter per radian N · mm/rad		112.984 8	pound inch per radian	lb$_f$ · in/rad
Plane Angle	degree		57.295 780[b]	radian	rad
Stress, Modulus of Elasticity	⎰ pascal Pa (N/m²) ⎱ kilopascal kPa ⎰ megapascal MPa		6894.757 3[c] ⎱ 6.894 757 3 ⎰ 0.006 894 757 3 ⎱	pound per square inch lb$_f$/in²	(= psi)
Density of Material e.g. for Steel	kilogram per cubic meter kg/m³ 7850 kg/m³		27 679.90	pound per cubic inch ~0.283	lb$_m$/in³ lb$_m$/in³
Acceleration "g" due to Gravity (by International Agreement)	9.806 650 m/s² 9.806 650 m/s²		0.3048§ 0.0254§	32.174 386.09	ft/s² in/s²
Natural Frequency (Hz = cycles/s) $\dfrac{1}{2\pi}\sqrt{\dfrac{g}{\delta}}$ where δ = static deflection = load/spring rate	$f = \sqrt{0.248/\delta \text{ (m)}}$ $= \sqrt{248/\delta \text{ (mm)}}$			$f = \sqrt{9.78/\delta \text{ (in)}}$	

[a] 4.448 221 6 = 0.453 592 37 · 9.806 650

[b] 57.295 780 = 180/π

[c] 6894.757 3 $= \dfrac{4.448\ 221\ 6}{0.000\ 645\ 16}$

Part 4

Incorporating Pneumatic Springs in Vehicle Suspension Designs

SAE HS 1576

SPRING COMMITTEE

E. C. Oldfield (Chairman), Hendrickson Spring
K. Campbell, Ontario, Canada
D. Curtin, General Motors Corp.
B. E. Eden, NI Industries
M. Glass, Marmon Group, Inc.
L. Godfrey, Associated Spring
R. S. Graham, Rockwell Int'l.
R. A. Gray, Troy, MI
D. J. Hayes, Redford, MI
P. W. Hegwood, Jr., GMC Delco Prods.
J. V. Hepke, GMC
E. H. Judd, Associated Spring Co.
J. F. Kelly, Marmon Group Inc.
M. Lea, GKN Composites
D. J. Leonard, Firestone Tire & Rubber Co.
J. Marsland, Chrysler Corp.
W. T. Mayers, Peterson Amer. Corp.
J. E. Mutzner, GMC
W. C. Offutt, Key Bellevilles Inc.
J. P. Orlando, General Motors Corp.
R. L. Orndorff, Jr., BF Goodrich Co.
W. Platko, General Motors Corp.
G. R. Schmidt, Jr., Moog Automotive Inc.
A. Schremmer, Assoc. Spring-Barnes Group
G. A. Schremmer, Schnorr Corp.
K. E. Siler, Ford Motor Co.
R. W. Siorek, U.S. Army Tank-Auto Command
A. R. Solomon, Analytical Engrg. & Res. Inc.
M. C. Turkish, Valve Gear Design Associates
F. J. Waksmundzki, Eaton Corp.

PNEUMATIC SPRING SUBCOMMITTEE

D. J. Leonard (Chairman and Sponsor), Firestone Tire & Rubber Co.
T. E. Burkley, Akron, OH
E. L. Harrod, General Motors Corp.
R. E. Houser, Grumman Allied Industries, Inc.
J. R. Hughlett, Ford Motor Co.
W. S. Locke, Knoll International Inc.
J. M. Mann, Navistar Int'l. Trans. Corp.
R. L. Orndorff, Jr., BF Goodrich Co.
W. C. Pierce, Lear Siegler Inc.
C. Wreford, Chrysler Corp.
M-C. Yew, Rochester, MI
W. J. Young, Freightliner Corp.

ACKNOWLEDGEMENT

It should be noted that this paper by the subcommittee was primarily directed under the chairmanship of Mr. Thomas A. Bank, retired from Firestone Tire and Rubber Company in 1982, while Mr. Bernhard Sterne of Bernhard Sterne Associates, provided many valuable editorial comments during the final compilation of this document. For the proofing prior to release, Mr. C. William Grepp, also retired from Firestone Tire and Rubber Company in 1982, is greatfully acknowledged for his efforts in making this paper as useful as possible to the design engineer.

TABLE OF CONTENTS

Chapter 1

Introduction and History

This manual has been prepared to assist the engineer and the designer to have a better understanding of the basic principles, types, and uses of pneumatic springs. In addition, it has been designed to serve as a useful guide in the selection of pneumatic springs for specific applications. The need for such a manual has been dictated by the popularity and growth of pneumatic spring applications.

The use of pneumatic springs in commercial applications has been a relatively recent development, but the idea of such a spring is not a new concept.

The earliest available records of a practical approach to pneumatic springs is a patent granted to John Lewis in February 1847. Prior to 1910, Benjamin Bell was engaged in experimental work on sleeve-type pneumatic springs with various piston shapes. The work and ideas put forth by these two gentlemen were made possible through the combined efforts of an Irishman and an Englishman some 300 years before that time.

Robert Boyle, the Irishman, published a paper in 1660, "New Experiments—Touching the Spring of Air", setting forth the law, "absolute pressures and volumes are in reciprocal proportion when temperature remains constant". Eighteen years later, Robert Hooke, the Englishman, formulated the relationship between force and elongation of elastic solid materials.

The first serious work toward adapting the pneumatic spring to the automotive industry was carried out by the Firestone Tire and Rubber Co. in the early 1930's. By 1935, experimental Buick and Plymouth cars were equipped with pneumatic springs. These were soon followed by installations on other vehicles, such as Studebaker, Chrysler, Ford, Lincoln, etc. In spite of this interest, the pneumatic spring was not adopted in the late 1930's primarily because of costs and the tremendous improvements being made with steel springs and suspensions in general.

In 1938, General Motors Corp. became interested in a new suspension with pneumatic springs for its buses. Working with Firestone, the first buses were tested in 1944 and first production was realized in 1953. This breakthrough triggered the growth and development of the pneumatic spring into the many new fields and applications that are in use today. In addition, new concepts in pneumatic springs have evolved to meet the requirements of these new and changing needs.

During the mid-50's Goodyear developed and patented the rolling lobe type air spring. The rolling lobe air spring load-deflection characteristics can be greatly influenced by piston contour.

The pneumatic spring has been able to make inroads into uses formerly reserved for more conventional and better known springs because of some unique characteristics and versatility. The more common and better known advantages of pneumatic springs are:

1. Controllable spring rate
2. Adjustable load capacity
3. Simplicity of height control
4. Reduction of friction
5. Nearly constant frequency with respect to load variations

Why and how pneumatic springs can offer these advantages and meet the needs of modern industry will be described in the following chapters which cover: basic principles, types, design problems, and special uses of pneumatic springs.

Chapter 2

Basic Principles of Pneumatic Springs

1. General Discussion

The pneumatic spring is basically a column of confined gas in a container designed to utilize the pressure of the gas as the force medium of the spring. The compressibility of the gas provides the desired elasticity for suspension use.

The pneumatic spring's ability to support a mass depends upon its effective area, which is a nominal area found by dividing the load supported by the spring by its gas pressure at any given position. The effective area is a function of deflection. Whether it remains constant, increases, or decreases is governed by the design of the spring and its components. The spring rate is the result of change in effective area and the change in gas pressure as the spring is deflected. The gas pressure varies with the speed and magnitude of deflection; for a unit of deflection, the pressure and, therefore, the spring rate will be different for isothermal, adiabatic, or polytropic processes.

Pneumatic springs provide an adjustable spring rate, adjustable load carrying ability, simplicity of height control, and a low friction action. Pneumatic springs are adaptable for light or heavy suspension applications.

2. Compression Processes

For a specific spring design, the minimum pneumatic spring rate occurs under isothermal compression conditions and the maximum spring rate occurs with adiabatic compression. The polytropic rate varies between the isothermal and the adiabatic. The isothermal rate results when all the heat of gas compression escapes so that the gas remains at a constant temperature. The isothermal rate is approached when the spring is deflected very slowly to allow time for the heat to escape, the gas temperature remains constant, and the gas pressure rise is minimum.

Adiabatic rate occurs when all the heat of compression is retained within the gas. This condition is approached during rapid spring deflection when there is insufficient time for the heat to be dissipated. The higher temperature of the gas results in a higher gas pressure and, therefore, a higher spring rate.

When the heat of compression is partially retained within the gas, a polytropic rate results. This occurs during most normal spring deflections and produces neither isothermal nor adiabatic rates, although in normal use it is much closer to the adiabatic situation.

3. Constant Volume Pneumatic Springs

Pneumatic springs which maintain a relatively constant volume at a given operating height regardless of static load or gas pressure are referred to as constant volume pneumatic springs and are the more common type in use at this time. At a given height, the load-carrying ability and the spring rate are varied by changing the pressure of the confined gas. With this type of spring, an external source of compressed gas is needed to maintain the spring height as the load on the spring is changed. The natural vibration frequency of the constant volume pneumatic spring remains more uniform with changes in load than does the natural frequency of the constant mass pneumatic spring.

4. Constant Mass Pneumatic Springs

Pneumatic springs which use a fixed mass of gas as the elastic medium are constant mass pneumatic springs. A given amount of gas is sealed in the system and remains constant for all conditions of load or deflection.

As the load on a constant mass pneumatic spring is increased, the gas volume is reduced and the spring rate increases. Conversely, when the load on the spring is reduced, the gas expands and the increased gas volume results in a reduced spring rate. Thus, the natural vibration frequency of a suspension system using a constant mass pneumatic spring increases as the load on the system increases.

Pneumatic springs which are not connected to a gas source with height control or other valve arrangements, often called "locked in" systems, are classed as constant mass pneumatic springs, as are most hydro-pneumatic springs.

5. Basic Cylinder and Piston Springs

A. Cylinders with pistons can be used as pneumatic springs but they have several major drawbacks:

(1) Sliding friction transmits significant forces through the spring. Short impulses are especially detrimental.
(2) It is difficult to maintain zero gas leakage past the piston and rod seals for the desired life of the unit.
(3) Clevises are required at the top and bottom for most mountings.
(4) The effective area cannot be manipulated.
(5) The piston and rod guide present wear problems.

B. An advantage is that high operating pressures may be used, and the unit can combine load-carrying and damping functions.

6. Reinforced Flexible Member Springs

The use of a reinforced flexible member in conjunction with rigid structures overcomes many of the above deficiencies. However, careful design is required to prevent high local stresses and severe fluctuations in stress which result in poor life. The flexible member structure carries only a portion of the developed spring force, with the remainder being transmitted directly through the gas column to the rigid supporting members. With some designs the circumferential stress created by the internal pressure, not the direct load stress, is the principal stress the flexible member encounters. Gauge pressures with currently regularly used materials are generally limited to 700 kPa for 2-ply and 1200 kPa for 4-ply reinforcement. Some severe operational conditions limit this still further.

7. Flexible Member Operating Life Considerations

Pneumatic spring designs which have the lowest maximum stress and low stress variation with cycling will achieve the best durability. Durability is also directly coupled with imposed stresses which are the result of the suspension design.

In applications encountering repeated severe stresses, the design maximum static pressure should be less than one-third the normal burst pressure at the static design height of the spring. For moderate and light-duty service, the maximum pressure may be increased up to one-half the normal burst pressure. However, more conservative operating pressures will generally result in increased life.

8. Operation With Only Slight Internal Pressure

To maintain their correct shape, pneumatic springs must have at least slight positive internal pressure under all deflection conditions. This means that springs with long rebound must have higher design position pressure than springs with short rebound. Failure to meet this requirement may cause girdle hoops to slip out of proper position with multiple convolution designs. Reversible sleeve springs may pinch extra folds between the piston and the top mounting metals, resulting in rupture of the flexible member. Generally, 70 kPa minimum design height gauge pressure will prevent operation troubles. In some cases, half this gauge pressure will suffice, but in a few special cases up to 140 kPa may be required.

9. Basic Calculation Considerations

The basic gas laws apply quite well for design calculations of general characteristics in the pressure, temperature, and frequency ranges normally used by pneumatic springs. In addition to these factors, an effective area varying with deflection must frequently be considered. This can be accomplished by changing the size or shape of the piston.

It may also be necessary to take into account the fact that frequently a nonproportional change in volume occurs with changes in deflection.

10. Design Balance

Factors that provide the most efficient design from a theoretical standpoint, must be weighed against factors that deteriorate spring life. Thus, each operating situation must be evaluated on its overall requirements, and compromises must frequently be made. Fortunately, there is generally a variety of ways to attain the desired results, and good overall performance is usually attainable.

11. Spring Characteristic Features Peculiar to Constant Volume Pneumatic Springs

Effective static deflection is determined by the dynamic rate at the static design position. It can be shown graphically by drawing a line tangent to the dynamic load-deflection curve at the static design position and extending it through the zero load line, then measuring the distance back to the static design position. (See Fig. 2.1.) Natural frequency is directly related to the effective static deflection.

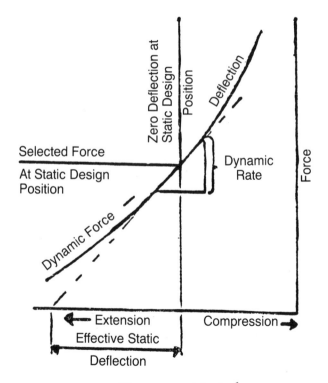

Fig. 2.1—Effective static deflection[*]

*All drawings courtesy of Mr. Greg Van Meveren.

4.4

Rates are generally considered to vary in direct proportion to force. Usually the natural frequency of the system stays reasonably constant throughout the normal force range.

System natural frequencies are variable and are determined by spring design, spring volume, and the gas law processes involved.

Fig. 2.2 shows the effect of piston shape on the effective area (A_e) and on the dynamic spring force (F_d) curves versus spring position. The lowest of the four sketches shows the effect of a large diameter flexible member combined with a small diameter piston. This variation must be done with care because of possible adverse effect on the service life of a pneumatic spring assembly.

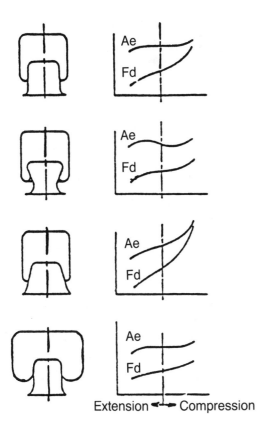

Fig. 2.2—Characteristic variations due to piston shapes and flexible member size

12. Gas Law Processes

A. Definitions — Units

The mass of a vehicle and of its cargo is measured in kilograms (kg) and is usually called "weight"; this mass, less the unsprung mass, acts upon the suspension springs as a load, or more accurately as a vertical downward force F (now designated 'force of gravity'), equaling mass times acceleration of gravity and measured in newtons (N = kg × 9.806650 m/s^2).

With a pneumatic suspension spring, this load or force is supported by a force which is developed as the product of gas gauge pressure (that is, pressure above atmospheric pressure) and an effective area within the flexible member of the spring.

In this manual the force is measured in newtons (N), the gas pressure is measured in kilopascals (kPa), the effective area is measured in square millimeters (mm^2). The relationship between these three values is

$$1 \text{ N} = 1 \text{ kPa} \times 10^3 \text{ mm}^2*$$

The pressure of the atmosphere at sea level is in balance with a 760 mm column of mercury at 0°C; it equals 101.32 kPa. The sum of the atmospheric pressure and the gauge pressure is known as absolute pressure. The fundamental gas laws deal with this absolute pressure.

B. Vertical Supporting Force

The supporting force (F) is created as the product of gas gauge pressure and effective area:

$$F = A_e \times P_g$$

Effective area (A_e) can be found directly when the force and pressure are known. Then it is the result of dividing force by pressure.

C. Constant Pressure, No Gas Flow, Constant Effective Area

$$\text{With Constant Pressure} \quad \frac{V_1}{T_1} = \frac{V_2}{T_2} \text{ or } \frac{V_1}{V_2} = \frac{T_1}{T_2}$$

where: T = absolute temperature
V = total pneumatic spring and working volume

These relationships affect the pneumatic spring system when the system is at rest and only temperature changes occur. Dynamically, the only way to maintain constant pressure is in combination with infinite volume and thus is not generally useful.

*The SAE Manuals on Metal Springs use MPa (= 10^6 Pa) as the SI unit for stress etc. In these other Manuals the relationship between the three values is therefore 1 N = 1 MPa × 1 mm^2.

D. Constant Pressure, Constant Temperature, Gas Flow, Varying Displacement, Varying Effective Area, Varying Volume

With varying effective area, volume is a function of displacement. However, force is still the product of pressure and effective area for all attainable displacements.

Charts prepared from test data using a number of specified pressures are very useful since data in this form allow manipulation of design height position, volume, and polytropic exponents. Data in less general form is not as versatile.

E. Constant Volume, Non-Flow Process

This process (see Fig. 2.3) can be shown as:

$$\frac{P_1}{T_1} = \frac{P_2}{T_2}$$

where: P = absolute pressure
T = absolute temperature

From a practical standpoint, with true gases this is a desirable but unattainable process because of the nature of the flexible member. Extremely large volumes may allow an approximation of the process and there are ways to obtain small ratios of volume change to total volume with feasible total volumes. (An example, interconnecting springs with only one spring undergoing a volume change at a time.)

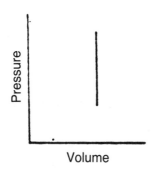

Fig. 2.3—Constant volume, non-flow process

F. Constant Temperature, Non-Flow Process (Isothermal Process)

$$P_1 \times V_1 = P_2 \times V_2$$

where: P = absolute pressure
V = total volume
PV = constant

This process (see Fig. 2.4) must be taken into account when examining the static stability of systems. It determines the practical limit of low rate operation. Under isothermal conditions, the spring rate must be appreciably in the positive range.

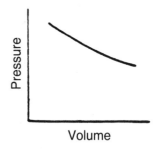

Fig. 2.4—Constant temperature non-flow process (isothermal process)

G. Adiabatic, Non-Flow Process:
This is defined as a process with no heat transferred to or from the working fluid (see Fig. 2.5).

$$\frac{T_2}{T_1} = \left(\frac{P_2}{P_1}\right)^{(\gamma-1)/\gamma} = \left(\frac{V_1}{V_2}\right)^{\gamma-1}$$

where: V = absolute temperature
P = absolute pressure
V = volume
γ = adiabatic exponent

This is a theoretical process; in practice it is not attainable with pneumatic springs. However, for rapid deflections it is closely approached.

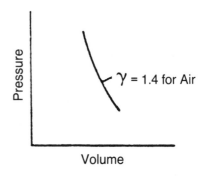

Fig. 2.5—Adiabatic non-flow process

4.6

H. Polytropic, Non-Flow Process

$$\frac{T_2}{T_1} = \left(\frac{P_2}{P_1}\right)^{(n-1)/n} = \left(\frac{V_1}{V_2}\right)^{n-1} \quad \text{or} \quad \frac{P_2}{P_1} = \left(\frac{V_1}{V_2}\right)^{n}$$

where: T = absolute temperature
       P = absolute pressure
       V = total volume
       n = polytropic exponent

This is the general case where the terms pressure, volume and temperature all vary (see Fig. 2.6).

Pneumatic springs operate in the full range from nearly isothermal to almost adiabatic. A generally acceptable value for n is 1.38 when the natural frequency of the system is being determined.

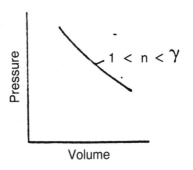

Fig. 2.6—Normal pneumatic spring operation situation

Chapter 3

Types of Pneumatic Springs (See J511a for Definition of Terms)

1. Bellows Types

Load-deflection characteristics in this type are determined by the position relationship of the various components more than by the physical dimensions and shapes of the components. Beads are generally the same diameter and, in conjunction with girdle hoops for multiple convolution springs, restrict the expansion diameter of the convolution. No bead passes through another bead or girdle hoop.

A. Single convolution bellows type consists of two beads and one convolution.

B. Double convolution bellows consist of two beads, one girdle hoop, and two convolutions (see Figs. 3.1 and 3.2).

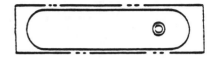

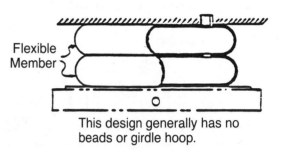

This design generally has no beads or girdle hoop.

Fig. 3.2—Two convolution bellows, oblong section

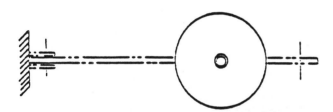

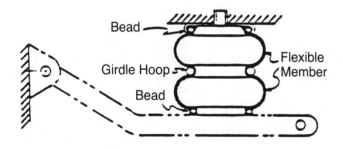

Fig. 3.1—Two convolution bellows, circular section

C. Triple convolution bellows consist of two beads, two girdle hoops, and three convolutions.

D. More convolutions are generally unuseable because of convolution instability.

2. Piston Types

Load-deflection characteristics are largely determined by the shape and physical characteristics of the rigid components. These springs, in their normal operation position, have one or both ends or beads positioned well inside the outer flexible member. Support for the small bead in the form of a contoured piston may be used to control the load-deflection characteristics by varying the effective area to a considerable degree (see Figs. 3.3 and 3.4).

A. Reversible Diaphragm

The flexible member is restrained and controlled by rigid structures attached to the small and large beads. In this type, the small bead or a portion of the piston passes axially through the flexible member and usually through the large outer bead.

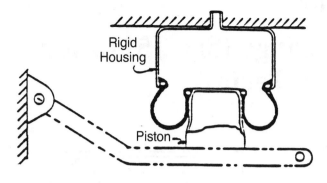

Fig. 3.3—Reversible diaphragm

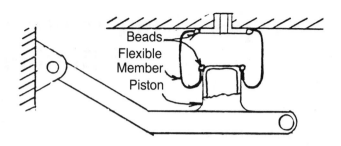

Fig. 3.5—Reversible sleeve type

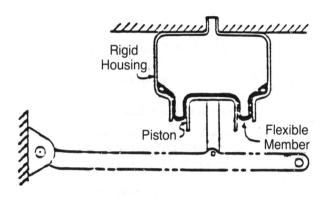

Fig. 3.4—Reversible diaphragm

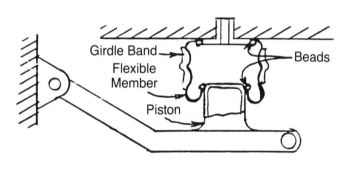

Fig. 3.6—Reversible sleeve type

The load-deflection characteristics are determined by the volume of the rigid structures more than by the volume of the flexible member. The shape and related positions of the rigid structures are also important factors.

B. Reversible Sleeve

In this type, the outer flexible member may be self- or externally-supported (see Figs. 3.5 and 3.6).

In the self-supporting type, the flexible member is restrained in diameter by its own internal construction, and load-deflection characteristics are determined by the volume of the outer flexible member, plus the size and proportions of the small bead rigid structure to which the small bead is attached. The small bead is positioned well inside the outer flexible member but does not pass through the opposite bead. These designs generally have little or no lateral stability.

The flexible member in the externally supported type is restrained by the floating external rigid structure and the size and proportions of the small bead rigid structure. The load-deflection characteristics are determined by the volume of the outer flexible member. Again the shape and related positions of the rigid structures are also important factors.

1. Straight Sleeve Type Pneumatic Springs

The straight sleeve type pneumatic spring is a form of the reversible sleeve type spring. The flexible member is built and cured on a mandrel. The molded shape may be cylindrical or conical. A significant feature of the straight sleeve type flexible member is the elimination of the end beads in the cured product. Instead, the separate external beads are forced or shrunk in place as the sleeve is assembled and attached to its end closures. Various length requirements for a given size flexible member can be accomplished simply by cutting the sleeve to the desired length.

Straight sleeve type pneumatic springs for vehicular applications are in wide usage for air adjustable shock absorbers, air seats, and truck cab suspensions.

3. Hydro-Pneumatic Springs

Pneumatic springs which contain both a liquid and a gas in the spring system are hydro-pneumatic springs (see Fig. 3.7). The gas, which is the elastic medium for the system, may be in the accumulator separated from the liquid by a diaphragm; it may be in the accumulator, in the connecting lines, and in the spring assembly; or the liquid and gas may

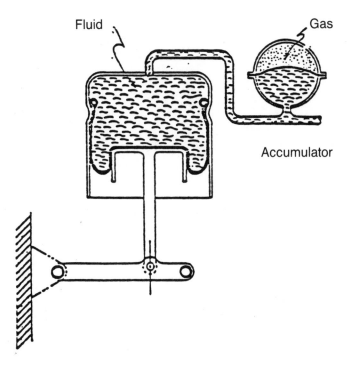

Fig. 3.7—Hydro-pneumatic spring

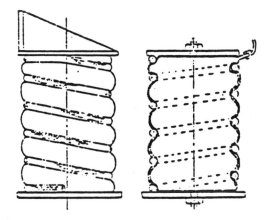

Fig. 3.8—Bladder type spring

pumping into a single strut type unit. The vertical motion of the vehicle provides the actuation for the pump.

4. Bladder Type Spring

This type utilizes no integral reinforcement. It relies on being contained within a restrictive structure, such as a coil spring, for its support (see Fig. 3.8).

5. Variations of Basic Designs

The bellows and reversible types may be blended in various ways to obtain desirable features of both.

Reversible sleeves and reversible diaphragms may also be blended if desired.

All types except the reversible self-supporting sleeve may be made oblong as well as round.

The addition of special shaped metals or other support members can produce desired lateral rates with the reversible sleeve type pneumatic springs.

be mixed as in the Oleo strut used for aircraft landing gears.

The usual form of the hydro-pneumatic spring is one in which the gas is sealed in part of the accumulator and is separated from the liquid by a flexible diaphragm. The liquid portion is in the cylinder or flexible member and in the connecting lines. Liquid for leveling the vehicle is supplied from a high-pressure source and controlled by sensing valves. The accumulator may be separate from the load transfer component of the system, or it may be attached directly to the unit and be an integral part of it. Vertical motion of the vehicle's axles or chassis causes displacement of the liquid, which in turn compresses or expands the gas in the accumulator containing the elastic medium. With this arrangement, liquid flow control valves may be incorporated in the system for damping purposes. This eliminates the need for separate hydraulic shock absorbers.

In another type of hydro-pneumatic spring, the gas or spring medium completely fills the cylinder, the connecting lines, and a portion of the accumulator. The vehicle height is regulated by admitting or releasing the fluid from its portion of the accumulator.

The "Oleo strut" suspension is a hydro-pneumatic spring which utilizes a mixture of gas and liquid in an extensible cylinder. The motion of collapsing or extending the strut compresses or expands the gas bubbles in the liquid to provide the spring action. Damping and shock absorption are usually provided by forcing the mixture through a controlled or uncontrolled restriction.

Several manufacturers produce hydro-pneumatic suspension units which incorporate automatic leveling and self-

Chapter 4

Pneumatic Spring Application Considerations on Commercial Vehicles

This chapter is devoted to bridging the gap between the theoretical characteristics of pneumatic springs and their functional application. Recognizing that the usual application of pneumatic springs will be in conjunction with other components making up a vehicle suspension, examples are presented here in terms of the pneumatic spring design for actual suspension systems. From this presentation, the designer should be able to determine the sequence of data accumulation, analysis, and calculation required to arrive at a suitable design within his particular parameters. He should also gain an understanding of how the pneumatic spring characteristics, discussed in earlier chapters, become pertinent to overall suspension design.

In addition to the design procedure examples and calculations, a section of this chapter presents generalized considerations for pneumatic spring suspension design which have evolved from the experience of suspension designers and pneumatic spring manufacturers.

The advantages of pneumatic suspension are not automatically attained. The entire suspension must be properly designed if it is to take full advantage of pneumatic spring capabilities and to compensate for their few limitations.

1. Reasons for Using Pneumatic Springs

A. To obtain better cargo protection through low spring rate and frequency.

B. To make a lighter vehicle possible, thus allowing an equivalent increase in cargo mass.

C. Pneumatic springs lend themselves to auxiliary axle pick-up when the axle is not required, thus providing operating economies and improved maneuverability.

D. Pneumatic springs optimize load distribution on multi-axle units.

E. If it is elected to maintain essentially the same vehicle structure as would be used with a higher rate and frequency suspension, then less damage to the vehicle from road shock will occur, and maintenance costs will be reduced.

F. Because pneumatic suspensions are normally leveled automatically, there is no change between the vehicle height at the "curb load" (that is, the load on the spring which is due to the mass of the vehicle without any payload) and at the "design load" (that is, the load on the spring which is due

to the mass of the vehicle plus the payload). This permits the cargo space to be designed with a higher top and a lower floor without interfering with the tires while still staying within maximum vehicle height limits. Also, a constant height can be maintained for use at loading docks and for trailer pick-ups.

G. As the cargo benefits from an improved ride, so does the vehicle or trailer for the same reason. Thus, the vehicle can be made lighter and still perform acceptably, again adding more cargo capability.

H. More comfortable ride is provided for vehicle occupants.

I. When used in conjunction with other types of suspensions, they make good axle load distribution possible.

J. They may make other functions possible, such as jacking, load shifting for increased traction, etc.

K. Low spring rate and frequency plus excellent load distribution protects the cargo, which in turn protects the roads and may make pneumatically sprung vehicles acceptable on roads where units of large mass are not now allowed.

2. Suspension Design Considerations

A. To design a pneumatic suspension system the designer must know, in addition to what is needed for the basic design, what will be required of the pneumatic spring and what is needed to fit the spring into the total system.

The system requirements that the spring is to supply are as follows:

1. Satisfactory mechanical operation over the full axle travel.
2. Support for the sprung mass with the available gas pressure.
3. The desired dynamic spring rate and system frequency throughout the sprung mass load range.
4. Desirable or at least satisfactory dynamic force characteristics throughout the full axle travel.

The pneumatic spring requirements that the suspension system must provide are as follows:

1. A space envelope that allows the spring to function properly at all lengths.

2. An operating environment that does not seriously affect spring life.

B. To take maximum advantage of these possibilities, many considerations should be kept in mind. There is a lot of reserve capability and versatility built into pneumatic springs. While some violations of the following application principles can be tolerated, it is best to design to obtain as many preferred conditions as possible for long, trouble-free pneumatic spring service. Pneumatic springs can be utilized in many geometries, but in designing to get the maximum benefit in one area, care should be taken not to create an unacceptable situation in another area. For example, good spring life and ultralow rate and frequency characteristics may be operating opposites with some designs.

The following are some things to strive for and some things to avoid:

C. The pneumatic spring design length should be established within the recommended design length range since life and performance characteristics may both be adversely affected if the springs operate continuously either above or below their design range.

D. For lowest spring rate and frequency for a particular suspension design, choose springs which keep the normal operating pressure within 80-100% of the normal rated value.

E. Moderate operating pressures (40-70% of rated pressure) will provide maximum life. It is also necessary to maintain some positive pressure under lightest load and full rebound conditions.

F. Low spring rates mean less control in vehicle roll; therefore, some auxiliary restoring force must be supplied. Leveling valves, shock absorbers, and shaped pistons may sometimes help in this situation but it is best to have a suspension design or linkage which supplies this restoring force. Some methods that have proved successful are:

1. Using a roll stabilizing bar connecting one suspension arm with the other.
2. Having rigid suspension arms with a rigid axle connected to the suspension arms with flexible mountings.
3. Using flexible suspension arms attached rigidly to the axle.
4. Keeping the roll moment as low as possible, consistent with other design considerations.
5. Utilizing a suspension design that has as high a roll center as practical.

G. Low spring rates produce more axle travel relative to the frame over irregular road surfaces. Thus, more axle travel is needed before cushioned stops, with their inherent high rates, come into operation. Bumpers should come into action smoothly. Cushioned rebound stops are recommended. Hydraulic shock absorbers or air damping may be needed to control vehicle and axle action.

H. To increase the pneumatic spring's load carrying capacity beyond what the spring normally provides, and gain improved vehicle roll control, place the spring on a trailing arm behind the axle. However, this arrangement will work the spring harder because of repeated longer travel required to provide desired axle motion.

I. For extra low suspension rate and frequency, place the pneumatic spring between the suspension arm pivot and the axle. This will generally provide good spring life if spring operating geometry and pressure are within proper design parameters.

J. Any operating condition that creates high stresses in the flexible wall of the spring will adversely affect the spring life. Examples are springs with high design operating pressures, with long compression deflections, and springs with severe misalignment between top and bottom mounting surfaces. The significant thing to remember is not just the high stresses, but the number of times the springs are subjected to these high stresses.

K. Rapid and repeated large changes in flexible member stresses will reduce life. Examples are springs which have portions of their structure subjected to repeated lateral motion because of excessive flexure of suspension components when cornering and springs which have small gas volume and undergo large deflections.

L. Life may be considerably improved if the alignment between the upper and lower mounting surfaces is balanced so that the maximum misalignment is held to a minimum. The adverse effect from misalignment varies with the design and style of the spring (see Chapter 7, Section 2).

M. Continued operating temperatures above 65°C are to be avoided for best life, since life is related to the spring's total heat history. Heat shields or a separation from the source of heat (such as the brake drums or exhaust systems) may need to be considered. An occasional problem associated with heat is caused by welding of parts adjacent to the pneumatic springs, causing local deterioration of the sidewall and subsequent failure.

N. Avoid situations where springs rub or nearly rub against anything. This also includes rubbing against themselves. Low operating pressures increase the chance that rubbing may occur between the outer inside wall adjacent to the top of the piston and the inner inside wall where the rubber fabric sleeve rolls over the top of the piston (see Fig. 4.1).

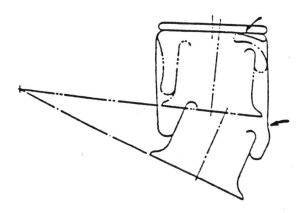

Fig. 4.1—Internal chafing

Verification of satisfactory radial clearance with other vehicle components must be made under all anticipated operating conditions. Remember that the springs will grow slightly with time and that suspensions have manufacturing tolerances. If the operating clearance of rolling sleeve springs appears to be a problem, keep in mind that the spring diameter is greatest at maximum pressure, which occurs when the pneumatic spring is in the fully compressed position. Considerable clearance may exist at the normal design height and even more clearance may be noted when the spring is in full rebound. Points of possible interference must be checked experimentally; there may be more clearance than anticipated—or possibly less. If the spring is used at considerably less than the normal maximum rated pressure, there will be extra clearance. If pneumatic springs are used above their normal rated pressure, they will have larger than rated diameters.

O. Avoid spring extensions which appreciably flex the spring wall, even locally, close to its attachment regions. Both extension and compression stops must be used to protect against the worst condition. If this is not controlled, the spring will be damaged resulting in reduced life.

P. Bumpers must deflect some distance before they will carry their rated load, and this should be taken into account when considering spring compression travel. In addition, internal bumpers are limited by volume considerations and are not generally designed for continuous ride use. They are meant to be used as dynamic compression bumpers and emergency ride springs only.

Q. Special shorter bumpers may be required if auxiliary axle pickup is used. Hence, they become only compression travel stops and are not auxiliary springs or true compression bumpers.

R. For ideal spring life, it is generally best to choose a spring with large volume or to use an auxiliary volume, rather than resorting to severely shaped pistons or other rigid components to obtain a desired low frequency operation.

This also reduces the difference between the adiabatic and isothermal rates and can produce very low dynamic rates with an isothermally stable system.

S. Isothermal instability occurs only on pneumatic springs of the reversible diaphragm type, or of the reversible sleeve type with hourglass-shaped pistons having an effective area curve with negative slope. An isothermally unstable system has three spring height positions that will carry the load at the same pressure. The center position is unstable so that if, for example, the operating gas cools, the system will settle to the lower stable position; but when this occurs, the height control valve will come into play and add gas which will bring the system back to the mid-position. Then, because the effective area is increasing as the load is raised, the added gas pressure will bring the system beyond the proper height to the upper stable position where again the effective area and pressure just balance the load. But now the height control valve operates to return the system to the center position, and an unacceptable cycling situation is created.

T. Road tar and sand, especially sand 3 mm across flats, is detrimental to spring life. In cases where the spring is in line with material thrown up by the tires, a protective shield should be placed in front of the spring assembly.

U. Effective Static Deflection

Static deflection of a steel spring is easily obtained by compressing the spring slowly in a laboratory, generating a load-deflection curve. This is used to obtain the spring rate which in many cases is approximately the same regardless of the speed of cycling. Then from this and the value of the load under study, the system frequency can be calculated.

An effective spring (or system) static deflection can also be obtained for a pneumatic spring but, several other factors must be considered. The speed of cycling must at least approximate the conditions being studied (anticipated system natural frequency or sudden shock inputs for example). When the spring operates through an arm system this must also be incorporated in the test setup if the resulting rate and frequency are to apply to the system. The test force—length trace must pass through the desired force condition at the desired spring static design length.

Having obtained such a trace, the effective static deflection can be found using the technique shown in Fig. 4.2. This value can then be converted to rate and frequency. But note that the resulting rate applies only in the tangency region of the curve. However, the frequency will be approximately true for a much wider band of operation.

This analysis method is not very practical in many instances and the following design procedure has been found to be more suitable and yields results that match normal field operation quite well.

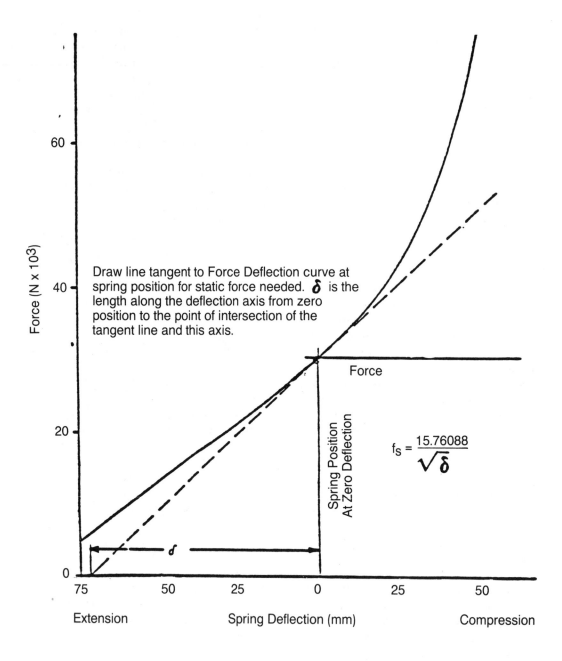

Draw line tangent to Force Deflection curve at spring position for static force needed. $\boldsymbol{\delta}$ is the length along the deflection axis from zero position to the point of intersection of the tangent line and this axis.

$$f_S = \frac{15.76088}{\sqrt{\delta}}$$

Fig. 4.2—Dynamic force deflection curve showing graphic method of determining effective static deflection
Example: $\delta = 72$ mm

3. Design Procedure

A. The initial step in specifying a pneumatic spring for a suspension design is to record the known and estimatable system parameters. To assist in this, a Data Record Form is provided in Appendix C from which copies can be made and be filled in for use in developing trial designs.

The form is self-explanatory, but reference to the glossary (Appendix B) may be helpful. The data compiled on the form is used in the selection of a trial spring design for the proposed application. The spring will function in this application if it meets or exceeds requirements stipulated on this form.

A preliminary check for a suitable spring can be made by applying the data record form values to Table 4.1. This can be useful in making a first trial choice as to type and size spring that could be used. The frequencies noted in the right hand column of this Table can be corrected to values which will be obtained with the proposed suspension by the formulas shown at the bottom of the Rate and Frequency Calculation Form (Appendix D) which converts from a direct acting spring to one with an arm ratio as required by the suspension design.

B. The calculations performed on the Rate and Frequency Calculation Form find the dynamic rate and system frequency for a sprung mass system which uses the trial spring and having an arm ratio of 1.0. These intermediate values are then converted to fit the trial suspension design with the correct arm ratio in the remaining steps.

If the results are not satisfactory, change any suitable specified item and recalculate the rate and frequency. It is recommended that only one thing at a time be changed insofar as possible so as to develop trends and the magnitude of change possible with that item.

It is frequently useful to alter the spring total volume as this has a very significant effect on the rate and frequency. If a different volume produces satisfactory results, another spring design having approximately this volume may be available. However, other design parameters may have to be changed to permit the new spring to be used. Other arm ratios may also be tried if the design allows but if nothing tried from any change or combination of changes proves satisfactory, consult the spring manufacturers as they may be able to put together combinations of spring components that will meet your requirements.

C. If a substantially lower natural frequency is needed, auxiliary volume must be added. This additional volume may be approximately determined by applying the Principle of Boyle's Law when the spring rate and volume are known, and the desired spring rate has been determined. This method applies only to springs having a constant effective area at the spring desired length.

$$V_2 = V_1 (R_1/R_2)$$

Where: V_2 = actual volume of spring and reservoir
V_1 = volume of spring
R_2 = desired rate of spring
R_1 = spring rate of present spring

Auxiliary volume to be added to the spring system is $V_2 - V_1$.

D. Any mass which has been suspended by the use of pneumatic springs will have from 1–6 deg of freedom and may have six natural frequencies. The suspended mass will tend to vibrate at one or more of these natural frequencies. For good isolation, all of the natural frequencies should be one-half or less of the forcing frequencies. In some applications, the existing frequencies pass through the natural frequency zone of the suspended mass and a temporary resonance exists. It is then necessary to include a viscous damping device in the system, such as a surface vehicle shock absorber.

4. Example of Application

The following example is typical of the procedure for selecting a suitable pneumatic spring for use in a new pneumatic suspension design:

A. Start by obtaining all the required information as shown in the *Data Record Form for Pneumatic Spring Application* (see Appendix C). After an initial selection has been made from Table 4.1, consult pneumatic spring manufacturers' catalogues to determine a more exactly defined model that your suspension design can accept. If the manufacturers' catalogues do not provide spring assemblies which meet the necessary requirements, contact the manufacturers directly, as they may have other models, not listed, which will meet your needs. Alternately they may be able to make modifications to available assemblies which will change them so as to meet your requirements.

B. In this example, it has been determined that a reversible sleeve type spring assembly XXXX may be suitable. To verify this, *Rate and Frequency Calculation Form for Pneumatic Spring Applications* may be used (see Appendix D). The volume, effective area, and pneumatic spring gas gauge pressure are obtained from the example spring force-length graph shown in Fig. 4.4. Careful scaling on these graphs is required if accurate results are to be obtained. A good procedure is to use dividers to find more precise values for volumes. Effective areas vary somewhat with changes in gas pressure, thus it is wise to pick a pneumatic spring force-length curve that is close to the anticipated pneumatic spring gas pressure. Using this curve, determine the force in N at the desired spring design length

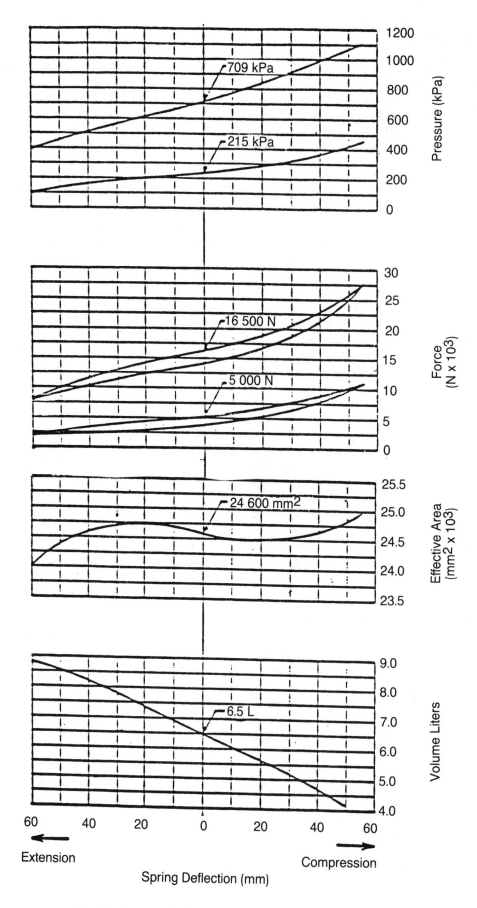

Fig. 4.3—An example of pneumatic spring dynamic characteristics

TABLE 4.1—SOME TYPICAL PNEUMATIC SPRING CHARACTERISTICS FOR SURFACE VEHICLE USE

| Mass Support Capability Range | Effective Area | Nominal Operating Diameter | Normal Design Limits | | | Useful Travel | Spring Mass System Natural Frequency Direct Acting Without Aux. Volume Without Internal Bumper |
| | | | Normal Max. Extended Length | Static Design Position Length Range | Normal Min. Compressed Length | | |
kg/Spring	mm^2	mm	mm	mm	mm	mm	Hz
BELLOWS TYPE							
1 Convolution							
420 - 1 430	17 400	250	150	127 - 110	50	100	2.9
950 - 3 000	38 700	320	164	127 - 110	50	114	2.6
100 - 4 920	63 600	378	166	127 - 110	50	116	2.4
2 Convolution							
200 - 1 000	11 600	210	275	240 - 200	75	200	2.0
340 - 1 600	17 100	244	275	240 - 200	75	200	2.1
910 - 3 300	40 000	320	305	240 - 200	75	230	1.9
450 - 5 000	61 600	376	305	240 - 200	75	230	1.7
130 - 6 800	83 200	434	312	240 - 200	82	230	1.7
500 - 13 600	177 400	570	324	240 - 200	82	242	1.7
3 Convolution							
500 - 5 100	62 600	376	450	330 - 240	110	340	1.4
350 - 7 500	94 100	460	450	330 - 240	110	340	1.4
350 - 10 500	134 800	515	460	330 - 240	110	350	1.4
REVERSIBLE SLEEVE TYPE							
68 - 154	2 260	76	295	200	117	178	1.3 - 1.7
80 - 340	4 840	102	267	178 - 105	102	165	0.9 - 2.0
136 - 467	6 770	127	247	165	102	145	1.4 - 1.5
635 - 1 270	19 000	203	405	292	140	265	1.0
560 - 1 460	20 800	220	508 - 375	355 - 203	203 - 127	305 - 248	1.1 - 1.5
794 - 1 815	23 100	229	368	254	140	228	1.2
180 - 2 500	34 200	254	407	292	147	260	1.2
590 - 3 220	46 500	280	406 - 381	292 - 270	180 - 147	234 - 226	1.3 - 1.5
225 - 3 265	46 500	315	667 - 444	457 - 178	216 - 108	451 - 336	1.1 - 1.6
360 - 3 580	49 000	305	660 - 341	414 - 184	218 - 117	442 - 224	1.0 - 1.7
452 - 3 580	55 500	330	635 - 343	414 - 229	219 - 107	416 - 236	1.1 - 1.4
680 - 4 000	71 000	356	914 - 544	457 - 305	356 - 150	457 - 394	1.1 - 1.3

DATA RECORD FORM FOR PNEUMATIC SPRING APPLICATION

1. Gross axle mass rating (GAMR) .... 10 000 kg

2. Axle unsprung mass .... 680 kg

3. Axle sprung mass (both sides) .... 9 320 kg

4. Axle sprung mass (at either side) .... 4 660 kg

5. Axle static support force per vehicle side (4660 × 9.806650) .... 45 699 N

6. Desired natural frequency of sprung mass .... 1.3 - 1.5 Hz

7. Maximum spring gas pressure .... 760 kPa

8. Spring arm (a_s) .... 762 mm

9. Wheel axle arm (a_w) .... 508 mm

10. Arm ratio (a_s/a_w) 1.500

11. Spring design length selected from a suitable Force-Length Graph   320   mm

12. Maximum axle compression travel required   100   mm

13. Resulting change in spring length from spring compression
deflection = axle compression travel $\times$ (a_s/a_w)   150   mm

14. Minimum spring length along centerline resulting   170   mm

15. Maximum axle extension travel required   105   mm

16. Resulting change in spring length from spring extension
deflection = axle extension travel $\times$ (a_s/a_w) . . . .   157.5   mm

17. Maximum spring length along centerline resulting . . . .   477.5   mm

18. Total change in spring length from full axle compression
and extension travel . . . .   307.5   mm

19. Smallest radial clearance limitation from spring centerline   165   mm

20. Spring design force required = (axle static support force
per vehicle side) / (a_s/a_w) . . . . 30 466   N

If an internal compression bumper is required and its operation is vital to the suspension performance, fill in the blanks in Items 21-23 and consult a pneumatic spring manufacturer for their information so that a suitable bumper can be provided. A bumper which can endure two times the spring design force may be recommended for satisfactory service.

21. Axle compression travel before bumper contact . . . .   83.3   mm

22. Spring compression deflection to bumper contact =
axle compression travel before bumper contact $\times$ (a_s/a_w)   125   mm

23. Minimum force bumper should provide . . . . 60 932   N
at a bumper compression distance of   25   mm

70 to 700 kPa

SPRING DESIGN LENGTH RANGE

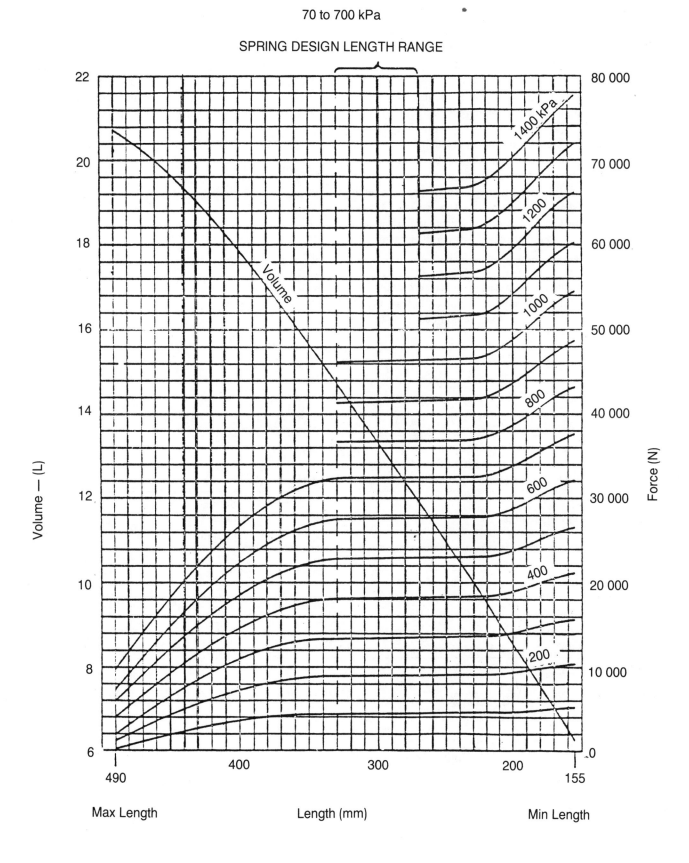

Fig. 4.4—Example spring force-length graph

precisely. After multiplying this value by 10^3, divide it by the pressure value given for that line in kPa as shown on the form to check out the effective area. Do the same for the lengths ±10 mm from the design position. Proceeding down this form will lead you to the answers defining the rate and frequency which the pneumatic spring provides to your design. These data include the gauge pressure at design position required—dynamic forces at ±10 mm—the dynamic spring rate at design position—the system's natural frequency for ±10 mm amplitude. As can be seen, it is easy to determine the effect of different loads, different arm ratios, and the addition of different auxiliary volumes on the suspension system characteristics. The results obtained should be satisfactory for most uses. If more precise answers are required, consult the pneumatic spring manufacturer for a more thorough evaluation.

C. The principle of comparing volumes at ±10 mm amplitude with volume at design position can be extended to any other position along the force-length curve, so long as the ratio developed is raised to the 1.38 power before multiplying with the effective area for that position. This can be used to create a dynamic spring force-length curve for the full stroke. At the same time, a dynamic gauge pressure curve can be drawn from the gauge pressures developed in the calculation process (see Rate and Frequency Calculation Form for Pneumatic Spring Applications).

D. Also, if desired, a dynamic spring rate curve can be drawn, but be sure to obtain the calculated values in $\frac{\text{N}}{\text{mm}}$.

EXAMPLE

Rate and Frequency Calculation Form for Pneumatic Spring Applications

Pneumatic Spring Selected: <u>XXXX</u> Spring Design Length <u>320</u> mm
Sprung Mass to Be Supported at Either Side <u>4 660</u> kg
Spring Force at Design Position Required F_s (for a_s/a_w = 1.5) <u>30 466</u> N

Spring Data and Calculations		+10 mm	Spring Design Length	−10 mm
Spring Length:		330 mm	320 mm	310 mm
Volume Data:				
Spring Volume	V_s	14.71 L	14.24 L	13.79 L
Auxiliary Volume	V_a	0.0 L	0.0 L	0.0 L
Bumper Volume	V_b	−0.714 L	−0.714 L	−0.714 L
Total Volume	V_t	13.996 L	13.526 L	13.076 L
Effective Area Data*:				
Effective Area $A_e = \dfrac{F_s\,(\text{N}) \times 10^3}{P_g\,(\text{kPa})}$		46 219 mm²	46 387 mm²	46 387 mm²
*Note: Using 700 kPa curve for A_e.				
Spring Dynamic Pressure Data:				
Gauge Pressure at Design Length $\quad P_g = \dfrac{\text{Spring Force (N)} \times 10^3}{A_e\,(\text{mm})^2}$			656.8 kPa	
+ Atmospheric Pressure $\quad P_{atm}$			+ 101.3 kPa	
= Absolute Pressure $\quad P_a$			758.1 kPa	

Spring Data and Calculations	+10 mm	Spring Design Length	−10 mm
Dynamic Press. = Absolute Press. @ +/− 10 mm $\quad P_a \left[\dfrac{V_t}{V_t \pm 10\ mm} \right]^{1.38}$	723.2 kPa		794.3 kPa
− Atmospheric Pressure $\quad P_{atm}$	−101.3 kPa		−101.3 kPa
Dynamic Gauge Pressure $\quad P_g \pm 10$ mm @ +/− 10 mm	621.9 kPa		693.0 kPa

Spring Dynamic Force Data:

Dynamic Force $\quad F_d = A_e \times 10^{-3} \times P_g \pm 10mm$ @ +/− 10 mm	28 743.6 N		32 146.2 N

Spring Dynamic Rate:

Dynamic Rate $\quad R_s \quad \dfrac{F_d - 10mm - F_d + 10mm}{20}$		170.13 N/mm	

Direct Acting System Frequency: *

Frequency $\quad (f_s) \quad = 15.76088 \sqrt{\dfrac{Spring\ Rate}{Spring\ Force}}$		1.178 Hz	

Calculated Dynamic Data for Suspension with

Arm Ratio $\quad = a_s/a_w$		1.500	
Wheel Force $\quad F_w = F_s\ (a_s/a_w)$		45 699 N	
Wheel Rate $\quad R_w = R_s\ (a_s/a_w)^2$		382.79 N/mm	
Actual Spring Mass Natural Frequency $\quad = f_s\ \sqrt{(a_s/a_w)}$		1.44 Hz	

$$* \quad from\ \frac{1}{2} \times \sqrt{\frac{g\ (m/s^2)}{Force\ (N)/Rate\ (N/mm)}}$$

$$= 0.15915494 \times \sqrt{\frac{9.806650\ (m/s^2) \times 10^3 \times Rate\ (N/m)}{Force\ (N)}}$$

$$= 0.15915494 \times 99.028531 \times \sqrt{\frac{Rate}{Force}}\ (H_z)$$

$$= 15.76088 \times \sqrt{\frac{Rate}{Force}}\ (H_z)$$

4.23

Chapter 5

Low Gamma Gas, Heat Sink, and Pneumatic Damping

1. Dynamic Response of Pneumatic Spring

The pneumatic spring rate expression is (see Fig. 5.1):

$$F_s = P_g A_e$$

$$\text{Rate} = \frac{dF_s}{dx} = P_g \frac{dA_e}{dx} + A_e \frac{dP_g}{dx}$$

$$PV^n = C \text{ (constant)}$$

$$\frac{dP_g}{dx} = \frac{nP_a}{V}\frac{dV}{dx} = \frac{nP_a A_e}{V}$$

$$\text{Rate} = \frac{nP_a A_e^2}{V} + P_g \frac{dA_e}{dx}$$

where: A_e = effective area
$\quad$ n = polytropic process exponent
$\quad$ P_a = absolute pressure
$\quad$ P_g = gauge pressure
$\quad$ V = spring volume
$\quad$ x = spring deflection

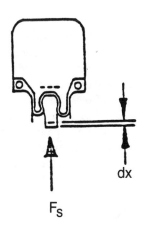

Fig. 5.1—Derivation of pneumatic spring rate

n approaches 1 at low frequency, and it approaches γ (γ = c_p/c_v, often referred to as k) at high frequency. Obviously, the lower the γ, the lower will be the dynamic rate; also, the less will be the difference between dynamic and static rate. n approaches 1 at low frequency when there is sufficient heat exchange between the spring medium and its environment. As the heat exchange characteristics are a function of the geometric configuration of the spring container, piston, flexible member, etc., there is not a clear demarcation that can be pinpointed as the frequency division between isothermal (n = 1) and adiabatic process (n = γ). However, it is reasonably safe to say that most of the truck or car pneumatic springs have ride and roll frequencies close to or higher than 1 Hz and operate in a nearly adiabatic process with a γ value very close to 1.38.

2. Low Gamma Gas

Strictly from an academic point of view without paying attention to practicality, it is obvious that any gas with a γ lower than air is desirable, because it enables us to obtain the same dynamic rate with a smaller volume. If sulfur hexafluoride with a γ of 1.09 is used in lieu of air with a γ of 1.38, we can expect a 20% reduction in spring rate.

Proper care must be exercised to assure that the gas selected does not condense within the design pressure and temperature range. Otherwise, $P_1 V_1^\gamma = P_2 V_2^\gamma$ relationship will no longer hold, the dynamic rate or the pressure-volume function will depend upon the condensation rate and other thermodynamic properties of the two-phase mixture. Besides thermodynamic properties, chemical and physical properties of the gas, such as toxicity, chemical decomposition, thermal decomposition, compatibility with the spring liner, and permeability, should also be considered. The additional cost of the gas, the price of using a closed system, and the availability of the gas in remote territory should also be weighed against the benefit of low gamma gas.

3. Heat Sink

A pneumatic suspension designer will sometimes experience difficulty in finding space for a sufficiently large tank to provide the desired ride softness. Incorporation of a heat sink, which can absorb and give off heat rapidly to lower the n value, seems to offer another alternative. The effectiveness of a heat sink depends upon its ability to absorb and give off heat rapidly.

The thermal capacity of the heat sink must be high as compared with the heat variation accompanying compression and expansion of the pneumatic spring. Furthermore, since the heat transfer between the air and the heat sink must be rapid as compared to the design ride frequency, the exposed area must be as large as possible and also must be in close proximity to substantiate all of the working gas.

Various materials in various forms, from aluminum balls to polyester fibers, have been used for heat sinks; fine fibers seem to offer the best combination in exposed areas and proximity. Because metallic fibers are subject to corrosion and fatigue due to continuous spring deflection, their peak effectiveness is relatively short.

Experiments in the early passenger car pneumatic spring development found that fine polyester fibers are one of the most effective heat sinks for that type of application. A test was conducted on a reversible diaphragm (see Fig. 5.2) designed for the rear suspension of a full size passenger car. A (238 mm) stroke at 1 Hz was used with a normal standing height gauge pressure of 689 kPa and a normal volume of 3.933 L. Without any heat sink, the measured rate was 14.2 N/mm. Various amounts of fine polyester fibers were weighed and packed into the test pneumatic spring chamber, the fixture was allowed to cycle for 3 min before the oscilloscope trace was recorded photographically. Fig. 5.3

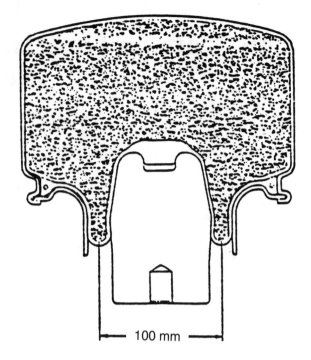

Fig. 5.2—Reversible diaphragm

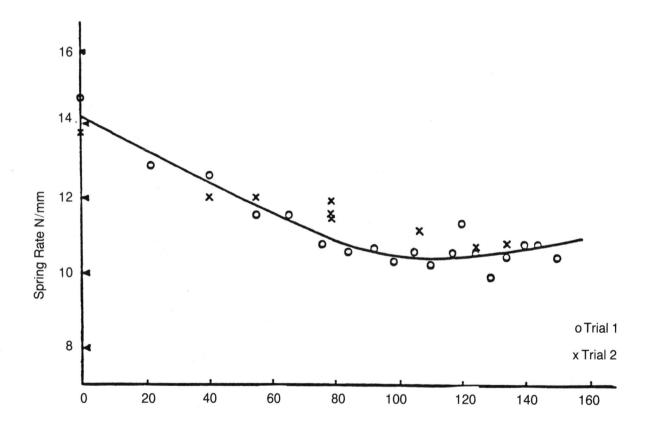

Mass of Polyester Fibers Added (Grams)

Fig. 5.3—Pneumatic spring with heat sink

shows a definite trend of the spring rate to decrease as the amount of polyester fiber inserted is increased up to an optimum point (about 110 g); a further increase of polyester fiber causes the spring rate to increase. At the optimum point, the spring rate read 10.3 N/mm, thus a 27% reduction of pneumatic spring rate was obtained, in this case, by heat sink. It was also found that dirty or matted fibers seemed to be nearly as effective as new fibers in reducing the dynamic pneumatic spring rate.

For springs with pressure higher than 700 kPa, it becomes increasingly difficult to pack the necessary amount of heat sink material without substantially affecting the working gas volume and the effective heat exchange area.

Another item which cannot be overemphasized is that any heat transfer requires time. A heat sink designed to be effective for normal ride frequency will not be equally effective at high velocity wheel impacts.

4. Pneumatic Damping

The subject of pneumatic damping has enticed suspension engineers since the dawn of pneumatic spring development. Many were obsessed by the thought of partitioning the pneumatic spring chamber, and forcing the air through proper orifices to obtain the necessary damping. Unfortunately, some of their acclaimed advantages were due to increasing spring rate rather than damping. It was indeed, for some cases, quite a high price to pay for the amount of damping obtained.

A damper is usually characterized as a linear damper with the damping force expressed as a function of velocity u and damping coefficient c:

$$F = -cu$$

Most pneumatic dampers are by no means linear dampers.* A convenient way to approximate the effect of nonlinear damper is by defining an equivalent linear damping coefficient c_e which will dissipate the same amount of energy per cycle as the nonlinear damper. For a linear damper, the work done per cycle is:

$$W_c = -\pi \, c\omega x_0^2$$

where: ω = circular frequency
x_0 = amplitude of the forced oscillation

The effective damping coefficient c_e of pneumatic damping is equal to:

$$C_e = \frac{\oint F_s dxs}{\pi \omega x_0^2}$$

where: F_s = Spring force
X_s = Spring deflection
ω = Circular frequency

An experiment was run with a reversible diaphragm type pneumatic spring designed for a passenger-car front suspension. The effective area of the pneumatic spring was 16 130 mm^2 and the normal pressure was 689 kPa. The normal volume was 4.883 L, and it was partitioned into two equal volumes. During the compression stroke, the air was forced from the lower chamber into the upper chamber through a small flapper valve. During the rebound stroke, the air was to flow through the 3.2 mm orifice only. The load was measured through a load cell link connected to the wheel spindle which had approximately 2:1 linkage ratio. It can be noted from Table 5.1 that the pneumatic damper has considerable hysteresis by paying the price of slightly higher compression spring rate and much higher expansion spring rate. If the frequency of the pneumatic spring is increased significantly, very little air will pass through the orifice and only the air in the inner chamber will be compressed and expanded. This will result in very high spring rate and very little damping.

TABLE 5.1—EXPERIMENTAL RESULTS OF AIR ORIFICE DAMPING

Frequency, Hz	Rate in Compression N/mm	Rate in Extension N/mm	Hysteresis at Static Design Position N
With pneumatic damper			
1	11.716	19.702	2,002
0.83	11.926	19.159	1,895
0.67	11.173	18.598	1,824
0.50	11.488	17.513	1,690
0.14	10.945	9.299	667
0.03	9.842	8.756	578
With hydraulic shock absorber			
1	10.945	10.070	890
0.83	10.945	10.945	827
0.67	11.488	11.488	756
0.50	11.488	11.488	667
0.03	9.842	7.671	623

*For the derivation of equivalent linear damping, refer to:

1. S. Timoshenko, D. H. Young, W. Weaver
   "VIBRATION PROBLEMS IN ENGINEERING"
   John Wiley & Sons, 1974

2. B. W. Anderson
   "THE ANALYSIS AND DESIGN OF PNEUMATIC SYSTEMS"
   John Wiley & Sons, 1967

Chapter 6

Auxiliary Equipment

This chapter describes some of the functions which are performed by auxiliary equipment. The topics covered include pneumatic control systems, jounce control devices, rebound control devices, roll control systems, and air supply systems. A resourceful designer will find a vast range of auxiliary equipment available which can be combined to perform specific unique functions required by a particular application.

1. Pneumatic Control Systems

The discussion in this section is limited to commonly used pneumatic control systems. Many other functions can be performed by pneumatic control devices. The scope of this section is also limited to the application of these systems to vehicles.

A. Height Control Systems

Height control systems are commonly used to automatically control the dimension between the axle and the frame of a vehicle.

There are two types of valves generally used: one which incorporates a mechanical time delay and another which does not. The unit which incorporates a time delay of 3-20 seconds has proved to be considerably more durable. With the time delay type valve, the suspension system operates effectively as a closed system—except during an event causing height variations which lasts longer than the time delay period. Such events may include long curves, adding or removing load, or stopping on uneven terrain. The use of height control valves with mechanical time delay decreases the number of times the mechanism in the valve is actuated and thus the wear on the valve and the amount of gas expended from the vehicle supply.

There are two common ways of piping vehicle height control systems. One method is to mount one height control valve on each side of the vehicle. Here, all the pneumatic springs on each side of the vehicle are connected together and are controlled by the valve on that side of the vehicle. However, the height control valve must be located so that it can effectively control the vehicle height at all axles in the system. Otherwise, more height control valves will be required for separate systems. Another method of connection uses only one valve and all the pneumatic springs on the vehicle are connected together. The single-valve system allows gas to be transferred from the pneumatic spring on one side of the vehicle to the pneumatic springs on the other side of the vehicle with roll. This air transfer may be a disadvantage. With the single-valve system, the roll rate of the vehicle is very dependent upon components of the suspension other than the pneumatic springs. With the two-valve system, side-to-side pressure differences help provide roll stability.

One of the important benefits of pneumatic suspensions is realized by connecting all of the pneumatic springs on a given side of a multiaxle vehicle together. Such interconnecting provides excellent equalization of load from axle to axle under static conditions when all axles use the same configuration suspensions. If the suspensions are of different design, or they are on a tractor and semitrailer combination more than two valves are probably required. The accuracy of equalization obtained with pneumatic spring type units is very difficult to obtain with other types of suspensions. With this type of plumbing, another benefit of pneumatic suspensions is realized under dynamic conditions. Under dynamic conditions, each axle is in effect independently suspended. The axles act independently because the inter-connecting gas lines are generally small and the time duration of the dynamic input is short, which does not allow gas transfer and equalization to occur as in the static condition.

When a vehicle is completely pneumatically suspended, one of the axles—usually the steering axle—must have a single height control valve system. If four valves are used, terrain or structural variations will cause three of the pneumatic springs to support the vehicle. Such a condition will result in one of the pneumatic springs being deflated much of the time. Four valve systems are not recommended. There are, however, exceptions for some unusual cases.

Another system which is not recommended, but which is often tried, uses a height control valve connected to a mechanical suspension to control the load on an added nondriving axle on a truck or tractor. Because of the manner in which corrosion, temperature, and overloading affect mechanical spring suspensions, this system is seldom, if ever, successful.

Automatic height control systems are the most commonly applied types. These systems are reliable and, if a clean, dry gas supply is used, troublefree.

B. Manual Pressure Control Systems

With manual pressure control systems, the load on a particular axle is controlled. The effective area of the pneumatic spring, multiplied by the gas pressure applied to that spring, is proportional to the load on the axle. Pressure control systems are used where the pneumatic suspensions are used in conjunction with other types of suspensions. The load on an added non-driving axle on tractors or trucks is usually controlled by a pressure-type control system. Specifically, the most common system used to control the load on the nondriving added axle of a tractor is an operator-controlled, cab-mounted pressure regulator. The driver sets the pressure regulator to a pressure as indicated on a gauge in the cab of the vehicle to obtain the desired axle loading. This control system accommodates frame height variations caused by deflection of the associated mechanical spring suspension unit. In plumbing this system, both pneumatic springs are connected together and their pressure is controlled by the same regulator. This interconnection of pneumatic springs is not a major disadvantage in this type of application, since the roll stability requirement for the vehicle is substantially determined by the mechanical suspensions used in conjunction with the pneumatic suspension.

Another system incorporates two or more preset regulators which are selected by a multiposition switch. This system is very useful when the loads carried are predetermined and repeated. A variation of this circuit incorporates a regulator which proportions load from one side of the vehicle to the other. This system is used, for example, on cement mixers. One side of the vehicle is more heavily loaded than the other when the barrel is rotating.

C. Pneumatic Switch (On/Off) Type Control Systems

Pneumatic on/off switches have many uses with pneumatic suspensions. The most common application is in the lifting of one axle of a vehicle. The switch is used to control the axle lift—that is, adding gas to the pneumatic lift spring and exhausting gas from the suspension pneumatic springs. Another use is raising and lowering the vehicle to provide a desired height. One switch on each side of the vehicle can be used to roll the vehicle from side to side. Switches controlling the gas pressure in the front and rear pneumatic springs of a vehicle will permit tipping the vehicle toward the front or rear. On/off switches can be operated by other than mechanical means. Pneumatically or electrically actuated switches are used in many of the systems described.

D. Load Measurement Systems

Load measurement is easily accomplished with pneumatic suspension systems. Some of the considerations in their use will be described here. Rolling sleeve springs provide a more constant effective area over a greater range of design heights than do convolution springs. Scaling systems are, therefore, usually more accurate when used with rolling sleeve springs. The unsprung mass of a suspension and axle must be considered in load-measuring systems. This is usually done by calibrating the load-measuring system on a legal weight scale after installation. Other variables, which may affect the accuracy of the scaling device, are suspension geometry and pinion angle variations.

There are three basic types of load-measuring systems. One system incorporates simply a pressure gauge attached to the pneumatic spring on each side of the vehicle. To read this system the operator must read each of the gauges, multiply by the appropriate factor for the suspension involved, and add that reading to a similar measurement from the opposite side of the vehicle. Through these calculations, the operator can obtain the load on the unit.

A refined system incorporates a valve with two reservoirs. This system samples the gas pressure from each side of the vehicle, averages the two pressures together, and displays the results on a gauge calibrated to read load. Once calibrated, this type of system is operated merely by pushing a button which actuates the valve.

A third system which has been used, but which is not recommended, incorporates valves which when actuated connect all of the pneumatic springs on the vehicle together. When the pressure equalizes, it is read and multipled by the appropriate factor for the suspensions involved. This system is not recommended because of the amount of gas wasted and the amount of time necessary to allow the spring pressures to equalize.

The use of scaling devices with pneumatic suspensions allows the operator to load the maximum amount of cargo without incurring the risk of overload during a drive from the loading site to a scale.

E. Modulated Brake Systems

Valves are available which allow proportioning of braking effort to load on pneumatically suspended axles. The proportioning is accomplished by modulating the gas pressure available to the braking system. The modulation is controlled by the pressure in the pneumatic suspension system which is proportional to the load on that system. Such systems are not in common use.

F. Automated Multifunction Control Systems

Automatic pneumatic controls can be incorporated to prevent the sudden rise of the rear of a tractor when it is being disconnected from a loaded trailer. These same systems usually prevent the pneumatic spring from folding in by maintaining equal pressure in all of the pneumatic springs when the vehicle is operated in a light condition. Actuation of the axle lift on trailer units or on the nondriving axle of a pneumatically suspended dead axle drive axle tandem can be accomplished automatically, as can load proportioning from axle to axle (which is desirable on certain vehicle combinations). The dumping of gas from the system automatically under certain conditions of vehicle usage is another system which can be incorporated. Many of these systems are triggered by a signal received when the trailer gas supply hose is disconnected or when the fifth-wheel latch is actuated.

Preventing the sudden rise of the rear of the tractor when it is being disconnected from a loaded trailer is accomplished by discharging the gas pressure in the system required to support the loaded trailer through a quick release type valve. After the high pressure required to support the loaded trailer is discharged, a regulated low-pressure circuit is actuated. This circuit provides a regulated pressure through the height control valves only high enough to allow suspension of the vehicle by the pneumatic springs once the unit is disconnected from the loaded trailer. This low-pressure circuit prevents resupply of the pneumatic springs with the high pressure required to support the loaded trailer and, thus, the jumping which occurs without such a system.

On units with a height control valve on each side of the vehicle, it is often extremely difficult to maintain proper pneumatic spring inflation when the vehicle is not loaded. By switching automatically to a single valve system when the vehicle is unloaded, the problems associated with uninflated pneumatic springs are avoided.

One of the most sophisticated automatic systems currently available is used on a tractor having a pneumatically suspended nondriving added axle and a drive axle on pneumatic suspensions. The nondriving axle remains in a retracted position until the load on the drive axle reaches a predetermined load level. The nondriving axle is automatically lowered at the preset load level and any additional load above the preset level is applied to the non-powered axle. When the load on both axles reaches a second preset level, any additional load is applied equally to each axle of the tandem. As load is removed from the vehicle, the load on the nondriving axle decreases first. When the load on the combination of the nondriving and driving axle decreases to a predetermined amount which can be supported on the driving axle, the nondriving axle is automatically lifted. This system is completely automatic and requires no operator control.

A manually operated overriding system is often provided on units with automatic pneumatic control systems to allow a load to be transferred from a nondriving to a driving axle. Load transfer can be used to provide increased tractive effort when needed. Most traction control systems automatically revert to the original loading proportions when the operator releases the actuating control.

2. Compression Control Devices

The low spring rates obtained with pneumatic springs may permit very large amounts of axle travel without the buildup of sufficient force to prevent the axle from reaching its limit of travel. To avoid the abrupt "bottoming out" at the end of jounce travel, pneumatic springs often contain compression bumpers. Separate external stops are also sometimes used. Internal stops are usually made of rubber with a tailored compression rate featuring a soft entry and rapid buildup in load capacity. These compression bumpers are customarily sized to carry a load exceeding the pneumatic spring capacity without damage to the bumper. More rigid external stops are often po... utilized only after some c... occurred. External stops... used to prevent excessive lo... The internal stops, in the even... suspension system, allow the vehic... ice area without damage to the pneu... components.

3. Extension Control Devices

A pneumatic spring is usually intended to carry a lo... only a compressive direction and may be damaged by e... tending it beyond its normal fully extended position. Because of this characteristic, some means of controlling the motion of the axle in the rebound mode is required. The amount of axle travel in the rebound mode must be controlled by some component of the suspension other than the pneumatic spring. This function is most often accomplished by the use of hydraulic shock absorbers. Other mechanical linkages such as chains, cables, or links are also used.

4. Roll Control Systems

The low spring rates inherent with pneumatic springs, combined with a narrow spring base of trucks and buses with dual tires, impose the requirement for control of vehicle roll on corners and crowned roadways on some auxiliary device or system. Three types of systems can be used.

A. Roll control can be provided by mechanical, antiroll stabilizing linkages which are independent or which are integral parts of the major suspension system. In automotive systems, this is usually accomplished by an auxiliary bar loaded torsionally. On large highway truck, tractor, and trailer applications, this function is performed as an integral part of the mechanical characteristics of the suspension system.

B. A hydraulic system incorporating two double-acting cylinders with interconnecting lines between the top of a cylinder on one side and the bottom of a cylinder on the opposite side, with their stems connected to the suspension or axle, can be used to prevent or control roll.

C. A third method would be to control the gas pressure in each of the springs with a pneumatic control system with response fast enough to react to roll mode inputs. This, in effect, would be a very fast-acting height control valve system. It would have to be much faster than currently used systems.

5. Gas Supply Systems

Engine-operated piston-type compressors are the principal method of obtaining compressed gas for use in vehicle pneumatic suspension systems. For vehicles which do not

ine driven gas compressors available, electrically
d 12V gas compressors are available. Reliable units
ing pressures of 620 kPa (or greater) similar to that
ded by engine-driven compressor units are available.
se units will supply sufficient gas to accommodate most
eumatic suspension systems. Vacuum-powered units are
lso available. These units generally supply gas only at very
low pressures and in small quantities.

Vehicle pneumatic suspension systems, like all auxiliary pneumatic systems, are required by law to be isolated from the brake system. The isolation is usually accomplished by a valve called a brake protection valve which maintains a preset gauge pressure, usually 450 kPa, in the brake system. The brake protection valve is installed at the connection of the suspension system to the main supply system.

Chapter 7
Physical Application Considerations

Inasmuch as the inherently versatile use of pneumatic springs in suspension leads the designer to try new and different approaches in his designs, it is felt that tips as to what is known to work best with regard to the pneumatic springs themselves should be made available. Some of the things which are known to affect the life of the pneumatic springs adversely are also presented.

The basic concept in all the desirable mounting geometries is to minimize the peak cord or bond stresses. Also, pinching and rubbing reduce spring life and should be avoided.

1. Desirable Pneumatic Spring Mounting Geometries

Suspension basic style and geometry are not considered here.

A. Reversible Sleeve and Reversible Diaphragm Types

Figs. 7.1 and 7.2 show that the piston in a fully compressed position has the same centerline as the upper mounting plate.

A line drawn through the pivot point and perpendicular to the piston centerline cuts the piston between one quarter and one half the way down from the top of the piston. The range of positions is shown in Fig. 7.3.

A workable relationship between the spring top mounting plate and the piston should be such that when the spring is in the fully compressed position the top of the piston is at approximately 25% of the maximum horizontal displacement, as shown in Fig. 7.3. Also, the maximum horizontal displacement away from the arm pivot should be less than the maximum displacement toward the arm pivot.

B. Convoluted Types

For maximum compression stroke have mounting plates parallel in fully compressed position, as shown in Fig. 7.4. Be sure that a line drawn through the pivot point and perpendicular to the pneumatic spring assembly centerline also divides the pneumatic spring in half.

The mounting plates may be parallel in the design position if adequate compression stroke can be obtained. But the previous statement always applies.

The centerline of the bottom mounting plate and the top mounting plate should coincide in the fully compressed position.

C. Parallelogram Suspension Linkage

The centerline of the piston should move equal distances fore and aft of the top plate centerline when the piston is moved through its full travel. The normal design position is shown in Fig. 7.5.

D. Multi-Axle Pneumatic System

With a multi-axle arrangement, it is possible to connect all the springs on each side of the vehicle or trailer to a common gas supply after it has been through the height control valve (see Fig. 7.6). It is best to have at least 25.4 mm and preferably 32.0 mm diameter tubing connecting all springs in such a system. The advantages are that as an individual spring flexes all the other springs act as reservoirs, thus reducing the pressure change and the dynamic spring rate of that spring. The somewhat increased pressure in the other springs has only a mild effect on their spring rate and the shock to the vehicle is damped, reduced, and distributed over a greater area. This arrangement does not affect vehicle roll stability.

2. Undesirable Pneumatic Spring Mounting Geometries

The examples mentioned in this section are extreme cases.

Pneumatic springs are capable of long operation with much abuse, but poor operating geometries will result in earlier failures than springs used with good geometries. Rather small changes in operating geometries on long stroke life tests can show 4-10 times life improvement with improved geometries. Field operations will probably not show as dramatic life improvements because of fewer long strokes in service, but certainly the trend toward improvement will be there.

Avoid centerlines offset as is illustrated in Fig. 7.7. The rolling sleeve type is less sensitive to misalignment than the convoluted type.

Fig. 7.8 shows piston centerlines improperly related to arm pivot. There will be increased wear at the upper mounting plate and on the sleeve where it rolls on the piston, plus the possibility of internal rubbing.

The lower mounting designs shown in Fig. 7.9 are unstable and unnecessary if other proper design criteria are employed.

Fig. 7.10 shows a pneumatic spring under low pressure conditions which may cause it to buckle and fold when compressed and may be damaged in the fully compressed position. It will wear internally, causing reduced life.

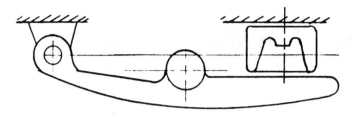

Fig. 7.1—Reversible sleeve fully compressed

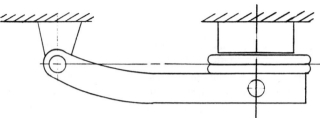

Fig. 7.4—Suspension arm with convoluted pneumatic spring

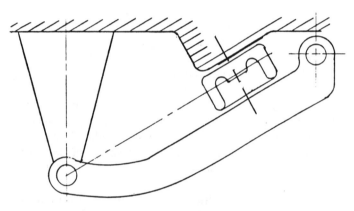

Fig. 7.2—Reversible sleeve relationship between pivot and spring

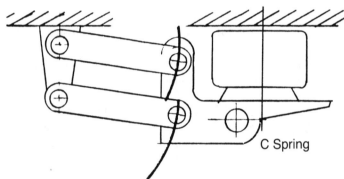

C Spring

Fig. 7.5—Parallelogram suspension linkage

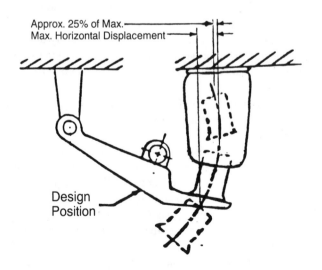

Approx. 25% of Max.
Max. Horizontal Displacement

Design
Position

Fig. 7.3—Reversible sleeve full travel

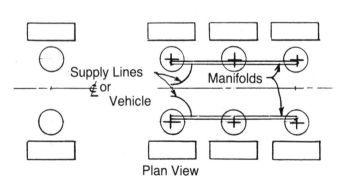

Supply Lines
₵ or
Vehicle

Manifolds

Plan View

Fig. 7.6—Multi-axle pneumatic system

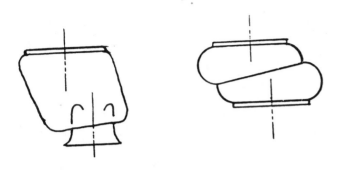

Fig. 7.7—Pneumatic springs with offset centerlines

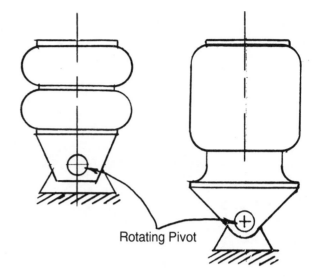

Rotating Pivot

Fig. 7.9—Unstable mounting

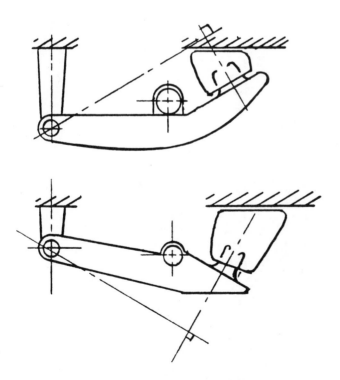

Fig. 7.8—Incorrect piston mounting angle

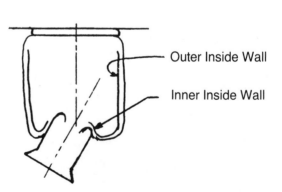

Outer Inside Wall

Inner Inside Wall

Fig. 7.10—Low pressure and internal chafing

Appendix A

SI Units and Symbols for Terms Used

	SI UNIT	SYMBOL
Absolute Pressure	kPa	P_a
Absolute Temperature	K	T
Adiabatic Exponent		γ
Amplitude of Forced Oscillation	mm	x_o
Arm Ratio		a_s/a_w
Atmospheric Pressure	kPa	P_{atm}
Auxiliary Volume	L	V_a
Axle Travel	mm	x_a
Bumper Volume	L	V_b
Centerline		C_L
Circular Frequency	rad/s	ω
Damping Coefficient	N/m/s	c
Dynamic Spring Force	N	F_d
Effective Area	mm^2	A_e
Effective Static Deflection	mm	δ
Effective Viscous Damping Coefficient	N/m/s	c_e
Force	N	F
Gauge Pressure	kPa	P_g
Length	mm	ℓ
Polytropic Exponent		n
Spring Deflection	mm	x_s
Spring Force	N	F_s
Spring Arm	mm	a_s

Spring Rate	N/mm	R_s
Spring Volume	L	V_s
Natural Frequency of Sprung Mass	Hz	f_s
Total Volume	L	V_t
Velocity	m/s	v
Volume	L	V
Wheel Axle Arm	mm	a_w
Wheel Force	N	F_w
Wheel Rate	N/mm	R_w

Appendix B

Glossary of Unique Terms Used in This Manual

Actual Sprung Mass Natural Frequency: The freely oscillating frequency of the sprung mass with the selected suspension design. The end result of the frequency calculations after the suspension arm ratio is applied to the direct acting sprung mass natural frequency.

Actual Wheel Rate: The end result of the spring rate calculations after the suspension arm ratio is applied to the spring rate.

Amplitude of Forced Oscillation: The distance from the spring static design length to either extremity of an oscillation.

Arm Ratio: The spring arm distance from the attachment to the frame divided by the wheel axle arm distance from the attachment to the frame. (a_s/a_w)

Auxiliary Volume: All volume added to the spring system beyond that included in the basic spring at the selected spring static design length as shown on the static force-length graph for that spring.

Axle Design Position: The location of the axle in its normal static design location relative to the supported structure.

Axle Gross Mass Rating: The axle's share of the vehicle's design rated capacity at the base of the tires.

Axle Sprung Mass: Axle gross mass less the axle unsprung mass.

Axle Travel: The distance the axle can move away from the static design position. It can be in the extension or compression direction, but must be labeled properly.

Axle Unsprung Mass: The mass of all the suspension and running gear components not supported by the pneumatic springs.

Bumper Volume: The space taken up inside the pneumatic spring assembly by the selected bumper which reduces the volume of the spring system by that amount.

Decay Trace: A line drawn on chart paper of amplitude versus time for a test run on a laboratory setup equivalent to the suspension design with suitable suspended mass. The system is force cycled to a large amplitude and then uncoupled so that the oscillations die out naturally creating a record of the nature of the decay.

Deflection: The amount of the increase or decrease in the spring length from the spring static design length. It must be properly labeled as extension or compression.

Design Conditions: The motionless (static), position or condition, of the vehicle and all its components as stipulated in the design.

Direct Acting Sprung Mass Natural Frequency: The result of the rate and frequency calculations incorporating data recorded on the data record form (Appendix C) and values selected from the spring force-length graph (Fig. 4.4). It is valid for the suspension design only if the arm ratio is 1.0. It is a necessary step in obtaining the suspension rate and frequency.

Effective Area: The effective area can be calculated for any spring length by dividing the spring output force by the gauge pressure at that length.

Effective Static Deflection: Obtained by extending a line tangent to the dynamic load deflection curve at the selected spring static design position (Figs. 2.1 and 4.2) down to the zero force line, and measuring the horizontal distance along this line separating this intersection and the vertical line defining the spring static design position.

Effective Viscous Damping Coefficient: A value found by evaluating decay traces from laboratory tests run at the natural frequency of the spring mass system with a specified volume configuration and a selected orifice size and location.

Flexible Member: The rubber fabric member between the top and bottom attachment points to other spring assembly components.

Force-Length Curve: A line generated by plotting the force generated throughout the full stroke of the spring while maintaining a constant pressure.

Height: A vertical distance between two points.

Height Control Valve: A unit which controls the distance between a selected location on the sprung mass and the axle or other unsprung suspension component.

Internal Bumper: A rubber unit normally used as an emergency spring, but may also be designed for use as a cushioning stop for extreme input forces to the pneumatic spring. It can also be designed for use as a maximum spring compression displacement stop for springs used with an axle pickup feature.

Load: The vertical downward force of gravity imposed by the mass above at the location being studied.

Maximum Spring Gas Pressure: The supply line pressure supplied to the spring which is anticipated in the normal operation of the vehicle. It is used to establish the load supporting capability of the spring and suspension system.

Normal Design Position: The location and configuration selected by the suspension designer for the suspension components relationships to the sprung mass and the ground with the vehicle at rest.

Normal Design Position Length: The spring length between the top and bottom mounting surfaces measured along the spring centerline with the spring located in its normal design position.

Orifice Area: The cross sectional area of an orifice opening separating the basic spring volume from the remaining spring system volume.

Pneumatic Spring Assembly: Any flexible member, normally combined with rigid components, which when pressurized with a suitable gas will support a mass with resiliency when the unit is subjected to vibration and shock inputs.

Radial Clearance: The minimum distance from the spring wall to nearby vehicle components. This must allow for all extreme pneumatic spring dynamic conditions.

Spring Arm: The perpendicular distance from the spring centerline, or an extension of it, to the arm attachment to the frame. (a_s)

Spring Deflection: The compression or extension distance occurring measured from the selected spring static design length.

Spring Design Force: The spring force required to support that location's share of the vehicle gross axle mass rated load.

Spring Design Length: The distance between top and bottom mounting surfaces measured along the longitudinal centerline when the spring is in its normal design position. This length is selected after consideration of the spring operating characteristics and the suspension design requirements.

Spring Design Length Range: A spring length range designated on the force-length graph which produces the spring's best overall operating characteristics in normal operation.

Spring Design Length Volume: The volume indicated at the intersection of the volume-length curve and the spring static design length vertical line as shown on the spring force-length graph.

Spring Force: The force generated by the pneumatic spring in the appropriate direction and which passes through the center of the effective area so as to oppose the applied force of the sprung mass.

Spring Force-Length Graph: A graph which provides static force at selected pressures and volume values. The data generally covers any usable spring length. The information is used in calculating vehicle rate and frequency characteristics.

Spring Length: The distance between the top and bottom mounting surfaces measured along the axial centerline of the spring.

Spring Minimum Length: A spring length limit which if shortened could seriously reduce the working life of the unit. This is the shortest length shown on the force-length chart and applies to any portion of the top and bottom mounting surfaces of the unit. It is not necessarily the length measured at the longitudinal center line of the spring.

Spring Rate: The difference in dynamic force resulting from a spring + and − amplitude of 10 mm divided by 20 so as to obtain an average rate in N/mm for this range.

Spring Rate Calculation Amplitude (usually + and − 10 mm): Used in the calculation of the suspension system characteristics and is instrumental in producing a typical oscillating frequency for the suspension design.

Spring Volume: The volume inside the flexible member and connecting inside spaces in the top and bottom attached components.

Spring Wall: The rubber fabric member between the upper and lower attachment points to other components.

Sprung Mass: All the vehicle mass which is supported by the suspension system.

Stroke: The total distance the spring length changes starting at one extremity of an oscillation to the opposite extremity of the oscillation.

Volume-Length Curve: A line generated by plotting the spring volume (without reservoir or bumper) generated throughout the full stroke of the spring.

Wheel-Axle Arm: The horizontal distance from a vertical line drawn through the axle centerline to the arm attachment to the frame. (a_w)

Wheel Force: The force acting at the wheel contact point and the ground resulting from the suspended mass acting through the various suspension system components.

Appendix C

Data Record Form for Pneumatic Spring Applications

1. Gross axle mass rating (GAMR) .... —————— kg

2. Axle unsprung mass .... —————— kg

3. Axle sprung mass (both sides) .... —————— kg

4. Axle sprung mass (at either side) .... —————— kg

5. Axle static support force per vehicle side (mass (Kg) $\times$ 9.806650) .... —————— N

6. Desired natural frequency of sprung mass .... —————— Hz

7. Maximum spring gas pressure .... —————— kPa

8. Spring arm (a_s) .... —————— mm

9. Wheel axle arm (a_w) .... —————— mm

10. Arm ratio (a_s/a_w) .... ——————

11. Spring design length selected from a suitable
    force-length graph .... —————— mm

12. Maximum axle compression travel required .... —————— mm

13. Resulting change in spring length from spring compression
    deflection = axle compression travel $\times$ (a_s/a_w) .... —————— mm

14. Minimum spring length along centerline resulting .... —————— mm

15. Maximum axle extension travel required .... —————— mm

16. Resulting change in spring length from spring extension
    deflection = axle extension travel $\times$ (a_s/a_w) .... —————— mm

17. Maximum spring length along centerline resulting .... —————— mm

18. Total change in spring length from full axle compression
    and extension travel .... —————— mm

19. Smallest radial clearance limitation from spring centerline .... —————— mm

20. Spring design force required = (axle static support force
    per vehicle side) / (a_s/a_w) .... —————— N

If an internal compression bumper is required and its operation is vital to the suspension performance, fill in the blanks in Items 21—23 and consult a pneumatic spring manufacturer for their information so that a suitable bumper can be provided. A bumper which can endure two times the spring design force may be recommended for satisfactory service.

21. Axle compression travel before bumper contact . . . . _____ mm

22. Spring compression deflection to bumper contact =
    axle compression travel before bumper contact $\times$ (a_s/a_w) . . . . _____ mm

23. Minimum force bumper should provide . . . . _____ N
    at a bumper compression distance of . . . . _____ mm

Appendix D

Rate and Frequency Calculation Form for Pneumatic Spring Applications

Pneumatic Spring Selected: _____ Spring Design Length ___ mm
Sprung Mass to Be Supported at Either Side _____ kg
Spring Force at Design Position required F_s (for a_s/a_w = ___) _____ N

Spring Data and Calculations		+10 mm	Spring Design Length	−10 mm
Spring Length:		mm	mm	mm

Volume Data:

Spring Volume	V_s	_____ L	_____ L	_____ L
Auxiliary Volume	$+ V_a$	_____ L	_____ L	_____ L
Bumper Volume	$- V_b$	_____ L	_____ L	_____ L
Total Volume	$= V_t$	_____ L	_____ L	_____ L

Effective Area Data:

Effective Area $A_e = \dfrac{F_s\ (N) \times 10^3}{P_g\ (kPa)}$ _____ mm^2 _____ mm^2 _____ mm^2

Spring Dynamic Pressure Data:

Gauge Pressure at Design Length $P_g = \dfrac{\text{Spring Force (N)} \times 10^3}{A_e\ (mm)^2}$ _____ kPa

+ Atmospheric Pressure P_{atm} _____ kPa

= Absolute Pressure P_a _____ kPa

Dynamic Press. = Absolute Press. @ +/− 10 mm $P_a \left[\dfrac{V_t}{V_t \pm 10\ mm} \right]^{1.38}$ _____ kPa _____ kPa

− Atmospheric Pressure $- P_{atm}$ _____ kPa _____ kPa

Dynamic Gauge Pressure @ +/− 10 mm $= P_g$ _____ kPa _____ kPa

Spring Dynamic Force Data:

Dynamic Force @ +/− 10 mm $F_d = A_e \times 10^{-3} \times P_g \pm 10mm$ _____ N _____ N

Spring Dynamic Rate:

Dynamic Rate $R_s = \dfrac{(F_d - 10mm) - (F_d + 10mm)}{20}$ _____ N/mm

Direct Acting System Frequency: *

Frequency (f_s) $= 15.76088 \sqrt{\dfrac{\text{Spring Rate}}{\text{Spring Force}}}$ _____ Hz

Calculated Dynamic Data for Suspension with

Arm Ratio $\qquad = a_s/a_w$ _____

Wheel Force $\qquad F_w = F_s \,(a_s/a_w)$ _____ N

Wheel Rate $\qquad R_w = R_s \,(a_s/a_w)^2$ _____ N/mm

Actual Spring Mass
Natural Frequency $\qquad = f_s \,\sqrt{(a_s/a_w)}$ _____ Hz

$$* \quad \text{from} \quad \frac{1}{2\pi} \times \sqrt{\frac{g\ (m/s^2)}{\text{Force (N)/Rate (N/mm)}}}$$

$$= 0.15915494 \times \sqrt{\frac{9.806650\ (m/s^2) \times 10^3 \times \text{Rate (N/m)}}{\text{Force (N)}}}$$

$$= 0.15915494 \times 99.028531 \times \sqrt{\frac{\text{Rate}}{\text{Force}}}\ (H_z)$$

$$= 15.76088 \times \sqrt{\frac{\text{Rate}}{\text{Force}}}\ (H_z)$$

Part 5

Design and Manufacture of Coned Disk Springs (Belleville Springs) and Spring Washers

SAE HS 1582

TABLE OF CONTENTS

Chapter 1

Introduction

Coned disk springs, also known as Belleville Springs, have a long history. In 1835, one Timothy Hackworth, in England, received acknowledgment for an application in a safety valve. A Frenchman, Julian F. Belleville, secured a British patent in the 1860's for a particular application. At that time, these springs were mainly used in the buffer parts of railway rolling stock and for recoil mechanisms of guns, etc.

Whenever space is limited, particularly in the presence of high forces, the use of coned disk springs will be of advantage. Linear, regressive, and even progressive load-deflection characteristics can be obtained by varying the basic dimensions or by stacking. A large number of standard off-the-shelf sizes is available, so that custom sizes may not be required.

For these reasons, coned disk springs are used today in virtually all branches of engineering with new applications surfacing all the time.

Chapter 2
Definition and Representation

1. Definition

Disk springs have the shape of conical washers (shells of truncated cones), with normally rectangular cross sections. Their geometry is defined by the outside diameter De, the inside diameter Di, the thickness t, and the overall height L, usually written in the form De × Di × T, L = ... mm. A disk spring with De = 50 mm, Di = 25.4 mm, T = 3 mm, and L = 3.6 mm for example can, therefore, be designated as Disk Spring 50 × 25.4 × 3, L = 3.6 mm. For further description, material, surface finish, etc., have to be added. The overall height L is taken as T + H, that is, thickness + dish height. A mathematically more correct relationship is $L = H + T \cos \alpha$. In almost all cases it can be approximated $\cos \alpha = 1$, because the cone angle α is rather small, therefore, L = H + T is of sufficient accuracy.

2. Representation

Fig. 2.1 shows a single disk spring with standard dimensioning. Fig. 2.2 shows a stack and the corresponding load-deflection diagram (characteristic).

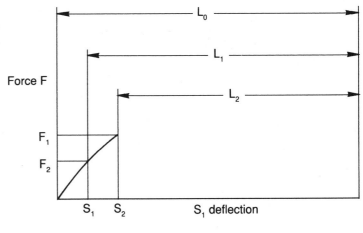

Fig. 2.1—Disk spring cross section

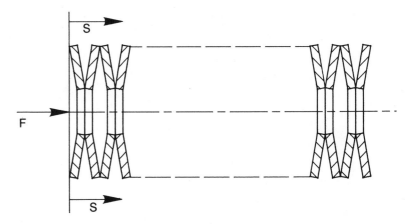

Fig. 2.2—Load-deflection diagram (characteristic) of a disk spring stack

5.3

Chapter 3
Nomenclature, Units

C   = Spring Rate, N/mm (Newton/millimeter)

D   = Diameter; mm
           (Indices indicate different locations)

E   = Modulus of Elasticity; MPa($=$N/mm^2)

F   = Load, Force; N (Newton)
           (Indices indicate specific applications)

H   = Dish (Free) Height; mm

HRC   = Hardness, Rockwell C-scale

i   = Number of Disk Springs in Series

K   = Constant
           (Indices indicate specific variations)

L   = Overall Height of Disk Spring; mm
           (Indices indicate specific cases)

N   = Number of Load Cycles

n   = Number of Disk Springs in Parallel

R   = Diameter Ratio (D_e/D_i)

S   = Deflections; mm
           (Indices indicate specific applications)

T   = Material Thickness; mm

W_1, W_2 = Width of Tongues
           (Indices indicate locations)

W   = Energy Storage Capacity; N·mm

Z   = Number of Tongues

α = Cone Angle of Unloaded Disk Spring

σ = Stress; MPa (N/mm^2)
           (Indices indicate locations or applications)

μ = Poisson's Number ($\mu = 0.3$ for most spring materials)

Chapter 4

Calculation and Formulae, Single Disk

1. Theory and Limitations

Several methods of calculating forces and stresses for given dimensions exist, some extremely complicated, others of limited accuracy.

The theory almost universally used today is the "Elastic Coned Disk"—method based on Almen and Laszlo (Ref. 2, 3).

Major assumptions of this theory are:

(a) Small initial cone angle (Fig. 4.1), so that the usual mathematical simplifications for small angles apply.

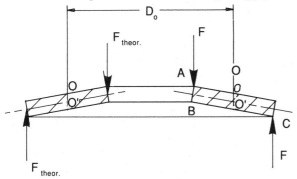

Fig. 4.1—Theoretical cross section of a disk spring

(b) Rectangular cross-section without radii, over the whole range of deflection.

(c) No stresses in radial direction.

(d) Load application "inside" edges (Fig. 4.1).

(e) Fully elastic behaviour of the material during deflection.

These assumptions result in a "center of rotation," point O′ in Fig. 4.1. Its location is defined as:

$$D_o = \frac{D_e - D_i}{\ln(D_e/D_i)} \qquad (1)$$

The theory gives good results for disk springs with common proportions (D_e/D_i = 1.3 to 2.5, H/T up to 1.5), compare Chapter 7, Section 3. The most noticeable deviations are:

(1) Measured loads at the beginning of the stroke are usually smaller than calculated. This may be due to some minor distortions from heat-treatment etc. Calculated loads in the first half of the deflection should be taken as a guide only and not toleranced closely. Loads at deflections of more than 80 to 85% of full deflection will be higher than

calculated due to a bottom-out effect, see Fig. 4.2.

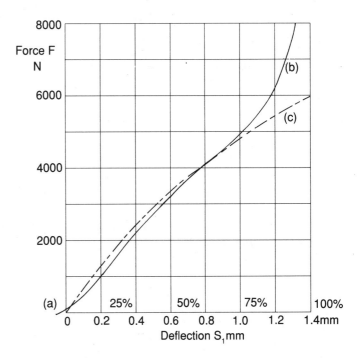

Fig. 4.2—Calculated (c) and measured characteristic of a disk spring 50 × 25.4 × 2 mm.
  (a)   'Flat' position at beginning of deflection
  (b)   'Bottom-out' effect

(2) Most disk springs have to be preset in order to avoid setting under working loads. A complicated state of residual stresses results from this presetting. These residual stresses are generally of the opposite direction when compared to the calculated stresses under deflection. The stress equations in Section 4B, therefore, result in stress differentials only, the "true" stress is the calculated stress minus the residual stress at a given point. Because the magnitude of the residual stresses is generally not known, a correction for the "true" stress is not possible. The presence of residual stresses explains the seemingly very high allowable stresses (Chapter 7).

Springs made from pre-hardened strip do not always conform closely to the theory for reasons not yet satisfactorily explained.

2. Single Disk, Usual Force Application

The equations in this chapter are valid for forces (loads) applied at points A and C, Fig. 4.1.

TABLE 4.1—VALUES FOR CONSTANTS K_1, K_2, K_3.
(These are rounded figures and should not be used for very accurate calculations.)

R	1.4	1.6	1.8	2.0	2.2	2.4	2.6	2.8	3.0	3.2	3.4	3.6	3.8	4.0
K_1	0.46	0.57	0.65	0.69	0.73	0.75	0.77	0.78	0.79	0.79	0.80	0.80	0.80	0.80
K_2	1.07	1.12	1.17	1.22	1.26	1.31	1.35	1.39	1.43	1.46	1.50	1.54	1.57	1.60
K_3	1.14	1.22	1.30	1.38	1.45	1.53	1.60	1.67	1.74	1.81	1.87	1.94	2.00	2.01

A. Force—Deflection

$$F = \frac{4 \cdot E}{1 - \mu^2} \cdot \frac{S \cdot T^3}{K_1 \cdot D_e^2} \left[1 + \left(\frac{H}{T} - \frac{S}{T} \right) \left(\frac{H}{T} - \frac{S}{2T} \right) \right] \quad (2a)$$

K_1 is a constant based on the diameter ratio $R = D_e/D_i$, see Section E.

The above equation simplifies for $S = H$, that is, spring at flat position to,

$$F = F_H = \frac{4 \cdot E}{1 - \mu^2} \cdot \frac{H \cdot T^3}{K_1 \cdot D_e^2} \quad (2b)$$

B. Stress—Deflection

Stresses are calculated in circumferential (tangential) direction. Each 'point' of the cross-section is under a different stress. Of practical importance are only the stresses at points O, A, B, C (Fig. 4.1).

$$\sigma_0 = \frac{3}{\pi} \cdot \frac{4 \cdot E}{1 - \mu^2} \cdot \frac{T \cdot S}{K_1 \cdot D_e^2} \quad (3)$$

$$\sigma_A = \frac{4E}{1 - \mu^2} \cdot \frac{T \cdot S}{K_1 \cdot D_e^2} \left[K_3 + K_2 \left(\frac{H}{T} - \frac{S}{2T} \right) \right] \quad (4)$$

$$\sigma_B = \frac{4E}{1 - \mu^2} \cdot \frac{T \cdot S}{K_1 \cdot D_e^2} \left[K_3 - K_2 \left(\frac{H}{T} - \frac{S}{2T} \right) \right] \quad (5)$$

$$\sigma_C = \frac{4E}{1 - \mu^2}$$

$$\cdot \frac{T \cdot S}{K_1 \cdot R \cdot D_e^2} \left[K_3 + (2K_3 - K_2) \left(\frac{H}{T} - \frac{S}{2T} \right) \right] \quad (6)$$

K_1, K_2, K_3 are constants based on the diameter ratio $R = D_e/D_i$, see Section E.

Negative results indicate compressive stresses; positive results indicate tensile stresses.

C. Spring Rate

The spring rate at any deflection S is given by:

$$C = \frac{dF}{dS} = \frac{4 \cdot E}{1 - \mu^2} \cdot \frac{T^3}{K_1 \cdot D_e^2} \left[\left(\frac{H}{T} \right)^2 - 3 \frac{H}{T} \cdot \frac{S}{T} + \frac{3}{2} \left(\frac{S}{T} \right)^2 + 1 \right] \quad (7)$$

Due to the non-linear nature of the spring characteristic, the rate C is not constant over the deflection range.

D. Energy Storage Capacity

$$W = \int_0^S F \cdot ds = \frac{2E}{1 - \mu^2} \cdot \frac{T^5}{K_1 \cdot D_e^2} \cdot \left(\frac{S}{T} \right)^2 \left[1 + \left(\frac{H}{T} - \frac{S}{2T} \right)^2 \right] \quad (8)$$

If the energy storage capacity between two deflections S_1 and S_2 is wanted, use equation (8) for S_1 and S_2 individually. The difference $W_2 - W_1$ is the desired result.

E. Constants K_1, K_2, K_3

With $R = D_e/D_i$, these constants are:

$$K_1' = \frac{6}{\pi \cdot \ln R} \left[\frac{(R-1)^2}{R^2} \right] \quad (9a)$$

or

$$K_1'' = \frac{1}{\pi} \cdot \frac{\left(\frac{R-1}{R^2} \right)^2}{\frac{R+1}{R-1} - \frac{2}{\ln R}} \quad (9b)$$

Either K_1' or K_1'' can be used, that is, $K_1 \cong K_1' \cong K_1''$. The results are the same for all practical purposes (Ref. 2). Further,

$$K_2 = \frac{6}{\pi \cdot \ln R} \left(\frac{R-1}{\ln R} - 1 \right) \quad (10)$$

and

$$K_3 = \frac{6}{\pi \cdot \ln R} \cdot \frac{R-1}{2} = \frac{3}{\pi \cdot \ln R} (R-1) \quad (11)$$

For some values, see Table 4.1

F. Load Acting Inside Edges

If disk springs are loaded inside the edges (Fig. 4.3), equation (2a) and (2b) can still be used. With the force F and deflection S from the usual loading at the edges, (Section 2), F_L and S_L can be calculated as:

$$F_L = F \cdot \frac{(D_e - D_i)}{(D_{Le} - D_{Li})} \quad (12a)$$

and

$$S_L = S \cdot \frac{(D_{Le} - D_{Li})}{(D_e - D_i)} \quad (12b)$$

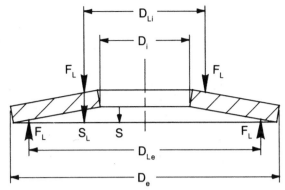

Fig. 4.3—Disk spring loaded inside the edges

It is apparent that such an arrangement will provide higher forces and smaller deflections when compared to a standard loading condition.

Similar reasoning can be applied when springs have very large radii or contact bearing flats (Section G).

G. Contact Bearing Flats

Frequently, larger size disk springs are provided with bearing flats. Often such springs are made interchangeable with springs of regular, rectangular cross-sections. In this case, the free height L has to remain unchanged. In order to match the loads as closely as possible, the thickness has to be reduced to T', Fig. 4.4. As a result, the cone angles will change slightly, with $\alpha_2 > \alpha_1$, Fig. 4.4.

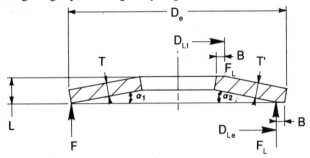

Fig. 4.4—Disk spring cross section, regular (left), with bearing flats (right)
B. Width of flats

A spring with flats can be calculated by using equations (2a), (3) to (6), substituting T' for T and H' for H. H' is the disk height before the flats are machined. A common flat width of $B \cong D_e/150$, H' can be calculated as:

$$H' \cong L - 0.9\,T' \qquad (13)$$

In addition, loads and deflections should be corrected with equations (12a) and (12b), Section F, because the loads will be acting inside the edges by the size of the flat width B.

The results of such calculations, particularly the stresses, should be taken as a guide only.

For standardized sizes with flats, see manufacturers' catalogs (Ref. 4, 5, 6).

H. Initially Flat Disk Spring

The initially flat disk spring of constant thickness (Fig. 4.5) is of little practical interest, mainly because of the need for special loading supports.

Equations (2a), (3) to (6) with H = 0 can be used for calculations.

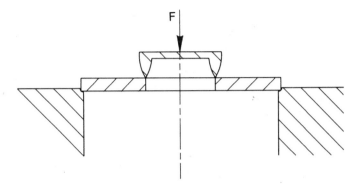

Fig. 4.5—Cross section of an initially flat disk spring

Chapter 5
Single Disk, Load-Deflection Diagrams

Disk springs are non-linear. The ratio H/T determines the actual shape of the load-deflection diagram (Fig. 5.1). As can be seen, disk springs with H/T - ratios smaller than 0.6 are practically linear over a wide deflection range. The diameter ratio $R = D_e/D_i$ has no influence on the shape of the curves.

Fig. 5.1 shows, in a non-dimensional representation, these curves up to 100% deflection (S/H = 1), that is, up to the flat, bottomed out state. Loading beyond flat is possible with specifically shaped supports. The characteristics for these large deflections will be point-symmetrical with respect to the flat-load point, as long as the spring works within an acceptable stress range.

Force F/F$_H$

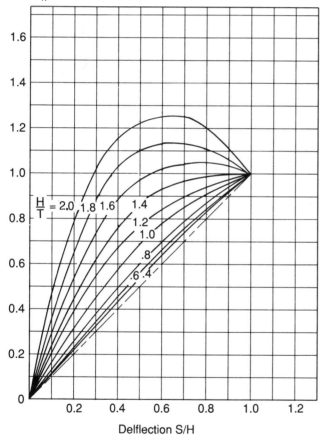

Delflection S/H

Fig. 5.1—Calculated characteristics (nondimensional) for disk springs with different ratios H/T. (Compare also Fig. 4.2) (F$_H$: load at flat position)

Chapter 6
Design Considerations

1. Combination of Single Disks (Stacking)

Disk springs can be stacked in different configurations. This may increase the load, the deflection, or both:

Parallel Stacking: This will increase the load proportional to the number of springs in parallel, Fig. 6.1. The deflection will remain unchanged.
(Friction between disks in parallel is treated in Section 6.5.)
$F_{total} = n \cdot F$ (n: number of springs in parallel)
$S_{total} = S$
$L_o = L + (n - 1)$ T unloaded height of parallel stack

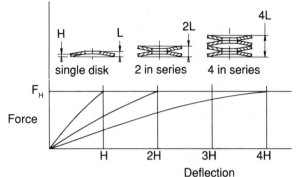

Fig. 6.2—Series stacking of disk springs.
(F_H: load at flat, single disk)

Parallel-Series Combination:
$F_{total} = n \cdot F$ total Force
$S_{total} = i \cdot S$ total deflection
$L_o = i [L + (n - 1)$ T] unloaded height

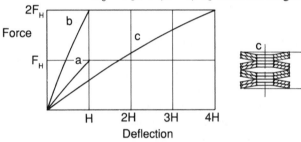

Fig. 6.3—Parallel/series stacking of disk springs
(F_H: Load at flat, single disk)
(a)   Single disk (n = 1)
(b)   Two disks in parallel (n = 2)
(c)   n = 2, i = 4

Another combination is shown in Fig. 6.4, resulting in a progressive characteristic. Fig. 6.5 shows a preassembled stack with n = 3, i = 25.

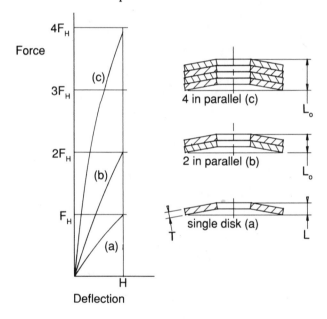

Fig. 6.1—Parallel stacking of disk springs.
(F_H: load at flat, single disk)

Series Stacking: Disk springs in series will result in increased deflection, proportional to the number of disks, Fig. 6.2.
$F_{total} = F$
$S_{total} = i \cdot S$ (i: number of disks in series)
$L_o = i \cdot L$ unloaded height of stack in series

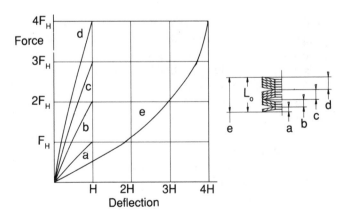

Fig. 6.4—Stacking for progressive characteristic
(F_H: load at flat, single disk)
(a)   Single disk n = 1
(b)   n = 2
(c)   n = 3
(d)   n = 4
(e)   combination

Fig. 6.5—Preassembled disk spring stack
n = 3, i = 25

For more examples of stacking, see Ref. 4.

Springs with ratios H/T > 1.4 show a force maximum before flat. If such springs are combined in stacks, individual disks may 'snap through', even before the nominal force maximum is reached, due to tolerance variations of individual disks. For this reason, practically all standard sizes are limited to H/T < 1.3.

Not all disk springs can be flattened out without set because of stress limitations.

2. Stack Length

Long stacks may show uneven deflection of the individual disks under load. This effect is more noticeable when disks with a relatively large ratio H/T are used. Typically, the disks close to the moving end will show higher than proportional deflection (with fatigue life consequences, Chapter 7, Section 2), and disks at the non-moving end may appear to undergo very little or no deflection, see Section 4. This is caused mainly by friction between the disks and the guide element (see also Sections 3 and 4). As a rule, the total length of a stack should not exceed 2 to 3 times its outside diameter.

Long stacks also tend to "buckle" under load. Guides prevent large sideways movements, but resulting forces perpendicular to the direction of the load may cause additional frictional hysteresis (Section 6). Flat washers can be used to break up a long stack. Such washers should operate with very low guide clearance and have to be thick enough to prevent cocking.

3. Guiding and Clearances, Lubrication

Disk spring stacks require guides, either on the inside or on the outside.

The diametrical clearance between springs and guides should be about 1% of the respective diameter, possibly as low as 0.5% for springs with very low H/T - ratios. Springs with standard rectangular cross-sections (Fig. 1.1) and small radii will not change their diameters perceptibly when loaded. The inside diameter may even get larger and the outside diameter smaller under load, unless springs with very high H/T ratios are considered.

Very small clearances are desirable when springs are assembled on rotating shafts in order to minimize imbalances. Small clearances will also facilitate alignment (Section 4).

The guides and the end thrust faces should be smooth and harder than the springs, for example, 55 HRC or more. This is particularly important for repetitive loading. For static or quasistatic loading (Chapter 7, Section 1), hardening of guides may not be necessary.

Practical considerations also determine if the outside or the inside diameter of the first and last spring in a stack should contact the supporting face.

Lubrication should be provided for repetitively loaded stacks, both between individual springs in parallel and between springs and guides. Extreme pressure, anti-seize lubricants are used for this purpose.

4. Alignment

The alignment of springs or spring stacks with respect to the guide should be as perfect as possible. Insufficient alignment may be the cause of uneven deflection (Section 2) or increased frictional hysteresis (Section 6). If assembled and preloaded stacks are accessible, alignment may be possible with a straight edge or by the use of a rubber mallet. If a stack is not accessible, onetime loading to flat will improve alignment.

5. Loading Beyond Flat

Sometimes it is desirable to load springs beyond flat, in order to increase the travel per disk or to utilize the near zero-spring rate in this deflection range for springs with larger ratios H/T (see Fig. 5.1 and Chapter 5).

Fig. 6.6 shows a possible arrangement, using guide rings. Advantages are the larger deflection range, possibly with a near zero spring rate and the inherent self-centering feature. Disadvantages are the increase in unloaded height and the higher costs of such an arrangement.

Depending on the actual dimensions of the guide rings, the springs will be supported inside the edges as they become "flat". For load corrections, necessary beyond this point, the method described in Chapter 4, Section F can be used.

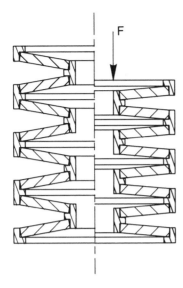

Fig. 6.6—Disk spring stack for loading beyond flat

Most standard stock sizes can safely be taken to flat position, but may be overstressed when taken beyond flat, resulting in a more or less pronounced set.

6. Frictional Hysteresis

Friction results when disk springs slide along the guide under load. In long stacks this may lead to uneven deflection (Section 2), so that the last disks, that is, the non-moving end of the stack, may see little or no deflection at all.

If a single spring is loaded between flat plates (Fig. 6.7), the upper inside edge of the spring will move towards the inside, the lower outer edge towards the outside, resulting in frictional forces F_F opposite to the direction of the movement. Single disks in series will, at the contacting edges, also move slightly with respect to one another, if they are not perfectly aligned. This motion will also cause some friction. Disk springs in parallel stacking will move opposite to one another over the whole contact area, when under a load, also resulting in friction.

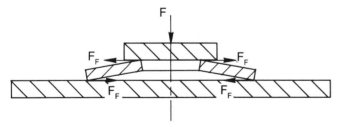

Fig. 6.7—Single disk spring, loaded between flat plates

All these mechanisms will cause frictional hysteresis, Fig. 6.8. The more springs in parallel stacking are used, the more pronounced is the frictional hysteresis.

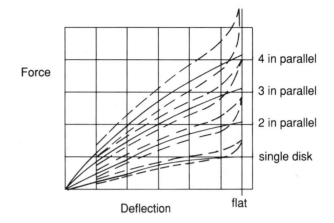

Fig. 6.8—Hysteresis of disk spring stacks.
(Also shown: typical bottom-out effect)

Table 6.1 shows typical figures for the frictional hysteresis. The higher numbers usually apply for dry, the lower ones for lubricated surfaces. The additional influence of stack length, alignment and clearance on frictional hysteresis should be determined experimentally for specific applications.

TABLE 6.1—TYPICAL PERCENTAGES OF FRICTIONAL HYSTERESIS IN DISK SPRINGS

single disk	±2 to 3%
2 in parallel	±4 to 6%
3 in parallel	±6 to 9%
4 in parallel	±8 to 12%
5 in parallel	±10 to 15%

7. Presetting, Recovery, Creep and Relaxation

Most disk springs are designed to a high stress level when loaded flat (Chapter 7, Section 1). Such springs are preset after heat treatment. The ensuing loss of height should bring the spring to the design height. This procedure results in a complicated state of residual stresses in the cross-section.

If preset springs are left unloaded for some time, such as in inventory, the height will increase slightly. This phenomenon is called recovery. The increased height may cause noticeable force increases when loaded, but one or a few loadings to flat should restore the initial height.

Steady loads at high stress levels may cause creep or relaxation, particularly at elevated temperatures. Creep is the term used when the springs are held under constant load and the loaded spring height gets smaller. Relaxation is a reduction of the spring force if the springs are held in a constant length arrangement. If such springs are unloaded, the change of the unloaded height will be a measure for both creep and relaxation.

Chapter 7
Design Stresses

1. Static and Quasi-Static Loading

Such a loading exists in disk springs
  a) under constant load
  b) under occasional load changes with longer intervals. The total number of such load changes can be up to 1000, according to some sources even up to 10 000.

Disk springs for static loads are normally designed that they may be pressed flat, at room temperature, so that the free height L (after unloading) stays within the normal tolerance range (see Chapter 10, Section C).

The critical stress for designing statically loaded springs is σ_A (equation (4), Chapter 4, Section B).

The values given in Table 7.1 should be taken as a guide only. There are several variables, such as the actual hardness, the material thickness and the diameter ratio $R = D_e/D_i$, which have some influence on the safe stress level.

TABLE 7.1—MAXIMUM PERMISSIBLE STRESS σ_A FOR DISK SPRINGS IN STATIC APPLICATIONS

Material	Percent of Tensile Strength	
	Not Preset	Preset Springs
Carbon or Alloy Steel	120	250
Stainless Steels	95	160

A much simpler method of stress calculation is based on stress σ_0 (equation (3), Chapter 4, Section B). As long as σ_0 does not exceed the tensile strength of the chosen material, the spring can be manufactured without taking an undue set under static or quasi-static loading. These springs are usually preset by the manufacturer, and so the conditions as discussed in Chapter 6, Section 7 will apply.

2. Repetitive Loading (Fatigue)

Repetitive or cyclic loading is defined as repeated loading between a preload deflection S_1 and a maximum deflection S_2, with corresponding stresses σ_1 and σ_2. Tensile stresses are the cause of fatigue failures, and the critical stress in disk springs is the larger of σ_B (equation (5)) or σ_C (equation (6), Chapter 4, Section B). Fig. 7.1 will help to decide which of the two stresses should be calculated. For springs falling into the shaded area of Fig. 7.1, calculations should be done for both σ_B and σ_C; the lower figure for the fatigue life N applies.

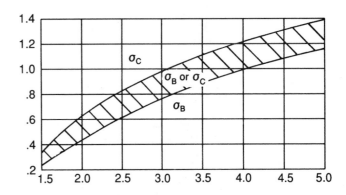

Fig. 7.1—Critical stress for calculating fatigue failures of disk springs

The modified Goodman diagrams (Figs. 7.1a, b, c) show the fatigue strength for various material thicknesses. Groups 1, 2 and 3 are further defined in Chapter 10, Section 1. These Goodman diagrams allow, if σ_1 and σ_2 are known, an estimate of the number of cycles until failure with a statistical survival rate of 99%. These diagrams are valid for stacks that are evenly deflected (up to 6 springs/stack). Long stacks (comp. Chapter 6, Section 2), substandard materials, surface imperfections such as scratches, nicks, corrosion and die breaks, will result in lower fatigue lives. The dashed lines imply some uncertainty, and readings in that range should be taken as a guide only.

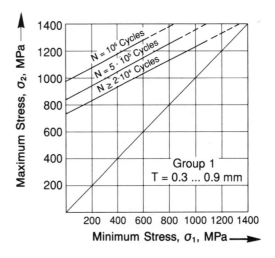

Fig. 7.1a—Modified Goodman diagram for disk springs

5.17

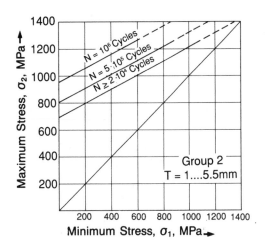

Fig. 7.1b—Modified Goodman diagram for disk springs
(For example, see Chapter 9, Section 2)

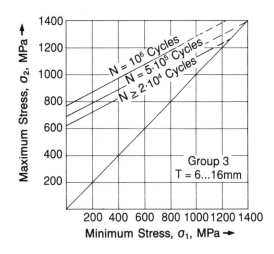

Fig. 7.1c—Modified Goodman diagram for disk springs

3. "Ideal" Shape of Disk Springs

Many applications require the unique load-deflection relationship typical of a specific H/T—ratio. In other cases, the diameters are fixed due to a particular envelope into which the spring has to fit.

If none of these conditions exist, the springs can be designed for best utilization of the design volume, that is, lowest weight per storable energy. Such springs have ratios $D_e/D_i = 1.7$ to 2.2 and H/T = 0.4 to 0.8. Similar reasoning will lead to preferable preload deflections of 20% to 50% of H for disk springs under repetitive loading (Ref. 8).

4. Minimum Preload

Most disk springs are preset and, therefore, are under high residual tensile stresses in the vicinity of the upper inner edge (Fig. 4.1, point A). If such springs are subject to cyclic loading without being preloaded, radial cracks may appear in this location. Normally, such cracks will not extend beyond a certain size and, therefore, do not constitute a fatigue failure as defined in Section 2. Nevertheless, such cracks should be avoided, which can be done by a preload deflection of at least 15% to 20% of H.

Improved results may be achieved with ground and polished surfaces, a very expensive process, or with shot-peening (see Chapter 8, Section D).

If the fatigue life of a given disk spring arrangement has to be known more accurately, an actual fatigue test is recommended, which is particularly important for applications with changing amplitudes.

Chapter 8

Materials and Finishes

1. Spring Materials

Table 8.1 summarizes the most frequently used materials for disk springs.

TABLE 8.1—SPRING MATERIALS

Material	E-Modulus MPa·10³ 1)	Tensile Strenght, MPa 2)	Hardness Range, HRC 3)	Thickness Range, mm 4)	Temp. Range °C	Remarks
SAE 1050 to 1095	207	1400 to 1700	46-50	0.1 to 5	100°	Standard
SAE 6150	207	1400 to 1700	40-49	0.3 to 60	200°	Standard
SAE 30301, 30302	190	1200 to 1400		up to 2	−200° to −250°	Corrosion Resistant
SAE 51400 - series	207	1200 to 1400	38-46	0.3 to 6	400°	Corrosion & Temp. Resistent
17-7 PH	195	1200 to 1400		up to 4	300°	Corrosion Resistant
Ph-Bz	115	600 to 700		0.1 to 5	100°	Corrosion Resistant, Non Magn.
Be-Cu	130	1270 to 1450		up to 2	−200° to +200°	Corrosion Resistant
INCONEL 750	210	1000 to 1100		up to 5	600°	Corrosion Resistant, Non Magn.

1) At room temperature, average values.
2) At room temperature, exact value dependent on hardness and other variables.
3) Typical range: Higher values for the thinner materials and vice versa.
4) Larger thicknesses available (special mill orders).

Other materials are used for special purposes, consult with manufacturers.

SAE 1050 to 1095: This is a family of high carbon steels, mostly used in cold rolled form for relatively thin disk springs, sometimes processed in prehardened form, but see Chapter 4, Section 1. Usually SAE 1060 to 1075 materials are preferred due to their balance of strength and durability. The final selection is often dictated by availability and should be left to the manufacturer.

SAE 6150: This material is used in cold rolled, hot rolled and forged form, for springs with medium and large thicknesses (1 and over), due to its even hardenability across the thick-ness and a reduced tendency to relaxation when compared to the standard carbon spring steels.

SAE 30301 30302: These austenitic Cr-Ni-Steels are not hardenable by heat treatment. Spring properties, that is, high tensile strength, etc., are obtained by cold working only. This restricts the useful thickness to approximately 3 mm maximum; the strength decreasing with increasing thickness.

These materials have excellent corrosion resisting properties. It should be noted that they will be somewhat magnetic due to the necessary cold working.

SAE 51400-series: This group consists of Cr-Alloys, which are ferritic at room temperature but become austenitic at elevated temperatures and can be rapidly cooled to produce a hard, martensitic structure.

These materials have a good corrosion resistance and can be used at temperatures up to 400°C. Both modulus of elasticity and tensile strength will be reduced as the temperature increases, Ref. 4.

There is some danger of delayed brittle fractures, that is, stress corrosion, under certain circumstances.

17-7 PH: This precipitation hardening material has an austenitic/ferritic structure. In order to achieve spring properties, the hardness has to be rather high, again with some danger of stress-corrosion.

2. Surfaces

The selection of the surface is very important and is dictated by the ultimate use of the spring. Initially, it is given by the material selection. Cold-rolled surfaces are associated with thin materials (up to 1 mm thickness). Hot-rolled surfaces are typical for thicknesses over 1 mm, where the last rolling operation is frequently done cold. Forged surfaces appear in very heavy cross-sections. Both hot-rolled and forged surfaces are frequently machined to size to minimize stress concentration effects. The diameters may be machined to remove die breaks in all springs with thicknesses over 1 mm.

A. Blank, Oiled

After the manufacturing of the disk springs, the surface may show discoloration typical to heat treatment. Such a surface, even if cleaned mechanically, will rust quickly (C-, Cr-V-steels). A slight oil film will prevent such rusting for a limited time. This surface condition should only be specified, if additional surface protection is considered upon receipt.

B. Zinc Phosphate, Black Oxide

Black oxide is a uniform black coating for ferrous steels. It is mostly decorative affording limited corrosion protection.

Zinc phosphate is gray to black in color and usually includes a supplementary oil treatment. It is widely used in disk springs for mild corrosion protection and indefinite shelf life in normal industrial surroundings. Typical coat thickness is 0.01 mm.

C. Plating for Corrosion Protection

The effect of a corrosive environment is difficult to predict with certainty.

Use of corrosion resistant (CRES) materials afford the most reliable protection; however, this is often costly and sometimes impractical. Protective coatings provide, in many cases, alternatives with sufficient protection. Either galvanic or mechanical methods of plating are used (Ref. 9, 10):

Galvanic Plating (Electroplating) should be avoided for hardened spring steels because these are inherently susceptible to hydrogen embrittlement. A subsequent baking operation is essential. The danger of hydrogen embrittlement exists also, if parts are acid-cleaned before plating.

Mechanical Plating (Ref. 10) provides an effective protection with practically no danger of hydrogen embrittlement. (Caution: acid cleaning (pickling) before plating may increase susceptibility to hydrogen embrittlement). This method is recommended for disk springs, which typically have a high hardness and high residual stresses, and do not tangle in the process, so that all surfaces are fully accessible. The existing plating equipment usually restricts the size of the springs to be plated to a maximum of 100 to 150 mm outside diameter.

These coatings are of a sacrificial nature and include zinc and, to a lesser degree, tin and aluminum. They act as a barrier between base material and environment, but they continue their protective role even after they are scratched, cracked or nicked.

Cadmium should only be specified where absolutely necessary, due to its toxicity.

D. Shot Peening

Shot-peening is a cold-working process used primarily to increase the fatigue life and prevent stress corrosion cracking. The surface of the finished part is bombarded with round steel shot in special machines under fully-controlled conditions (Ref. 11).

Disk springs are frequently shot-peened. As a rule of thumb, the cycle life is increased by at least a factor of two, at a rather moderate cost. The process should be specified for repetitively loaded disk springs, if there is any doubt that the regular execution will meet the fatigue requirements.

Loading shot-peened disk springs to flat should be avoided, because some additional set may result.

E. Teflon Coating

Disk springs are sometimes Teflon coated. This reduces friction for disks stacked in parallel (Chapter 6, Sections 1 and 5). It is also used for color coding.

F. Other Coatings

Electroless Nickel: This is a relatively expensive method to protect disk springs. There is no danger of hydrogen embrittlement. The coatings have to be rather heavy in order to avoid pores. These nickel coats may come off when the spring is loaded.

Hot Dip Galvanizing: This is a non-electrolytic dipping process, typically in a zinc bath. The possible thickness of the final coat depends on the thickness of the base material (see ASTM 153). Relatively thick, pore free coatings can be obtained. Due to the elevated temperatures of this process, annealing takes place in the hardened spring with a subsequent loss of hardness.

Chapter 9

Design Examples

A typical design problem consists of selecting a disk spring for a given load or loads, deflection or deflection range, fatigue life, environment, etc., within dimensional limitations. The first choice should always be, for reasons of ready availability and cost, a standard size from a manufacturer's catalog (Ref. 4, 5, 6, 7). If no standard size is available, or cannot be modified to fit the application, a custom size has to be designed.

The following steps will lead to a solution:

1. Outside and/or inside diameters can be obtained from dimensional restrictions. If one of the two is given, the ratio D_e/D_i should be chosen to be approximately 2, (see Chapter 7, Section 3).

2. Select material (Table 8.1).

3. Select the general shape of the characteristic (Fig. 5.1), that is, the preferable ratio H/T.

4. Select working load as a percentage of load at flat. Determine approximate load at flat F_H again with the help of Fig. 5.1.

5. Calculate the thickness T, using Equation (2b) in the form:

$$T = \sqrt{D_e} \cdot \sqrt[4]{\frac{F_H \cdot K_1}{\left(\frac{H}{T}\right) \cdot \frac{4E}{(1 - \mu^2)}}} \qquad (12)$$

Adjust T to a standard thickness, according to the ones listed in manufacturers' catalogs or Table 10.1, Chapter 10, Section A; this will ensure availability.

6. Calculate stresses σ_O or σ_A from Equations (3) and (4), for full deflection S = H, and compare to the permissible stresses listed in Chapter 7, Section 1.

7. If stresses are too high, change assumptions and recalculate. For instance, half the load F_H, means two springs in parallel to satisfy the load requirements.

1. Static Loading

Problem: Design a disk spring to be used for preloading a seal. The spring diameters are dictated by the seal dimensions. D_e max = 173 mm, D_i min = 150 mm. The seal is to be preloaded with F = 1050 N ± 15%; the load should not fall below 800 N after the spring moves 0.7 mm due to the relaxation of the seal.

Solution: Use D_e = 173, D_i = 150 mm, that is, D_e/D_i = R = 1.153. This ratio is very low, but the diameters are dictated by the available space. Choose H/T = 1.4 (see Fig. 5.1), to keep load as constant as possible over a wide range of deflection. (Note: Snapping will not be a problem because single disk is used and no external forces will be applied).

Assume load at flat, F_H = 1050 N. Choose material SAE 6150 with E = 207 000 MPa (Table 8.1), and tensile strength 1400 MPa.

The use of Equation (12) with K_1 = 0.24 (Equations (9a) or (9b), Chapter 4, Section E), results in T = 1.56 mm, and H = 1.4 · T = 2.18 mm.

Choose next standard thickness T = 1.5 mm and H = 2.2 mm.

Check with Equation (2a): F_H = 954 N (load at flat with T = 1.5, H = 2.2 mm).

Stresses (Equations (3) and (4)): σ_O = 405 MPa, σ_A = 747 MPa

Both stresses are so low that this spring will not have to be preset, see Chapter 7, Section 1.

Assume that the maximum deflection S_2 of the spring is 80% of H, that is, 1.75 mm, which leads to:

$$S_2 = 1,75 \text{ mm} \quad \text{and} \quad F_2 = 960 \text{ N} .$$

A change of 0.70 mm will reduce the deflection to S_1 = 1.75 − 0.7 = 1.05 mm, with F_1 = 845 N.

Result: Disk Spring 173 × 150 × 1.5, L = H + T = 3.7 mm. The loads will be in the desired range. Due to the small D_e/D_i ratio of 1.153, the theoretical load application point is noticeably different from the theoretical point, compare Chapter 4, Section 1(d). This will result in a higher load than calculated, but possibly still within the desired limits. If the load is measured too high, one may consider to shift the deflection range lower or reduce the dish height H.

2. Repetitive Loading

Problem: A standard disk spring with the dimensions $D_e = 50.8$ mm, $D_i = 25.4$ mm, $T = 2.46$ mm, $H = 1.22$ mm ($L = H + T = 3.68$ mm) has a nominal load of $F_1 = 3043$ N at $S_1 = 0.35$ mm, and $F_2 = 5036$ N at $S_2 = 0.61$ mm. How many load cycles between S_1 and S_2 can be expected until fatigue failure occurs?

Solution: a) Find critical stress locations with $R = D_e/D_i = 50.8 / 25.4 = 2$ and $H/T = 1.22 / 2.46 = 0.5$. It follows from Fig. 7.1, Chapter 7, Section 2, that stress σ_B or σ_C is critical for fatigue failures.

b) From Equations (5) and (6), Chapter 4, Section B, we calculate:
at $S_1 = 0.35$ mm, $\sigma_{B1} = 379$ MPa, $\sigma_{C1} = 447$ MPa
at $S_2 = 0.61$ mm, $\alpha_{B2} = 710$ MPa, $\sigma_{C2} = 748$ MPa

c) The applicable Goodman diagram, Fig. 17.1b, Chapter 7, Section 2, checked for both σ_B and σ_C, will give $N \geqq 2 \times 10^6$, that is, infinite cycle life.

d) If the same spring is loaded from deflection $S_1 = 0.35$ mm to a higher deflection $S_2 = 0.9$ mm, we calculate $F_2 = 7081$ N (Equation (2a), Chapter 4, Section A), $\sigma_{B2} = 1129$ MPa, and a fatigue life of $N = 10^5$, see Fig. 7.1b.

Chapter 10

Manufacturing and Tolerances

1. Manufacturing Methods

For standardized disk springs, 3 different manufacturing groups are defined.

Group 1:   Springs with small material thicknesses (up to 1 mm). These are made from cold-rolled carbon steel strip, stamped and formed in one tool. Tumbling, heat-treatment and pre-setting (if necessary) are further major steps.

Group 2:   Material thickness between 1 and 6 mm. Either carbon or Cr-V-steels are used. These springs can be stamped to size, or have machined diameters which eliminate die breaks and, therefore, improve the fatigue life. Cr-V-steels (SAE 6150, Chapter 8) are preferred for these higher thicknesses because a more even grain structure results after heat treatment.

Group 3:   Material thickness over 6 mm. Made from hot-rolled plate or from forged blanks. Machined all over to remove all scale, etc. Only Cr-V steels, but no plain carbon steels should be used.

Special Grade: Custom designed disk springs often require different manufacturing approaches. If, for instance, the surface has to be ground for thickness tolerances below standard mill tolerances, flat rounds are stamped or flame cut etc., ground to thickness and then formed. It is frequently more advantageous to modify standard stock sizes, than to make a custom size spring, particularly when the production quantities are small.

2. Tolerances

A. Thickness Tolerances

TABLE 10.1—THICKNESS TOLERANCES OF DISK SPRINGS

Group	Nominal Thickness, mm	Tolerances, mm
1	0.2, 0.25, 0.3, 0.35, 0.4	±0.02
	0.5, 0.6, 0.7, 0.8	±0.03
	0.9, 1.0	±0.04
2	1.0, 1.1	±0.04
	1.25, 1.5, 1.75	±0.05
	2.0, 2.25, 2.5	±0.06
	2.7, 3.0, 3.5	±0.07
	4.0, 4.3, 5.5	±0.08
3	over 6	±0.08

For thicknesses in Group 1 and 2 of Table 10.1, the given tolerances are the standard mill tolerances for rolled steel. The tolerances for Group 3 springs result from machining.

Frequently the actual mean thickness is different from the mean nominal thickness, for instance 0.77 mm instead of 0.8 mm. The nominal thickness is used in the calculations, the actual thickness is used for manufacturing in order to get a good match between calculated and manufactured spring. This small thickness reduction is necessary because disk springs typically show small radii at the edges, either from tumbling (Group 1), or machining (Group 2). These radii will cause an increase in measured force; therefore, the thickness is reduced. In such cases, the tolerances of Table 10.1 apply to the actual mean thickness and not to the nominal thickness.

B. Diameter Tolerances

TABLE 10.2—DIAMETER TOLERANCES OF DISK SPRINGS

Dimensions, mm		Tolerances, mm	
Over	to	Outside Dia D_e	Inside Dia D_i
		minus only	plus only
3	6	−0.12	+0.12
6	10	−0.15	+0.15
10	18	−0.18	+0.18
18	30	−0.21	+0.21
30	50	−0.25	+0.25
50	80	−0.30	+0.30
80	120	−0.35	+0.35
120	180	−0.40	+0.40
180 & over		−0.46	+0.46

C. Overall Height Tolerances

TABLE 10.3—TOLERANCES FOR THE FREE HEIGHT, L, OF DISK SPRINGS

Group	Material Thickness, mm	Tolerances of L, mm
1	up to 1	+0.075, −0.025
2	1 to 2.25	+0.10, −0.05
	2.25 to 6	+0.15, −0.05
3	6 and over	±0.2

D. Force Tolerances

It is not possible to make a disk spring to conform to the theoretical characteristic over the whole deflection range, Chapter 4, Section 1. For this reason, it is customary to tolerance only the load at S = 0.75 H (75% deflection). It is better to measure the loaded height, as opposed to the actual deflection. The loaded height at 75% deflection is $L_1 = L - 0.75\ H$.

TABLE 10.4—LOAD TOLERANCES FOR SINGLE DISK SPRINGS (75% DEFLECTION)
(For definition of Group 1, 2, 3, see Section C.)

Group	Load Tolerances at $L_1 = L - 0.75\ H$
1	+25% − 7.5%
2	+15% − 7.5%
3	± 5%

Loads at deflections other than 75% deflections can also have tolerances. In such a case, the tolerance ranges given in Table 10.4 may not be applicable, see also Chapter 4, Section 1. Narrow load tolerances for more than one loaded height have to be avoided.

Table 10.4 applies to single disks during the loading stroke. Spring stacks usually show somewhat lower forces for given loaded heights when compared to single disks. Actual measurements have to be compensated for deflections in the testing equipment, a fact which is often overlooked and could lead to unjustified rejections.

E. Reference Dimensions

In order to allow the manufacturer to meet functional requirements, it is necessary to designate some dimensions as reference dimensions only. In Section A, it has been shown that the thickness T is often reduced. Table 10.5 shows the general relationship between functional requirements and reference dimensions.

TABLE 10.5—FUNCTIONAL REQUIREMENTS AND REFERENCE DIMENSIONS

Functional Requirements	Reference Dimension
One Spring Force at a Specific Loaded Stack Length	L, or L and T
Two Spring Forces at Specific Loaded Stack Lengths	L and T

Chapter 11
Standardization

There are several standardized disk spring series on the market today, listed in manufacturers' catalogs (Ref. 4, 5, 6, 7) and generally available from stock. Ref. 12 shows a MIL-Standard which contains a selection of standard stock sizes from the above mentioned catalogs.

Under the heading CONICAL SPRING WASHERS or SAFETY SPRING WASHERS (Ref. 4, 6, 14) one can find "disk springs" adapted for the use in bolted connections, such as bus bars. Such washers are usually not preset and will, therefore, take a substantial set when loaded for the first time.

An effort should be made to use standard sizes in new designs, in order to avoid high tooling and set-up costs and long lead times. If no standard size can be found for a particular application, standard sizes can sometimes be modified by enlarging the inside diameter and reducing the outside diameter or the thickness. This may be helpful at least in the prototype stage. In any case, custom sizes should be designed around standard material thicknesses or customary inside and outside diameters.

Chapter 12
Modified Disk Spring Shapes

In this chapter, several types of spring washers different from standard disk springs are discussed. Not all these shapes are directly related to the disk spring, but may compete for some applications.

1. The "Open" Disk Spring

Disk springs with a radial gap are called "open" disk springs, Fig. 12.1. They have a linear characteristic, Fig. 12.2. The load F_H at flat position is the same as if there were no gap. The shaded area in Fig. 12.2 indicates the influence of the closed shape versus the open shape.

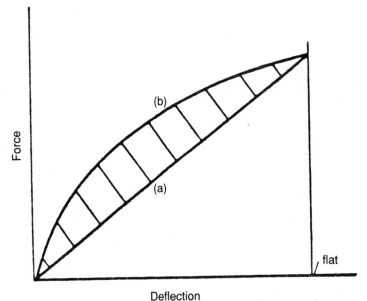

Fig. 12.2—(a) Characteristic of open disk spring
(b) Regular disk spring of same dimensions

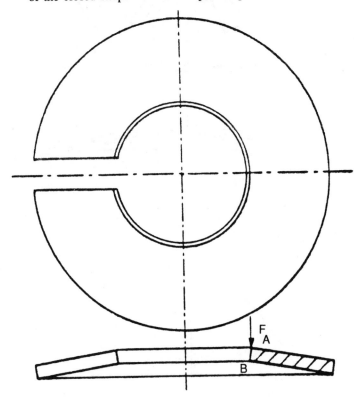

Fig. 12.1—The open disk spring

The load-deflection characteristic of the open disk spring can be calculated from Equation (2a), Chapter 4, Section 2, with the expression in the bracket equal to 1, that is:

$$F = \frac{4E}{1 - \mu^2} \cdot \frac{ST^3}{K_1 D_e^2} \qquad (13)$$

For values of K_1, see Chapter 4, Section E.

The open disk spring is equivalent to a curved beam, which is bent under a load.

Fig. 12.1 shows the locations A, B of the critical bending stresses, calculated as:

$$\sigma_A' = - \frac{2EST}{(1 - \mu^2)D_e^2} \cdot \frac{R^2}{(R - 1)} \qquad (14a)$$

and

$$\sigma_B' = + \frac{2EST}{(1 - \mu^2)D_e^2} \cdot \frac{R^2}{(R - 1)} \qquad (14b)$$

σ_A' is a compressive stress and is smaller than σ_A from Equation (4), Chapter 4, Section B. σ_B' is a tensile stress which is larger than σ_B from Equation (5), Chapter 4, Section B. Fatigue calculations can be made using Equation (14b) and the Goodman diagrams in Chapter 7, Section 2.

There are a few deliberate applications of the open disk

spring. Regular disk springs with a radial crack, such as caused by fatiguing, will keep operating with a characteristic according to Equation (13). Because the critical stress σ_B' is now larger than σ_B as initially calculated, a second fatigue failure will develop very quickly, splitting the disk in two pieces and, therefore, render it completely broken and unusable. This is of particular interest in a stack. It is more likely that the already cracked spring will develop a second crack before other disks in the stack will show first cracks.

2. Radially Tapered Disk Spring

Radially tapered disk springs may be defined as disk springs having a non-constant thickness. Such springs are sometimes used in order to achieve a more uniform stress-distribution. They are much more expensive to manufacture and to use and are, therefore, rarely used. It does not fall within the scope of this manual to treat the theoretical approaches. For the initially "flat" version, Fig. 12.3(a), see Ref. 3, for the initially 'coned' version, Fig. 12.3(b), see Ref. 4.

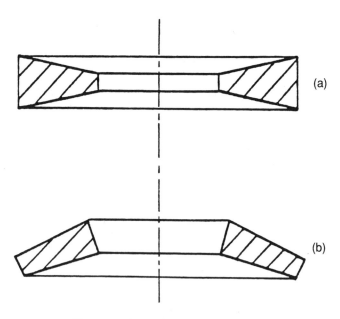

Fig. 12.3—Radially tapered disk spring
(a) Initially flat
(b) Initially coned

3. Slotted Disk Spring

Some disk springs are slotted, typically from the inside, Fig. 12.4. The actual disk spring portion is the closed ring section, defined with D_e, D_t, T and H, see Fig. 12.4(a). The "fingers" can be considered as levers (parallel to each other). This type spring will show a significantly larger deflection than the regular disk spring in a single disk. The closed ring section can be designed with a ratio H/T = 1.4, which gives a nearly zero-rate portion of the characteristic,

Fig. 12.4(b), see also Chapter 5, if loaded to the flat or near flat position. Often supported at the edges only, these springs can be loaded beyond flat and will, therefore, have an even larger deflection range with very small load variation, making them ideally suited for many clutch applications.

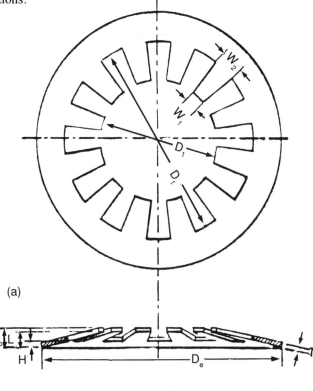

(a)

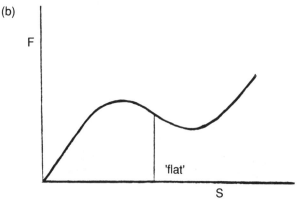

(b)

Fig. 12.4—(a) Slotted disk spring (no radii shown at the bottom of tongue)
(b) Typical load-deflection diagram

Theoretically, this spring can be treated as a standard disk spring with the load acting "outside" the inner diameter, compare Chapter 4, Section F. Such an approach would neglect the effect of the bending of the fingers or tongues. The following design equations (Ref. 13) apply when the bending of the tongues is taken into account and the load is applied at the edges.

Load F as a function of material, dimensions, deflection S_1 at the inside of the closed ring section:

$$F = \frac{4E}{(1 - \mu^2)} \cdot \frac{T^3 S_1}{K_1 D_e^2} \cdot \left[1 + \left(\frac{H}{T} - \frac{S_1}{T} \right)\left(\frac{H}{T} - \frac{S_1}{2T} \right) \right]$$

$$\times \left[\left(1 - \frac{D_t}{D_e} \right) \div \left(1 - \frac{D_i}{D_e} \right) \right] \qquad (15)$$

K_1 is again a factor (Chapter 4, Section E) depending on the diameter ratio, in this case $R = D_e/D_t$.

Deflection S_2 of the tongues only:

$$S_2 = C \cdot \frac{(D_t - D_i)^3(1 - \mu^2)}{2ET^3 \cdot W_2 Z} \cdot F \qquad (16)$$

Here, C is a constant depending on the ratio W_1/W_2 (Fig. 12.4(a)) see Table 12.1. Z is the number of tongues.

TABLE 12.1—CONSTANT C FOR VARIOUS RATIOS W_1/W_2

W_1/W_2	0.2	0.3	0.4	0.5	0.6	0.7	0.8	0.9	1.0
C	1.31	1.25	1.20	1.16	1.12	1.08	1.05	1.03	1.0

Total deflection:

$$S = S_1 \left[\left(1 - \frac{D_i}{D_e} \right) \div \left(1 - \frac{D_t}{D_e} \right) \right] + S_2 \qquad (17)$$

Stress calculations: For quasi-static and static loading, and for a check if a set can be expected, the compressive stress at the upper inner edge of the closed ring section is decisive. It corresponds closely to the stress at the same location of the regular disk spring, Equation (4).

$$\sigma_A = - \frac{4E}{(1 - \mu^2)} \cdot \frac{TS}{K_1 D_e^2} \left[K_3 + K_2 \left(\frac{H}{T} - \frac{S_1}{2T} \right) \right] \qquad (18)$$

The constants K_1, K_2, K_3 are functions of $R = D_e/D_t$, see Chapter 4, Section E. As long as the value of σ_A is safely below the yield stress of the chosen material, there will be no set. The deflection S_1 is larger than H, if the spring is loaded beyond its flat position. At the bottom of the cutouts, stress concentration factors may apply.

For repetitive loading, the critical stress is the stress σ_C at the lower outer edge:

$$\sigma_C = \frac{4E}{(1 - \mu^2)} \cdot \frac{T \cdot S}{K_1 \cdot R \cdot D_e^2}$$

$$\times \left[K_3 + (2K_3 - K_2)\left(\frac{H}{T} - \frac{S_1}{2T} \right) \right] \qquad (19)$$

This stress is the maximum tensile stress, and can be used to determine the fatigue life the same way as described in Chapter 7, Section 2.

4. Serrated Spring Washers

Serrated spring washers, also known as Serrated Safety Washers, Fig. 12.5, are fastening elements used under the head of a bolt or under a nut. They are dimensionally designed to be useful for socket head screws. Due to their basic conical disk spring shape, they add to the elasticity of a bolted joint. This will reduce the possibility of a loss in clamping force, an effect that can also be achieved with unserrated disk springs, see Chapter 11 and Ref. 14. The serrations will bite into the surface of the clamped and clamping parts, thus impending back-off. This feature has been found particularly useful when vibrations are present. For a list of standardized stock sizes, see Ref. 4.

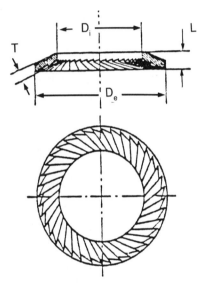

Fig. 12.5—Serrated spring washer (safety spring washer)

Chapter 13

Spring Washers

Disk springs as well as special spring washers exert thrust loads in axial directions. Whereas disk springs are distinguished by their conical shape, the washers discussed here are not conical, but bent in a specific way so that the state of stress is primarily bending. These spring washers are usually less predictable and associated with higher load tolerances when compared to the disk spring. They also do not show the degressive load-deflection behaviour of the disk spring, but rather have a linear or even progressive characteristic.

1. Curved Washer

Curved washers (Fig. 13.1) exert a relatively light thrust load and are often used to absorb axial end play. The spring rate is approximately linear up to 75% deflection, then the rate increases and gets considerably larger than calculated; see also Ref. 7.

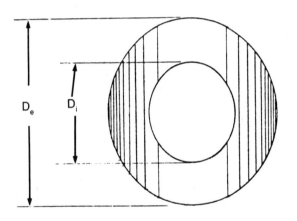

Fig. 13.1—Typical curved spring washer

Design equations for curved washers are based on the equations for simple bending beams, but with an empirical correction factor K_4, Fig. 13.2.

The equation for the load is:

$$F = \frac{4 \cdot E \cdot T^3 \cdot S}{D_e^2 \cdot K_4} \qquad (20)$$

The critical stress is calculated as:

$$\sigma = \frac{1.5 \cdot F}{T^2} \cdot K_4 \qquad (21)$$

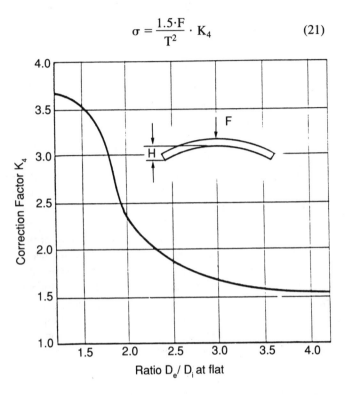

Fig. 13.2—Empirical correction factor K_4 for curved spring washers

2. Wave Washer

Wave spring washers, Fig. 13.3, are used to exert moderate thrust loads when space is limited, similar to thin disk springs with relatively large inside diameters. The rate is linear between 20% and 80% of the deflection. Washers that are round in the free position go out-of-round when loaded

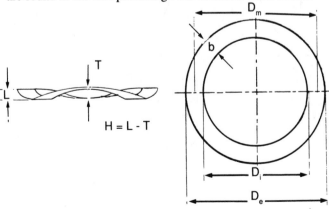

Fig. 13.3—Typical wave spring washer
T: material thickness

and vice versa. Generally, a ratio of $D_m/b = 8$ gives a well-balanced design. When this ratio is substantially lower than 8, a disk spring is preferable.

The number N of the waves can be 3 or more and can be selected on the basis of the desired spring properties.

The load equation (for approximate results) is:

$$F = \frac{E \cdot b \cdot T^3 N^4 D_e}{2.40 \cdot D_m{}^3 \cdot D_i} \qquad (22)$$

Critical stress is given by:

$$\sigma = \frac{3\ F \cdot D_m \cdot \pi}{4b\ T^2\ N^2} \qquad (23)$$

D_e is the outside diameter in the free position. It will change slightly under load.

For more details and a list of stock sizes, see Ref. 7.

3. Finger Washers

Finger washers, Fig. 13.4, combine the flexibility of the curved washers and the distributed load points of the wave washers. Load, deflection and stress are approximated by assuming that the fingers are cantilever beams. Finger washers are made and tested to prove the design. They are used in static applications such as applying an axial load to ball bearings to reduce noise and vibration, and are competing in this field with special disk spring designs. For more details and stock sizes, see Ref. 7.

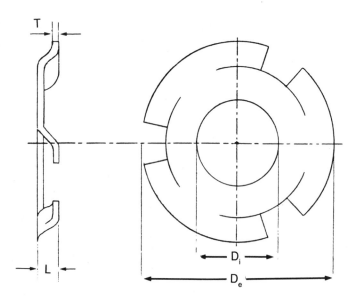

Fig. 13.4—Typical finger spring washer

Chapter 14

References

1. SAE J773b, Conical Spring Washers.

2. Almen and Laszlo, "The Uniform-Section Disk Spring", Trans. ASME (1936), pp. 305-314.

3. Wahl, A. M., "Mechanical Springs", Penton Publ. Co., 2nd printing 1949. Reprinted 1982 by SMI (Spring Manufacturers Institute).

4. SCHNORR DISC SPRING HANDBOOK, 11th Ed., 1981, and other literature from Schnorr Corp., Woodside, NY.

5. Catalogs from E. C. Styberg Eng. Co., Racine, WI.

6. Catalogs from Key Bellevilles Co., Leechburg, PA.

7. DESIGN HANDBOOK, 1981 ed., and other catalogs, Associated Spring, Bristol, CT.

8. Schremmer, G., "Endurance Strength and Optimum Dimensions of Belleville Springs", ASME publ. 68 WA/De 9 (1968).

9. QQ-P-1466, Federal Specification, "Plating, Cadmium (Electro-deposited)".

10. ASTM B 695-82, "Coatings of Zinc, Mechanically Deposited on Iron or Steel".

11. Metal Improvement Co., "Shot Peening Applications", 6th Ed. (1980).

12. Military Specification, MIL-W-12133/1.

13. Schremmer, G., "The Slotted Conical Disc Spring", ASME paper 72 WA/DE9 (1972).

14. Schremmer, G., "How to Keep Bolted Joints Tight", Machine Design, Sept. 21, 1972.

APPENDIX
CONVERSION TABLE

To Convert from SI Unit to U.S. Customary Unit, Divide by the <u>Factor</u>
To Convert from U.S. Customary Unit to SI Unit, Multiply by the <u>Factor</u>

Quantity	SI Unit		Factor (° = Exact)	U.S. Customary Unit	
Length	kilometer	km	1.609 344°	mile	
	meter	m	0.304 8°	foot	ft
	millimeter	mm	25.4°	inch	in
Area	square millimeter	mm²	645.16°	square inch	in²
Volume	cubic millimeter	mm³	16 387.064°	cubic inch	in³
	cubic millimeter	mm³	3 785 412.	gallon	gal (U.S.)
	liter (=10⁶mm³)	L	3.785 412	gallon	gal (U.S.)
Area Moment of Inertia	millimeter to the fourth power	mm⁴	416 231.425 6°	inch to the fourth power	in⁴
Mass	kilogram	kg	0.453 592 4	pound-mass	lb$_m$
Force (or Load)	Newton	N	4.448 222[a]	pound-force	lb$_f$
Elastic Energy, Work	Joule	J (= N·m = kN·mm)	0.112 984 8	pound inch	lb$_f$.in
Bending Moment, Torque	Newton millimeter	N·mm (= kN·m)	112.984 8	pound inch	lb$_f$.in
Linear Spring Rate	Newton per mm	N/mm (= kN/m)	0.175 126 8	pound per inch	lb$_f$/in
Torsional Spring Rate	Newton millimeter per radian	N·mm/rad	112.984 8	pound inch per radian	lb$_f$.in/rad
Plane Angle	degree	deg	57.295 780[b]	radian	rad
Stress, Modulus of Elasticity	pascal	Pa (N/m²)	6 894.757[c]	pound per	
	kilopascal	kPa	6.894 757	square inch	lb$_f$/in² (= psi)
	megapascal	MPa	0.006 894 757		
Density of Material	kilogram per cubic meter	kg/m³	27 679.90[d]	pound per cubic inch	lb$_m$/in³
for example, for Steel		7850 kg/m³		~0.2836	lb$_m$/in³
Acceleration of Gravity "g"—adopted 1901 by Internatl. Committee on Weights Measures		9.806 650 m/s²	0.3048°	32.174 05	ft/s²
		9.806 650 m/s²	0.0254°	386.089	in/s²
Natural Frequency "f"	cycles per second	Hz			

$$f = \frac{1}{2\pi} \sqrt{\frac{B}{S}}$$

$$f = \sqrt{0.2484/S \ (m)}$$
$$f = \sqrt{248.4/S \ (mm)}$$

$$f = \sqrt{9.760/S \ (in)}$$

where S
= static deflection

[a]4,448 222 = 0.453 592 37 · 9.806 650

[b]57.295 780 = 180/π

[c]6894.757 $= \dfrac{4.448\ 221\ 5}{0.000\ 645\ 16}$

[d]27 679.90 $= \dfrac{0.453\ 592\ 37}{0.000\ 016\ 387\ 064}$

Index